Methods in Neurosciences

Volume 19

Ion Channels of Excitable Cells

Methods in Neurosciences

Editor-in-Chief
P. Michael Conn

1994

Methods in Neurosciences

Volume 19
Ion Channels of Excitable Cells

Edited by

Toshio Narahashi

Department of Pharmacology
Northwestern University Medical School
Chicago, Illinois

ACADEMIC PRESS, INC.
A Division of Harcourt Brace & Company

San Diego   New York   Boston   London   Sydney   Tokyo   Toronto

Front cover photograph: A cartoon of an ion channel embedded in a lipid bilayer membrane. The channel comprises five subunits. The inset shows a cross-sectional view.

Academic Press, Inc.
525 B Street, Suite 1900, San Diego, California 92101-4495

United Kingdom Edition published by
Academic Press Limited
24–28 Oval Road, London NW1 7DX

International Standard Serial Number: 1043-9471

International Standard Book Number: 0-12-185287-3

PRINTED IN THE UNITED STATES OF AMERICA
94 95 96 97 98 99   E B   9 8 7 6 5 4 3 2 1

Table of Contents

Contributors to Volume 19

Article numbers are in parentheses following the names of contributors. Affiliations listed are current.

NORIO AKAIKE (8), Department of Physiology, Kyushu University Faculty of Medicine, Fukuoka 812, Japan

EDSON X. ALBUQUERQUE (7), Department of Pharmacology and Experimental Therapeutics, University of Maryland School of Medicine, Baltimore, Maryland 21201, and Laboratory of Molecular Pharmacology II, Institute of Biophysics "Carlos Chagas Filho," Federal University of Rio de Janeiro, Rio de Janeiro 21914, Brazil

SIMON ALFORD (18), Department of Physiology, Northwestern University Medical School, Chicago, Illinois 60611

MANICKAVASAGOM ALKONDON (7), Department of Pharmacology and Experimental Therapeutics, University of Maryland School of Medicine, Baltimore, Maryland 21201

KIMON J. ANGELIDES (17), Department of Cell Biology, Baylor College of Medicine, Houston, Texas 77030

RAMESH BANGALORE (10), Department of Physiology, University of Rochester, Rochester, New York 14642

NEWTON G. CASTRO (7), Department of Pharmacology and Experimental Therapeutics, University of Maryland School of Medicine, Baltimore, Maryland 21201, and Laboratory of Molecular Pharmacology II, Institute of Biophysics "Carlos Chagas Filho," Federal University of Rio de Janeiro, Rio de Janeiro 21914, Brazil

S. Y. CHIU (6), Department of Neurophysiology, University of Wisconsin—Madison, Madison, Wisconsin 53706

GRAHAM L. COLLINGRIDGE (18), Department of Pharmacology, The Medical School, University of Birmingham, Edgebaston, Birmingham B15 2TT, United Kingdom

JOHN A. DREWE (13), CoCensys Inc., Irvine, California 92718

JAMES EBERWINE (3), Department of Pharmacology, University of Pennsylvania Medical School, Philadelphia, Pennsylvania 19104

JERRY M. FARLEY (12), Department of Pharmacology and Toxicology, University of Mississippi Medical Center, Jackson, Mississippi 39216

ANTONIO V. FERRER-MONTIEL (19), Department of Biology, University of California, San Diego, La Jolla, California 92093

STEVEN R. GLAUM (18), Department of Pharmacological and Physiological Sciences, The University of Chicago, Chicago, Illinois 60637

ANNE GROVE (19), Department of Biology, University of California, San Diego, La Jolla, California 92093

NOBUTOSHI HARATA (8), Department of Physiology, Kyushu University Faculty of Medicine, Fukuoka 812, Japan

HALI A. HARTMANN (13), Department of Molecular Physiology and Biophysics, Baylor College of Medicine, Houston, Texas 77030

VICTOR HENZI (15), Department of Physiology and Cellular Biophysics and the Center for Neurobiology and Behavior, Columbia University, New York, New York 10032

BARRY W. HICKS (17), Department of Cell Biology, Baylor College of Medicine, Houston, Texas 77030

ROBERT S. KASS (10), Department of Physiology, University of Rochester, Rochester, New York 14642

GLENN E. KIRSCH (13), Department of Anesthesiology, Baylor College of Medicine, Houston, Texas 77030

HENRY A. LESTER (14), Division of Biology, California Institute of Technology, Pasadena, California 91125

AMY B. MACDERMOTT (15), Department of Physiology and Cellular Biophysics and the Center for Neurobiology and Behavior, Columbia University, New York, New York 10032

JOHN F. MACDONALD (16), Departments of Physiology and Pharmacology, University of Toronto, Toronto, Ontario, Canada M4Y 1R5

MAURICIO MONTAL (19), Department of Biology, University of California, San Diego, La Jolla, California 92093

JOSEPH G. MONTES (7), Department of Pharmacology and Experimental Therapeutics, University of Maryland School of Medicine, Baltimore, Maryland 21201

TOSHIO NARAHASHI (2), Department of Pharmacology, Northwestern University Medical School, Chicago, Illinois 60611

EDNA F. R. PEREIRA (7), Department of Pharmacology and Experimental Therapeutics, University of Maryland School of Medicine, Baltimore,

Maryland 21201, and Laboratory of Molecular Pharmacology II, Institute of Biophysics "Carlos Chagas Filho," Federal University of Rio de Janeiro, Rio de Janeiro 21914, Brazil

DONALD G. PURO (4), Departments of Ophthalmology and Physiology, The University of Michigan, Ann Arbor, Michigan 48105

FRED N. QUANDT (1), Department of Physiology and Multiple Sclerosis Research Center, Rush University, Chicago, Illinois 60612

MICHAEL W. QUICK (14), Division of Biology, California Institute of Technology, Pasadena, California 91125

MARTIN D. RAYNER (5), Department of Physiology, School of Medicine, University of Hawaii at Manoa, Honolulu, Hawaii 96822

DAVID B. REICHLING (15), Department of Anatomy, University of California, San Francisco, San Francisco, California 94143

DAVID J. ROSSI (18), Department of Physiology, Northwestern University Medical School, Chicago, Illinois 60611

MARY LOUISE ROY (2), Department of Neurology, Yale University School of Medicine, New Haven, Connecticut 06510

MICHAEL W. SALTER (16), Division of Neuroscience, Hospital for Sick Children, Toronto, Ontario, Canada M5G 1X8

MICHAEL C. SANGUINETTI (11), Division of Cardiology, University of Utah, Salt Lake City, Utah 84112

MICHAEL F. SHEETS (9), Department of Medicine and The Feinberg Cardiovascular Research Institute, Northwestern University Medical School, Chicago, Illinois 60611

P. SHRAGER (6), Department of Physiology, University of Rochester Medical Center, Rochester, New York 14642

N. TRAVERSE SLATER (18), Department of Physiology, Northwestern University Medical School, Chicago, Illinois 60611

JOHN G. STARKUS (5), Bekesy Laboratory of Neurobiology, Pacific Biomedical Research Center, University of Hawaii at Manoa, Honolulu, Hawaii 96822

D. JAMES SURMEIER (3), Department of Anatomy and Neurobiology, University of Tennessee, Memphis, Tennessee 38163

ROBERT E. TEN EICK (9), Department of Pharmacology and The Feinberg Cardiovascular Research Institute, Northwestern University Medical School, Chicago, Illinois 60611

LU-YANG WANG (16), Departments of Physiology and Pharmacology, University of Toronto, Toronto, Ontario, Canada M4Y 1R5

C. J. WILSON (3), Department of Anatomy and Neurobiology, University of Tennessee, Memphis, Tennessee 38163

Preface

Studies of ion channels have a long history. The first technical breakthrough was made by K.S. Cole who developed the voltage clamp technique in 1949, a technique extensively applied to squid giant axons by A.L. Hodgkin, A.F. Huxley, and B. Katz in their 1952 studies which clearly established the roles of sodium and potassium channels in nerve excitation. Although the original voltage clamp technique utilizing internal axial wire electrodes allowed extremely elegant and precise measurements of ion channel activity, it was not an easy technique to use at that time and, above all, its applicability was limited to certain preparations such as giant axons and nodes of Ranvier. Studies of neurotransmitter-activated channels were still difficult, and the application of voltage clamp techniques was limited mostly to those of muscle end plates and snail giant neurons. Thus, although a number of epoch-making developments and discoveries had been made through voltage clamp studies of various ion channels, progress was relatively slow, and the popularity of ion channel study was confined to certain groups of specialists.

Another quantum leap was made by E. Neher and B. Sakmann in 1976 when they successfully recorded single channel activity by using patch electrodes. Although the original patch-clamp technique was far from perfect, it was greatly improved in 1981 by the same group of investigators who developed gigaohm seal techniques, which broadened applicability to whole cell as well as single channel records. Since gigaohm patch-clamp techniques are applicable to practically any type of cell, the study of ion channels has become explosively popular. The techniques can now be used not only for neurons, but also for a variety of cells including, but not limited to, skeletal, smooth, and cardiac muscles, secretory cells, lymphocytes, and red blood cells.

The popularity of ion channel study was also aided by the development and application of at least two other techniques, i.e., molecular biology and imaging. Researchers of ion channels were quick to adopt molecular biology techniques for the determination of channel molecular structures. A variety of imaging techniques have also been used for the measurement of intracellular components that are controlled by ion channel activity. Today rapid progress is being made in the channel field by combining techniques such as patch clamp, molecular biology, and imaging. Because of the highly significant and widely recognized roles of ion channels in physiology, pathophysiology, pharmacology, and toxicology, the term "ion channel" has now become a household word in the biomedical sciences.

Since so many different ion channels and their subtypes have been discovered and are being studied in a variety of preparations, it was deemed highly useful to publish a volume devoted exclusively to ion channel methodology. Thus, this volume covers preparations and techniques for the study of various ion channels. Both voltage-gated and ligand-gated ion channels of neurons, axons, and cardiac and smooth muscles are included as are patch-clamp techniques as applied to different cells and other contemporary techniques as related to electrophysiology. These techniques comprise molecular biology, imaging, lipid bilayers, and channel expression in oocytes. Although it was impossible to cover all types of channels and methodologies currently in use, the chapters in this volume provide basic techniques which can be applied to other channels and cells with appropriate modifications. Other volumes of the *Methods in Neurosciences* series complement this one, particularly Volume 4, Electrophysiology and Microinjection.

TOSHIO NARAHASHI

Methods in Neurosciences

Section I

Neuronal Voltage-Gated Ion Channels: Electrophysiology

[1] Recording Sodium and Potassium Currents from Neuroblastoma Cells

Fred N. Quandt

Introduction

Numerous studies of neuronal cell lines grown in tissue culture have shown that they possess the functions of mammalian neurons *in vivo,* including electrical excitability (McMorris *et al.,* 1974). Voltage-clamp studies of neuroblastoma cells soon followed (Moolenaar and Spector, 1977, 1978, 1979; Kostyuk *et al.,* 1978). These studies indicated that neuroblastoma cells would be a useful preparation for studying voltage-dependent Na and K channels (see also the review by Spector, 1981). I have often used N1E-115 neuroblastoma to examine the pharmacology of Na and K channels (Quandt and Narahashi, 1982; Quandt *et al.,* 1985; Quandt, 1988b; Im and Quandt, 1992; Quandt and Im, 1992). The neurons have the advantage that they represent a convenient, homogeneous, and stable preparation for expression of these channels. Sodium and potassium currents can be easily measured from the whole cell, and single-channel analysis can be undertaken. In this article, I give the procedures which my laboratory has used for measuring these currents.

Cell Culturing

Neuroblastoma cells will grow under a wide variety of conditions. However, growth rates will vary. It is convenient to plate 1 ml of cells into a 75-cm^2 flask (Corning 25110-75) once a week, usually on Friday, and adjust conditions so that they become confluent in 7 days. When plated on Friday, the cells do not have to be fed until Monday. They are then fed every day until split on the next Friday. The growth rate appears to increase in proportion to the fetal bovine serum (FBS). We typically supplement the media with 5% FBS to maintain this growth rate.

Standard medium is composed of Dulbecco's modified Eagle's medium (Sigma D5523 or GIBCO 430-1600). Glucose is brought to 4.5 g/liter by the addition of 3.5 g/liter. Sodium bicarbonate is added in the amount of 3.7 g/liter and 4.25 g/liter HEPES is also added in order to maintain pH in the CO_2 incubator as well as in room air. We add 50 ml/liter of fetal bovine serum (Sigma F2138 or GIBCO 230-6140). Medium is then filtered with a 0.22 μm filter (Millipore Sterivex-GS, SVGSB1010) and stored in the refrigerator in 500-ml bottles. A total of 5 ml antibiotic–antimycotic solution (Sigma A9909 or GIBCO 600-5240) is added to an aliquot before its use.

Cells grow well for more than 40 to 50 passages (in this case, corresponding to the number of weeks in culture). With the high passage numbers, we find that cells clump and do not extend processes when exposed to a differentiating condition (defined below). We then culture new cells which were previously frozen. Although we have no specific information to indicate that the electrophysiological properties of the cells are altered at the higher passage numbers, this procedure avoids potential variability. Our primary stock of cells is preserved at passage number 20 to 25, so that cells are thawed twice per year.

To freeze cells, one flask is pelleted at 2000 rpm for 5 min and resuspended in 1 ml of growth medium to which dimethyl sulfoxide (DMSO) has been added at a concentration of 10% (v/v). This aliquot is then placed into a cryovial and placed into a styrofoam box. The box is placed into a −70°C freezer overnight. We find that this procedure gives a good rate of change of temperature for the freezing process. The vial is then stored in the liquid or vapor phase in a liquid nitrogen tank. Cells can be stored for many years.

Cells are thawed by quickly bringing them to 37°C via immersion of the cryovial into a water bath. The thawed cells are transferred to a 15-ml centrifuge tube and 10 ml of normal growth medium is added slowly in order to avoid osmotic shock. The resuspended cells are then placed into a 75-cm^2 flask. The solution should be changed the next day to remove the DMSO.

Prior to their use in experiments, cells are grown on 22-mm-diameter glass coverslips (Fisher 12-546-1) in Dulbecco's medium with 1 to 2.5% FBS and 1 to 2% DMSO. Three of the glass coverslips can be accommodated in 60-mm-diameter petri dishes (Falcon 1007). The coverslips are not routinely washed. The density of cells is not very critical, as long as the cells are not crowded. Cell growth is arrested on exposure to the altered medium, and many cells become differentiated after 3 days (see below).

Recording Techniques

Numerous techniques have been used in this laboratory to record membrane potentials and currents from neuroblastoma cells. The two-microelectrode recording technique was the first method used to voltage-clamp neuroblastoma cells (see Introduction). The technique is not optimal because rather large microelectrodes must be used (5 to 10 MΩ when filled with 3 M KCl) due to the large currents. Under these conditions we have found that the stability of the clamp is limited to 5 to 10 min, since KCl leaks out of the electrodes, causing the cells to swell.

The internal dialysis technique (Lee *et al.*, 1980; Kostyuk *et al.*, 1977) has been adapted to neuroblastoma cells (Kostyuk *et al.*, 1978; Huang *et al.*, 1982; Matsuki *et al.*, 1983; Quandt and Narahashi, 1984). This technique uses a suction pipette with a large opening at the tip, with continuous perfusion of the inside, and has major advantages. First, the clamp is very stable, often allowing recordings for an hour. Second, the pipette resistance is low, minimizing series resistance. Third, it allows the solution to be changed during an experiment. Although large neurons are required, N1E-115 cells attain the required size. Use of the patch-clamp technique (Hamill *et al.*, 1981) has superceded the suction pipette for measuring currents from the whole cell since it is more convenient and easier to switch to single-channel recording. Techniques are available to alter the internal pipette solution (Tang *et al.*, 1990). The series resistance tends to be higher in the whole-cell patch clamp than with a suction pipette. On occasion I have substituted two patch electrodes for the microelectrodes in the two-electrode voltage clamp in order to eliminate the series resistance. It is also possible to establish a whole-cell patch clamp and simultaneously record current through a patch of membrane with an independent patch pipette. This last configuration eliminates the current through the seal resistance and reduces the capacitative current for the membrane patch.

Whole-Cell Patch-Clamp Technique

Neuroblastoma cells can be readily patch-clamped following the general techniques which are given in Hamill *et al.* (1981). Whole-cell voltage clamp and single-channel recording from intact and excised membrane patches are rather routine.

Recording System

A coverslip on which cells are grown is placed into a plastic chamber, sometimes on a small drop of Vaseline to prevent movement. The glass can first be broken using a sharp-tipped tool to reduce the size of the chamber and maintain the cells not being used. The chamber is fixed to the mechanical stage on an inverted microscope. The cells are visualized with the microscope using long working distance, phase contrast objectives and a total magnification of 400 to 600×. The microscope power supply can be replaced with a DC power supply to reduce 60-cycle noise in recordings.

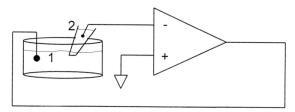

Fig. 1   Feedback circuit used to minimize changes in the bath potential. The potential of the bath is measured with a low-resistance pipette filled with saline. An Ag–AgCl electrode in the bath is connected to the output of a negative feedback circuit, shown here simplified, to maintain the bath potential at ground.

The patch pipette is maneuvered onto the cell using a Narashige three-dimensional hydraulic manipulator mounted on a second mechanical three-dimensional manipulator. The electrode is aimed at the cell at a 45° angle to the microscope stage. An anti-vibration air table will increase stability for whole-cell and intact patch recording situations. The controls for the hydraulic manipulator and the valve to the bath perfusion system are mounted off the air table to eliminate vibration following initiation of the clamp.

We employ a bath perfusion system in order to alter the electrolyte composition or add neuroactive agents. In addition, the temperature of the preparation is altered by cooling or heating the solution flowing into the bath using a Peltier device to cool polyethylene tubing. The temperature of the bath is measured using a thermocouple mounted close to the area of focus of the objective. Lower temperatures (10 to 15°C) are usually used to slow gating reactions for Na currents.

A simple oscilloscope and pulse generator is useful for measuring the pipette and seal (between the membrane and pipette) resistances. A computer-controlled pulse generator is routinely employed. Responses to step depolarizations are recorded using a computer with a 12-bit or greater analog-to-digital converter. Since capacitative currents are rather large, responses to four prepulses of 25% amplitude are usually summed to subtract these and linear leak currents. We record currents using a low gain to increase the range and prevent saturation. Software is used to increase the final gain of the record following subtraction.

Two electrodes are utilized in contact with the bath. These are shown in Fig. 1. The first is an Ag–AgCl pellet (1). This electrode is traditionally connected to the amplifier ground. In many experiments, the electrolyte composition of the bath is extensively altered and the junction potential across this electrode changes dramatically. A second, low-resistance pipette electrode (2) and feedback circuit is employed to clamp the bath to the appropriate voltage. In this arrangement, changes in the bath voltage are confined to the tip potential of electrode 2, yet the series resistance remains low. The time constant of the feedback is set to greater than 1 sec to minimize noise. Alternatively, electrode 2 can be connected to ground and electrode 1 eliminated if series resistance is not a concern. The patch pipette is mounted in an appropriate electrode holder and polyethylene tubing is used to apply pressure to the pipette. An Ag–AgCl pellet is used to connect the amplifier to the patch pipette having a high-resistance seal with the membrane. Compared to an Ag wire, the use of a pellet minimizes drift in the electrode potential when the pipette contains intracellular solutions.

Solutions

Our normal external saline is made of (in mM) 125 NaCl, 5.5 KCl, 3.0 CaCl$_2$, 0.8 MgCl$_2$, 20 HEPES, 25 dextrose. The pH is brought to 7.3 with NaOH. Normal internal solution consists of 150 potassium glutamate, 20 EGTA, 1 sodium HEPES, 20 HEPES. The pH is adjusted to 7.25 with KOH. In order to eliminate K currents, KCl in the external solution is replaced with CsCl. In addition, potassium glutamate in the internal solution is replaced with CsOH or N-methyl-D-glucamine. In this case, glutamic acid is added to titrate the pH to 7.25. CdCl$_2$ (2 mM) can be added to the external solution to block currents through Ca channels; however, Na and K currents will be reduced. The solution with which the pipette is filled, as well as the bath solution in which the seal between the pipette and the cell are formed, is filtered. We use a 0.45 μm filter (Gelman 47 mm supor-450, #60173) at the beginning of each day on which experiments are performed and periodically look for subsequent precipitation which can interfere with the formation of high-resistance seals.

Patch Electrode Construction

Electrodes are made by pulling 1.5- to 1.8-mm-diameter borosilicate glass capillaries (Kimax-51, 34500) using a two-stage puller (Narashige PP-83). It is rather difficult to adjust the heat on this puller, since small changes make great differences in the pipette size. We have found that measuring the AC voltage across the coil with a digital multimeter to an accuracy of 0.01 V both facilitates and quantifies the adjustment. A voltage stabilizer generally increases reproducibility between the pipettes, although the variability in glass diameter for the glass used is considerable. A relatively fast tapered pipette is made with a final opening diameter of 3 to 5 μm. This larger pipette is ideal for whole-cell clamp, as it has a very low resistance (about 1 MΩ)

and the resulting series resistance is low. A large opening facilitates the dialysis of the cell with the pipette solution. The seal resistance is adequate for this recording configuration.

The seal resistance is considerably higher and a consistent final pipette opening can be made if the electrodes are manually fire polished. Fire polishing is accomplished by heating the tip using a platinum ribbon under visual control. The ribbon should be narrow (75 to 100 μm) and should come to a sharp point to localize the heat. The ribbon and electrode are visualized with an inexpensive compound microscope using a magnification of 600×. The ribbon is fixed to the stage of the compound microscope and its position is adjusted relative to the objective with the stage and focus controls. The position of the electrode is altered with a micromanipulator. The heat from the platinum ribbon is adjusted with a variac in series with a step down transformer. The final AC rms voltage is then variable from 0 to 3 V. The opening of the pipette for the whole-cell clamp is typically at least 3 μm.

Selecting a Cell

Neuroblastoma cell geometries vary considerably in any culture. We usually use cells which have been grown to optimize differentiation (see below). The cells exist in three general configurations. These possibilities are shown in Fig. 2. First, many N1E-115 cells are bipolar, exhibiting long neurites (type a). Second, some cells are clustered (type b). Third, there is a population of cells which are small and round, isolated from other cells, and devoid of neurites (type c). Cells in configuration c are most ideal for whole-cell patch clamp. Cells which are not attached to other cells and have been exposed to differentiating solution for more than 3 days will have cell division arrested. In contrast, cells in clusters may or may not be dividing. In addition, these cells may be electrically in contact with an unknown number of other cells. Large cells with long processes are not isopotential with the cell body and will have a higher capacitance. Neuroblastoma cells with short processes may be relatively isopotential, since the density of Na channels is not greater in the neurites than in the cell body (Catterall, 1981). Further, cells with large-diameter short neurites have been found to be relatively space clamped (Grinvald *et al.*, 1981).

Establishing the Whole-Cell Clamp

Seals are obtained by maneuvering the pipette onto the cell under visual control. Because the cells are thin, it is useful to focus through the cell. This will ensure that the plane of focus is toward the top of the cell so first contact

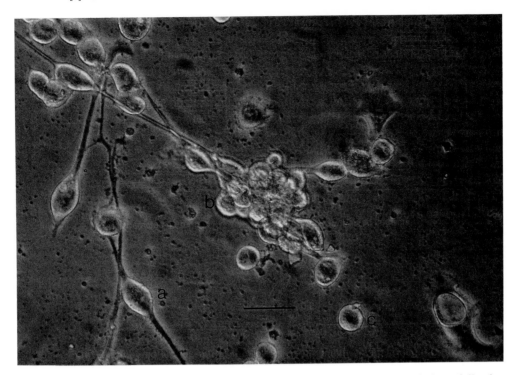

Fig. 2   N1E-115 neuroblastoma cells exhibit heterogeneous morphology following differentiation. Cells were photographed following growth in media containing 1.5% DMSO and reduced (2.5%) serum. Some cells are bipolar with long neurites (a), while other cells occur in clusters (b). Small cells which are isolated and devoid of neurites (c) are ideal for whole-cell clamp. The bar represents 50 μm.

between the cell and the pipette is observed. We apply a positive pressure to the pipette prior to contact with the cell to minimize dilution of the solution in the pipette with the bath solution. In addition, it is likely that the tip will be free of cellular debris. The resistance between the pipette and the bath is continuously monitored during this procedure by measuring the current in response to a 10-mV pulse across the pipette tip in order to ascertain the seal between the pipette and the cell. Seals are typically obtained with a 90% success rate. A pipette is only used in one attempt to obtain a seal.

The interior of the cell is usually made continuous with the pipette by an increase in negative pressure of the patch pipette. It is important that the electrode is tightly seated in the holder, otherwise the suction required to break the patch of membrane in contact with the pipette will move the

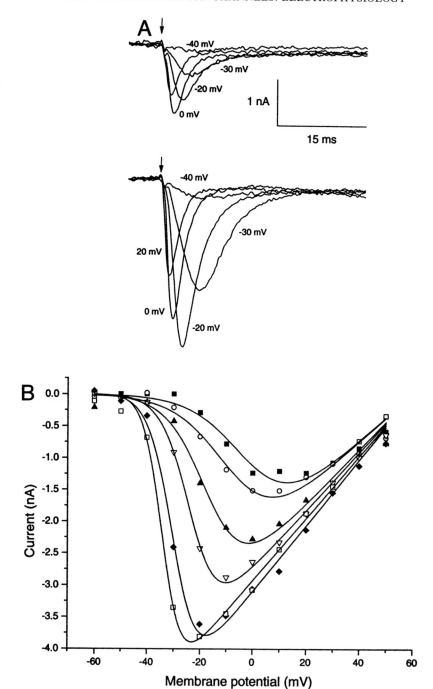

electrode. The capacitative current for the voltage pulse increases when the membrane under the pipette is ruptured and the holding potential is then applied. No pressure is applied to the back of the electrode during recording. However, the series resistance will be high if the membrane is not adequately broken. The status of the series resistance should be determined at this point to see that it does not decrease with application of negative pressure. The time constant of the decay in capacitative current following a voltage step will decrease with a decrease in series resistance.

Properties of Whole-Cell Currents

There are many voltage-dependent ion channels in the membranes of N1E-115 cells, including Na channels and at least three types of K channels (Quandt, 1988a). One of these is a large conductance Ca-activated channel. Two types of Ca channels are present (Narahashi *et al.*, 1987). Chloride channels have been reported in another neuroblastoma cell line (Bolotina *et al.*, 1987). However, we have not explicitly tested for their presence in N1E-115 cells. In addition to neurotransmitter-activated channels, a Ca-activated cation channel has also been reported (Yellen, 1982). There are also stretch-activated channels (Falk and Misler, 1989). There is a nonspecific cation background current which appears not to be gated, but exhibits outward rectification in normal Ca concentrations (Quandt, 1991).

Sodium currents typically increase with time after initiation of the whole-cell clamp rather than run down. This phenomenon should be taken into consideration when the effects of drugs are being tested. Figure 3 shows this phenomenon for N1E-115 cells. The amplitude of the Na current increases

FIG. 3   Time dependence of Na currents in whole-cell patch-clamped neuroblastoma cells. (A) Sodium currents recorded in response to various depolarizations, beginning at the arrow, are superimposed. (Top) Current was obtained 2 min after initiation of the clamp. (Bottom) After 10 min from the onset of the clamp. (B) Current–voltage curves were measured at various times following the initiation of whole-cell recording. The curves plot a Boltzman equation given by $I = (G/(1 + \exp((V - V_h)/k)))(V - E)$, where V is the voltage in mV, E is fixed at $+60$ mV, G is the Na conductance in mmho, and I is the current in nA. G varied in the equations from 37 to 52 mmho and k averaged -6.7. V_h varied systematically with time: 1 min, filled squares, $V_h = 0.2$ mV; 2 min, unfilled circles, -7.0 mV; 3 min, filled triangles, -16.2; 5 min, unfilled triangles, -23.3; 10 min, filled diamonds, -29.9; 15 min, unfilled squares, -34. No potassium external, N-methyl-D-glucamine internal solutions. Temperature 13.2°C.

following the initiation of the clamp. Gating parameters shift to more negative potentials with time. An example of the time course of this shift is shown in the figure. The causes of the alteration in Na current have not been investigated. However, numerous factors could be involved. The obvious effect would be the alteration in the composition in the cytoplasmic solution to one more favorable for operation of the channel. There could be an alteration in the surface potential of the internal membrane. There could also be an increase in the Na equilibrium potential with time, or a decrease in the outward K currents due to the substitution of K with N-methyl-D-glucamine. The potential of the patch pipette may also be drifting to some extent. However, this would appear to make the Na equilibrium potential less positive as well. Fernandez *et al.* (1984) attribute a similar shift, seen in GH_3 cells, to a time-dependent Donnan potential between the cell and the pipette.

It should be noted that Na currents can decrease, rather than increase, if the holding potential is less negative due to a corresponding shift in the availability of Na current to respond at any holding potential. A factor in the increase in the Na current is also due to recovery from slow inactivation upon application of a negative holding potential (see Quandt, 1987). The resting potential of neuroblastoma cells grown under standard conditions is rather low, typically -35 mV at 22°C. Most Na channels are unavailable to open following depolarization from -35 mV due to the process of slow inactivation. Sodium channels will recover from slow inactivation over a time course of minutes when the membrane potential is made more negative, increasing the Na current in response to a step depolarization. However, the voltage sensitivity of activation may be independent of slow inactivation.

Surface Area

It is often important to determine the average density of channels in the membrane of cells. To determine this parameter, the whole-cell current is divided by single-channel current and further divided by the cell surface area. Neuroblastoma cells have a surface which is much larger than what would be expected from the dimensions of the cell. Figure 4 shows the results of an experiment to determine the surface area. The cell was patch clamped in the whole-cell configuration. The patch pipette was used to record the membrane potential and inject a constant current pulse. The time constant for development of the potential was determined. The time constant (t) is given as $t = 1/rc$, where r is the membrane resistance and c is the membrane capacitance. r can be determined as $r = v/i$, where v is the voltage change generated by the application of current i. In the example experiment, $t =$

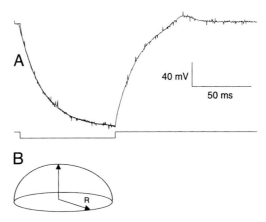

FIG. 4   Estimate of the area of the surface membrane from cell capacitance. (A) The membrane potential was measured from a neuroblastoma cell (top trace) while the constant current step was applied (bottom trace). The time course of charging of the cell membrane could be fit to a single exponential with a time constant of 22 msec. (B) The geometry used to calculate the expected surface area of the cell is shown.

22 ms, and $r = 175$ MΩ, giving $c = 1.25$ nF. Since cell membranes typically have a specific capacitance of 1 μF/cm^2, the cell would have an actual surface area of about 10^{-3} cm^2. This cell had an actual radius (R) of 12.5 μm. The maximum surface area (SA) would be given by the geometry shown in Fig. 4B, which considers a hemisphere with base. In this case, SA $= 2 \pi R^2 + \pi R^2$. For this cell, the predicted SA $= 1.5 \times 10^{-5}$ cm^2. The actual surface area is over 60$\times$ the area from a simple consideration of the geometry of the cell, suggesting that the surface membrane is highly invaginated.

Effects of Differentiation of Neuroblastoma Cells

My laboratory has routinely utilized cells grown under conditions which promote differentiation so as to reduce the variability between experiments. However, how these cells compare to dividing cells is not well known. I have used the whole-cell patch-clamp technique to investigate the changes in Na and K currents in N1E-115 cells following inhibition of cell division. Figure 5 shows the time course of cell growth with control media containing 5% serum, called growth media (GM), and in media with reduced serum and added DMSO, called differentiating media (DM). Similar results were obtained by Kimhi *et al.* (1976). Also shown is the number of cells which

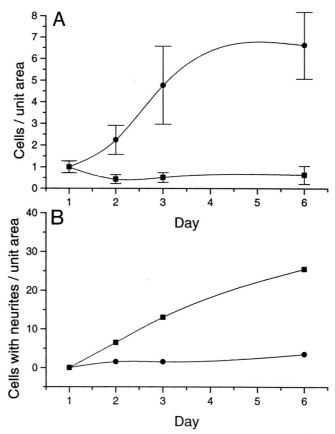

FIG. 5   Time course of growth and differentiation of neuroblastoma cells following plating. The density of cells was measured following initial plating, when grown in normal growth media (5% fetal bovine serum, circles) or in media having 1% serum and 2% DMSO (DM, squares). (A) The total cell count per unit area normalized to the maximum. (B) The number of cells exhibiting neurites. DM suppressed cell division and increased the fraction of cells exhibiting morphological differentiation.

bear neurites following plating of the cells under the two conditions. Note that cell division was effectively arrested in DM, although the number of cells which exhibit morphological differentiation increased.

Sodium and potassium currents measured from cells grown under each condition and having the standard morphology (type c in Fig. 2) and similar size (28 to 33 μm) were compared. Three differences were revealed. First, maximum Na currents were present in a larger portion of cells. Second, maximum K currents were larger in all cells. Third, the transient K current had a larger representation. Sodium current–voltage curves recorded from cells grown in DM showed a maximum Na current of 7.3 nA ± 3.0 nA (avg ± SD), $n = 6$, while cells grown in GM had an average of 4.4 nA ± 2.9 nA, $n = 6$. The Na current was then more variable for cells grown in GM and some cells had Na currents as large as that found for cells grown in DM.

Cells grown in DM also had a larger K current than those grown in GM. Figure 6A shows representative K currents grown under the two conditions. Note that the Na current is the same size in these cells, but the peak K current at any voltage is different. Potassium current of neurons grown in DM had a different waveform, since the current exhibited a component which inactivated rapidly following a step depolarization. N1E-115 cells have two voltage-dependent K channels having different rates of inactivation (Quandt, 1988a). The rapidly inactivating component of outward current was present in cells grown in DM. In Fig. 6A at the largest depolarization for the neuron grown in DM, approximately half of the current was associated with a decay having a time constant of 42 msec. The remaining current had a decay time constant of 1.3 sec. Only a long time constant is seen in the neuron grown in GM. The change in waveform can then be attributed to the increase in expression of a transient type of K channel after differentiation. Average current–voltage curves recorded from the two groups of cells are plotted in Fig. 6B. An increase in total K current upon differentiation has also been reported for N_2AB-1 neuroblastoma cells by Smith-Maxwell et al. (1991).

Single-Channel Recording

Single-channel recording can be accomplished from membrane patches on intact neurons, inside-out, or outside-out patches. The details of recording and electrode fabrication are similar to those for measuring currents from the whole cell. For single-channel recording, we use fire-polished pipettes with a reduced opening diameter of 1 μm. The electrodes are coated with red GLPT insulating varnish (GC electronics 10-9002) following fire polishing in order to reduce the capacitance between the pipette and the bath. In practice, this coating can reduce the rms noise at a bandwith of 1 kHz by a factor of at least 3. We paint the electrodes with an ordinary wire as close as possible to the tip, while looking under a magnifying light (Luxo). The varnish is easier to work with than Sylgard since it does not readily flow into the tip, and it becomes dry in air within a few minutes. Coating is not important in whole-cell clamp experiments, since currents are in the

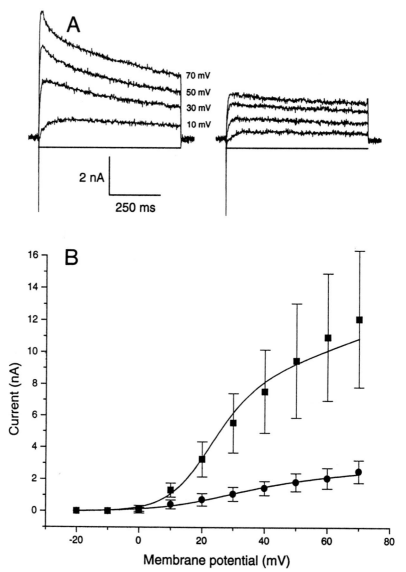

FIG. 6   Effect of differentiation on K currents in neuroblastoma cells. Whole-cell K currents were recorded from round, isolated cells having a diameter of 28 to 33 μm from matched cultures grown in normal growth media (GM) or in media containing 2% DMSO with 1% serum (DM). (A) Potassium currents are shown as recorded from cells grown in DM (left) or in GM (right). (B) Average current–voltage curves are shown for cells grown in DM (squares) and GM (circles). Data are included from five cells in each growth condition. The equations plot the Boltzman equation given in the legend to Fig. 3, with $E = -80$ mV, $G = 72$ mmho (DM) or 16 mmho (GM), $k = -7.1$ (DM) or -11.6 (GM), and $V_h = +21$ mV (DM) or $+24$ mV (GM). Normal saline external, potassium glutamate internal solutions. Temperature 22°C.

nanoampere range and noise generated by the pipette capacitance is relatively small.

Cells used for single-channel recording can theoretically be large and irregularly shaped with neurites. However, we have not investigated whether these cells vary in any systematic way from the cells used for the whole-cell clamp procedure. The pipette and bath typically contain normal saline for intact recording situations. For excised membranes, the concentration of the permeant ion can be increased to increase the size of single-channel currents. For inside-out excised patches of membranes, we obtain the seal between the pipette and the cell with normal saline in the bath. The bath solution is then changed to internal saline, and the membrane is rapidly separated from the cell. It should be noted that the bath potential can be altered during the change in solution, unless measures are taken (see above). For outside-out excised membranes, the pipette contains internal saline. After a seal is obtained, the membrane is ruptured under the pipette as done for recording from the whole cell. The holding current typically becomes more positive once the membrane is broken. The holding potential should then be made negative to bring the holding current to zero. The pipette is then slowly backed away from the cell. A strand of membrane can often be observed as it is extracted. Be sure to pull the electrode completely clear of the cell to eliminate any connection with the cell. Current in response to a 10-mV pulse can be monitored to ensure that the membrane reforms. We find that minimizing the holding current during this process increases the success of this procedure. Typically, outside-out membranes appear to be larger than inside-out membranes, judged from the number of channels for a given pipette tip diameter.

Membrane currents are filtered using an eighth-order Bessel filter (Frequency Devices 902 LPF). Low temperatures (5 to 20°C) are used to facilitate measurements of channel gating. Because single-channel analysis requires a large number of opening events for analysis, repetitive pulse stimuli are usually applied. Data for one patch can easily reach 5 megabytes. After first storing the data on a conventional hard disk, data are transferred to an optical disk drive which provides relatively low-cost archiving (e.g., Panasonic LF-5010/5014 for IBM compatibles). Membrane current can be recorded using a PCM-VCR recording system when continuous measurement is required. Sodium and potassium channels inactivate and the probability of opening is very small or zero during a maintained depolarization. However, Na channels will burst following removal of fast inactivation (Quandt, 1987).

One particular problem associated with recording single-channel currents is the rather large capacitive current following the depolarization and repolarization. This current can obscure the measurement of single-channel events. A number of techniques can be used to remove the capacitive

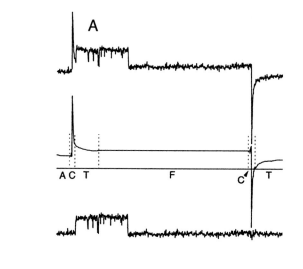

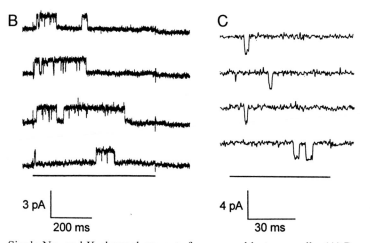

FIG. 7 Single Na- and K-channel currents from neuroblastoma cells. (A) Procedure for elimination of capacitative and time-independent currents. (Top) Original record. Capacitative currents mark the beginning and end of a depolarization. (Middle) A waveform for the capacitative and time-independent currents was obtained by averaging (A), tracing (T), or copying (C) segments of an original recording. (Bottom) The middle waveform was subtracted from the top record. (B) Typical single K-channel currents (upward) recorded from an inside-out patch are shown in response to a depolarization to +10 mV. Normal saline external, potassium glutamate internal solution. Temperature 14.6°C. (C) Typical single Na-channel currents are shown, as recorded from an inside-out patch in response to a depolarization to −60 mV from a holding potential of −120 mV. External solution (mM): 250 NaCl, 10 HEPES, 2 CaCl$_2$. Internal solution (mM): 230 cesium glutamate, 10 HEPES, 20 EGTA. Temperature 15.8°C.

current, including the subtraction of the average of traces not containing channel openings. Often these are not available. I have implemented a novel technique which does not require traces without openings. The technique involves using a mouse to trace the waveform. This tracing is used to construct a second waveform containing only capacitative and time-independent currents for purposes of subtraction. The procedure is diagrammed in Fig. 7A. One thousand twenty-four points of the 4K point recording are displayed on the computer monitor. The recording is traced using the mouse to obtain the values of the idealized waveform without opening events. Time segments of the waveform during the initial part of the capacitative transients can just be copied to the idealized waveform if no openings are present. One point on the waveform also can be used to fill a time segment, or the segment can be filled with an average value of a segment not having openings. The resulting 1K point tracing is then expanded to 4K by interpolation. Following subtraction it is sometimes necessary to correct the interpolated waveform by adding values which will bring the current for small time segments at the beginning of the transients to zero. The procedure is fast and reliable. Example single K- and Na-channel currents in response to step depolarizations are shown in Figs. 7B and 7C.

Acknowledgments

I thank Dr. T. Narahashi in whose laboratory my experiments using neuroblastoma cells were initiated when I was a postdoctoral fellow. Support for my studies using neuroblastoma cells has been provided by the Medical Research Council of Canada and the National Multiple Sclerosis Society (U.S.A.).

References

1. V. Bolotina, J. Borecky, V. Vlachova, M. Baudysova, and F. Vyskocil, *Neurosci. Lett.* **77**, 298–302 (1987).
2. W. A. Catterall, *J. Neurosci.* **1**, 777–783 (1981).
3. O. P. Hamill, A. Marty, E. Neher, B. Sakmann, and F. J. Sigworth, *Pfluegers Arch.* **391**, 85–100 (1981).
4. L-Y. Huang, N. Moran, and G. Ehernstein, *Proc. Natl. Acad. Sci. USA* **79**, 2082–2085 (1982).
5. A. Grinvald, W. N. Ross, and I. Farber, *Proc. Natl. Acad. Sci. USA* **73**, 2424–2428 (1981).
6. L. C. Falke and S. Misler, *Proc. Natl. Acad. Sci. USA* **86**, 3919–3923 (1989).
7. J. M. Fernandez, A. P. Fox, and S. Krasne, *J. Physiol.* **356**, 565–585 (1984).
8. W. B. Im and F. N. Quandt, *J. Membr. Biol.* **130**, 115–124 (1992).

9. Y. Kimhi, C. Palfrey, I. Spector, Y. Barak, and U. Z. Littauer, *Proc. Natl. Acad. Sci. USA* **73,** 462–466 (1976).

10. P. G. Kostyuk, O. A. Krishtal, and Y. A. Shakhovalov, *J. Physiol.* **270,** 545–568 (1977).

11. P. G. Kostyuk, O. A. Krishtal, V. I. Pidoplichko, and N. S. Veselovsky, *Neuroscience* **3,** 327–332 (1978).

12. K. S. Lee, N. Akaike, and A. M. Brown, *J. Neurosci. Methods* **2,** 51–78 (1980).

13. N. Matsuki, F. N. Quandt, R. E. Ten Eick, and J. Z. Yeh, *J. Pharmacol. Exp. Therap.* **228,** 523–529 (1983).

14. F. A. McMorris, P. G. Nelson, and H. Ruddle, Eds. *Neurosci. Res. Bull.* **11,** 412–536 (1974).

15. W. H. Moolenaar and I. Spector, *Science* **196,** 331–333 (1977).

16. W. H. Moolenaar and I. Spector, *J. Physiol.* **278,** 265–286 (1978).

17. W. H. Moolenaar and I. Spector, *J. Physiol.* **292,** 307–323 (1979).

18. T. Narahashi, A. Tsunoo, and M. Yoshii, *J. Physiol.* **383,** 231–249 (1987).

19. F. N. Quandt and T. Narahashi, *Proc. Natl. Acad. Sci. USA* **79,** 6732–6736 (1982).

20. F. N. Quandt and T. Narahashi, *Neuroscience* **13,** 249–262 (1984).

21. F. N. Quandt, J. Z. Yeh, and T. Narahashi, *Neurosci. Lett.* **54,** 77–83 (1985).

22. F. N. Quandt, *J. Physiol.* **392,** 563–585 (1987).

23. F. N. Quandt, *J. Physiol.* **395,** 401–418 (1988a).

24. F. N. Quandt, *Mol. Pharmacol.* **34,** 557–565 (1988b).

25. F. N. Quandt, *Soc. Neurosci. Abstracts* **17,** 1520 (1991).

26. F. N. Quandt and W. B. Im, *J. Pharmacol. Exp. Therap.* **260,** 1379–1385 (1992).

27. C. J. Smith-Maxwell, R. A. Eatcock, and T. Begenishich, *J. Neurobiol.* **21,** 71–77 (1991).

28. I. Spector, *in* Nelson P. G. and Lieberman, M. eds. "Excitable Cells in Tissue Culture," pp. 247–277, Plenum, New York, 1981.

29. J. M. Tang, J. Wang, F. N. Quandt, and R. S. Eisenberg, *Pfluegers Arch.* **416,** 347–350 (1990).

30. G. Yellen, *Nature* **296,** 357–359 (1982).

[2] Sodium Channels of Rat Dorsal Root Ganglion Neurons

Mary Louise Roy and Toshio Narahashi

Introduction

Dorsal root ganglion (DRG) neuron preparations isolated from avian, reptilian, amphibian, and mammalian species have been widely used by neuroscientists in their studies of a variety of research questions. Using mammalian DRG neurons, electrophysiologists and molecular biologists have studied voltage-dependent ion channels, including sodium (1–12), calcium (1, 13–16), chloride (17, 18) and potassium (19, 20), as well as ion channels activated by neurotransmitters such as GABA (17, 21–27), glycine (28), and adenosine triphosphate (ATP) (29). As the isolation and culture procedures of rat DRG neurons are relatively easy to master and can be tailored to various rat ages, this preparation is ideal for many different research projects.

Mammalian DRG neuronal cell bodies are clustered in connective tissue and embedded between the spinal column vertebrae on either side of the spinal cord. DRG neurons receive sensory input from mechanoreceptors, free nerve endings, nociceptors, and other peripheral sensory transducers. Information is subsequently transmitted by DRG axonal processes which give rise to a variety of ascending nerve tracts, such as the spinothalamic and proprioceptive tracts. The DRG axons ultimately make synaptic contact with nuclei of the dorsal column (30) or thalamus, at which point the information is transmitted to the cerebral cortex for further processing. Thus, DRG neurons serve as important centers of communication between the peripheral and central nervous systems. The study of the physiological properties of these ganglia, as well as their responses to pharmacological and toxicological agents, is of great medical relevance.

The aim of this chapter is twofold: (i) to outline basic procedures of rat DRG neuronal isolation and culture and (ii) to briefly discuss, by way of example, some of the biophysical data on the DRG neuronal sodium channel populations.

Dorsal Root Ganglion Neuronal Preparations

Two different DRG neuronal culture procedures have been utilized in our laboratory, each of which has been tailored to the specific interests of the investigator. The "acute" preparation, in which DRG neurons are used on the day of culture, has proved useful in our electrophysiological studies of rat DRG sodium channels, as the neuronal preparations have a smooth surface and spherical shape and are easily voltage clamped. The "primary" culture, in which DRG neurons are isolated and maintained in culture for experiments over a 2-week period, generates a population of DRG neurons with extensive dendritic and axonal processes and synapses. This preparation has successfully generated cells for researchers interested in neurotransmitter-activated ion channels such as those of GABA and glycine.

The preparations of both acutely dissociated and primary cultured rat DRG neurons require three major steps: (i) isolation of ganglia from an anesthetized Sprague–Dawley rat(s), (ii) enzymatic dissociation of neuronal cell bodies from encapsulating support tissue, and (iii) trituration and plating of isolated neurons onto poly-L-lysine-coated coverslips immersed in supplemented Dulbecco's modified Eagle medium (DMEM). Each of these steps is detailed in the following sections.

Materials and Solutions

The following sterile items are required for acute and primary DRG isolation, enzyme treatment, and plating:

> Five, 5-in. Pasteur pipettes.
> One 24-well cluster polystyrene culture plate (Cat No. 3047 (Falcon), Becton–Dickinson Labware, Lincoln Park, NJ).
> Twenty-four 12-mm glass coverslips (Cat No. 60-4888-12, PGC Scientifics, Gaithersburg, MD).
> One 35 × 10-mm plastic petri dish (Falcon 3801, Becton–Dickinson Labware, Oxnard, CA).
> One 15-ml polystyrene centrifuge tube (Cat No. 25310-15, Corning Incorporated, Corning, NY).
> One 50-ml plastic centrifuge tube (Cat No. 25339-50, Corning Incorporated, Corning, NY).
> One pair of dissecting scissors.
> One pair fine forceps.
> One pair general purpose forceps.
> Dissecting microscope.
> Dissecting pins.
> Dissection tray or Slygard-coated 60 × 15-mm petri dish (Falcon 1007, Becton–Dickinson Labware, Lincoln Park, NJ) suitable for dissecting pin insertion.
> Shaking water bath (36°C).

All solutions are to be prepared under a laminar flow hood to maximize sterile conditions. Both culture media and CMF-PBS solutions are sterile filtered using 250- or 500-ml Nalgene 0.22 μm membrane filter units. Enzyme solutions are not filtered prior to use.

Culture Media

Add 1.0 ml gentamycin (40 mg/ml) to 500 ml DMEM, containing D-glucose, L-glutamine, HEPES buffer, and sodium pyruvate, as prepared by GIBCO Laboratories (Cat No. 380-2320AJ, Grand Island, NY) and sterile filter.

Calcium- and Magnesium-Free Phosphate-Buffered Saline Solution (CMF-PBS)

Add 5 g D-glucose to 500 ml CMF-PBS and sterile filter.

Trypsin Solution

Add 10 mg of type XI trypsin from bovine pancreas (Cat No. T-1005, Sigma Chemical Company, St. Louis, MO) to 5 ml CMF-PBS with glucose, for a final concentration of 2.0 mg/ml.

Papain Solution

Add 0.8 ml papain (10 mg/ml) suspension (Boehringer–Mannheim GmbH, Germany) to 5 ml CMF-PBS with glucose. The solution should be initially cloudy, but will become clear and be suitable for DRG treatment about 10 min after placement into a 36°C water bath.

Media and Cluster Preparations

Prior to the dissection of the DRG from the rat, it is common practice to take the tissue culture 24-well cluster with poly-L-lysine-coated 12-mm coverslips (see Appendix), and to add 0.5 ml culture media to each well. The cluster is then placed into the incubator (5–10% CO_2, 100% humidity, 36–37°C), to allow for gas and temperature equilibration. For this reason, culture medium (50 ml) to be used in trituration is pipetted into a centrifuge tube and similarly placed in the incubator. Sterile techniques are strongly advised throughout the culture procedures.

Isolation of Dorsal Root Ganglion Neurons

On each experimental day of the acute DRG preparation, a single 3- to 12-day-old or adult Sprague–Dawley rat is anesthetized with sodium pentobarbital (ip, 0.4 ml at 3 mg/ml). The primary culture, which is done weekly, requires the use of four or five 1- to 2-day-old rat pups, which are similarly anesthetized.

1. After each rat exhibits no response to a firm pinch with forceps and/or breathing has ceased, a swath of skin 1 cm wide is cut from the tail to the base of the skull on the dorsal side of the animal and is moved away from the spinal column.

2. Using fine scissors, the ribs are cut through along either side of the spinal column. Once at the base of the skull, the spinal column is cut such that the cervical end of the vertebra is freed.

3. With general-use forceps, the column is lifted in an arch away from the body and the mesentery and kidneys attached to the ventral side of the column are carefully removed from the column as one moves posteriorly toward the tail.

4. After cutting through the lumbar vertebra near the tail, the vertebral column is removed and placed into a Sylgard-coated petri dish containing sterile CMF-PBS and glucose.

5. With the ventral side of the column up, a longitudinal cut is made through the ventral wall along the vertebral column, exposing the spinal cord.

6. The column is then turned over to expose the dorsal side, and a similar longitudinal cut is made, resulting in two hemisections.

7. Removal of the spinal cord with fine forceps allows the dorsal root ganglia to be visualized under a dissecting microscope (Fig. 1A).

8. Using fine forceps, the ganglia are carefully plucked from between the vertebrae of the spinal column and placed in a culture dish containing CMF-PBS (Fig. 1B). Every attempt is made to avoid shredding the connective tissue surrounding the neurons in order to avoid damage to the neurons. Separation of the vertebra allows greater accessibility to the ganglia.

Developmental Considerations

In younger animals (2- to 3-days old), the ganglia are fragile and look almost transparent. Younger animals typically yield fewer DRG cell body clusters, each of which is much more sensitive to enzymatic treatment and trituration than those in older animals, so the use of ganglia from two or more younger animals is recommended. Older animals (5- to 12-days old) typically have ganglia which are more tolerant to physical manipulation and appear opaque. However, the ganglia located in the lumbar region of these animals are often too embedded in the column for easy removal and thus are left out of the preparation. Long axonal processes are trimmed using fine scissors, as they interfere with trituration procedures. For acute electrophysiological recordings, neurons derived from one older animal are plated over six cluster wells. In the primary culture, neurons isolated from four to five pups are plated over 24 cluster wells.

Adult animal preparations are more difficult to work with, as the spinal column has fully calcified and is thus much harder to cut, and the embedded ganglia have developed extensive processes along the spinal cord.

Initial DRG isolation procedures may be performed on ice, but with practice, ice is no longer needed. The less amount of time spent between rat and culture media, and the fewer temperature changes the ganglia experience *in vitro,* the better the yield and health of the cultured neurons. However, in the primary culture isolations, which require multiple animals and thus more time, the PBS and ganglia are placed on ice, in hopes of enhancing neuronal survival.

Enzyme Treatment of Dorsal Root Ganglia

Once the ganglia are physically removed from the spinal column, it is necessary to enzymatically treat them as the neurons within the ganglia are surrounded by extensive connective tissue. This treatment typically involves the use of at least one protease, which is intended to weaken the overall structure of the connective tissue and thus ultimately to allow for the mechanical separation of the neurons from the tissue. A delicate balance is sought in this technique, as one hopes to weaken the encapsulating tissue without damaging the neurons within the tissue. Several different variables can be adjusted to achieve this balance, such as concentration of protease(s), temperature of incubation, or length of enzyme treatment. We have found that this latter variable can be easily and reproducibly tailored to the age of the rat used.

1. Using a sterile Pasteur pipette, the ganglia are transferred into a 15-ml graduated centrifuge tube containing either trypsin or papain enzyme solution.

2. Ganglia are incubated at 36°C in a shaking water bath for approximately 20 min.

3. After the enzyme treatment has been completed, the enzyme solution is reduced to a volume of 1 ml. To dilute the enzyme concentration to zero, the ganglia are washed three times with 14 ml of culture media which are initially placed in the incubator. Each wash entails addition of media via pipette, gentle shaking of capped centrifuge tube, the settling of ganglia to the bottom of the tube, and then the removal of supernatant with Pasteur

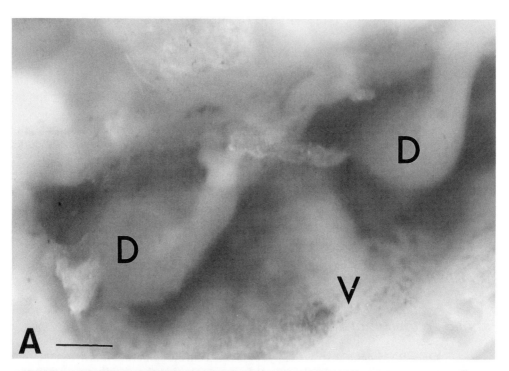

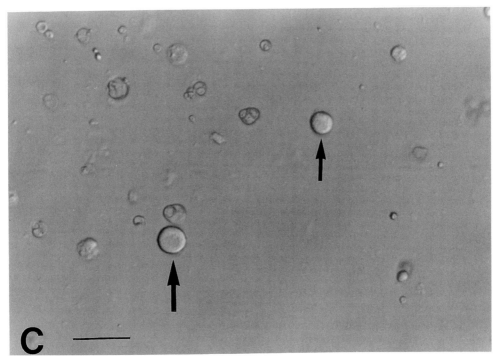

FIG. 1   Eight-day-old rat dorsal root ganglia observed at various stages of acute isolation. (A) Two ganglia with axonal projections are shown as they appear in the spinal column following removal of the spinal cord. (D, ganglia; V, vertebra; Calibration, 500 μm). (B) One isolated ganglion is shown, prior to enzyme treatment. Note the extensive processes radiating away from the spherical cluster of DRG neuronal cell bodies. Calibration, 500 μm. (C) Individual neurons are visualized following enzyme treatment and trituration. A large-diameter neuron is shown in the lower left (arrow), and a small-diameter neuron is in the upper right (arrow). Calibration, 50 μm.

pipette. Centrifugation of ganglia in the wash media (low speed, 1 min) has proved useful in primary culture, which generates a large number of ganglia, but is not used for the fewer ganglia used in the acute procedures.

4. After the third wash, the ganglia are suspended in 1 ml of DMEM for trituration.

As enzyme lots vary, incubation times vary, but the main reasons for caution regarding incubation times are the durability and size of the ganglia themselves. Younger animals (Days 2–4) produce very small and fragile ganglia, resulting in a lower yield of neurons and short incubation times (18–20 min). Older animals (Days 5–12) and adults produce larger and more robust ganglia, requiring longer incubation times (23–40 min) and resulting in a larger yield of neurons. A good indicator of the ideal treatment time is an observation of semitransparent ganglion connective tissue. It is deemed preferable to adjust treatment time rather than temperature.

Trituration and Plating of Dorsal Root Ganglion Neurons

The purpose of trituration is to gently free the neurons from surrounding extraneous support tissue. This is commonly accomplished by moving the ganglia in and out of a series of sterile Pasteur pipettes, the tips of which are heat polished to varying diameters. Good yields are achieved by slightly occluding the pipette tip during trituration by placing it on the rounded bottom of the 15-ml graduated centrifuge tube. The largest numbers of neurons are liberated by using multiple steps of trituration.

1. The first trituration is done using minimal occlusion of the pipette tip and is stopped as soon as the DMEM is cloudy to the eye, usually 10 times in and out of the pipette.
2. The resulting cloudy cell suspension is evenly dispersed into the prepared cluster dish wells containing 0.5 ml culture medium.
3. Culture medium is then added to the remaining ganglia, restoring the volume of the centrifuge tube to 1 ml.
4. Trituration continues with the pipette tip occluded such that ganglia are forced through a smaller opening of the pipette.
5. Again, the resulting cloudy cell suspension is dispersed into the cluster dish wells. This process is usually repeated at least once more, with increasing occlusion of the pipette tip.
6. After the final trituration, the remaining fragments of the ganglia are evenly distributed over the cluster wells. It is conceivable and likely that such tissues may provide the neurons with nutritive and beneficial substances, i.e., growth factors, which maintain or increase the health of the cultured cells.
7. Generally, 1 or 2 hr of incubation (5–10% CO_2, 100% humidity) are necessary to allow the isolated cells to settle and adhere to the coverslips (Fig. 1C).

Additional Considerations

In primary cultures and in acute preparations which are being maintained for more than 12 hr, the addition of fetal bovine serum (FBS) is necessary for neuronal survival. Serum is initially withheld from the culture procedures because the proteins within the serum may compete with the cells for binding sites on the poly-L-lysine-coated slips. Once the neurons are adhered to the coverslips, 0.25 ml of filtered FBS may be added to each well. Within 12 hr of such treatment, DRG processes are seen, such that by 3 days in culture, extensive synaptic networks are established.

Characteristics of Rat Dorsal Root Ganglion Sodium Channels

Tetrodotoxin (TTX) is a natural toxin which has proved effective in the biophysical characterization and molecular identification of sodium-channel populations in numerous preparations (31). In many neuronal and muscle tissues, nanomolar concentrations of TTX are sufficient to abolish all sodium-channel activity, while leaving other ion channels unaffected. In cardiac and denervated muscle tissues, however, micromolar concentrations of TTX are required to similarly block sodium current. TTX-sensitive (TTX-S) and TTX-resistant (TTX-R) sodium channels have been previously reported in mammalian DRG preparations (1–12). Using electrophysiological techniques, our research was designed to extend to the physiological and pharmacological characterizations of these types of sodium channels in the acutely dissociated DRG neurons.

Pipette Preparation

The whole-cell configuration of the patch-clamp technique is used to record whole-cell membrane currents (32). Pipette electrodes for whole-cell recordings are made from borosilicate glass capillary tubes, each with an inner diameter of 0.8–1.0 mm (Kimble Products, Vineland, NJ), by using a series of two pulls on a vertical microelectrode puller (PP-83, Narishige, Tokyo, Japan). Electrode tip resistances of 0.75–2 MΩ are ideal for these studies. Heat polishing of the electrode tip does not significantly enhance seal quality in these studies, so unpolished electrodes are generally used.

Solutions

The external and internal solutions used for electrophysiological recordings are specifically designed to record sodium-channel currents and minimize other ion-channel currents. External solution is composed of (in mM) NaCl, 75; KCl, 5; CaCl$_2$, 1.8; MgCl$_2$, 1; glucose, 55; N-2-hydroxyethylpiperazine-N-2-ethanesulfonic (HEPES) acid, 5.5; Na-HEPES, 4.5; tetramethylammonium chloride (TMA-Cl), 40; with a final pH of 7.4. Internal solution is composed of (in mM): CsF, 110; CsCl, 30; NaF, 10; MgCl$_2$, 2; HEPES-acid, 10; ethylene glycol bis(β-aminoethyl ether) N,N,N',N'-tetraacetic acid (EGTA), 2; Calpain Inhibitor II (Calbiochem, La Jolla, CA), 0.2; with the pH adjusted to 7.3 with 1 M NaOH. To minimize the osmotic stress placed on the neurons during experiments, the osmolarities of internal and external solutions are adjusted to match the osmolarity of the culture media (300 mOsm).

Sealing Procedures

For electrophysiological recordings, coverslips with DRG neurons are individually transferred to a 1.5-ml glass-bottomed Plexiglas perfusion chamber containing external solution which is mounted onto an inverted microscope stage. Each pipette is filled with internal solution, placed into a microelectrode holder, and mounted into the headstage (Axon Instruments, Foster City, CA). Using both gross (Narishige) and fine (Narishige, MO-1C3) micromanipulators, the pipette is placed onto the cell membrane. This positioning is verified visually through a microscope and electronically monitored as an increase in electrode tip resistance by applying for 20 msec 10-mV hyperpolarizing pulses. In many cases, a gigaseal is formed spontaneously between the pipette and cell membrane. When this is not the case, slight negative suction is used to increase seal quality.

Once a gigaseal is established for whole-cell recording, sharp negative suction results in breaking the membrane patch within the electrode, allowing for dialysis of the cell interior with internal solution, as well as the creation of a continuous electrical circuit between pipette and the whole-cell membrane. Using a gravity perfusion system, external solutions are cooled with a Peltier device to 13–18°C and subsequently applied to the chamber. The temperature does not fluctuate more than 1°C during the course of an experiment. All data are compensated for the liquid junction potential between internal and external solutions, which is approximately −5 mV. Series resistance compensation is accomplished by speeding up the decay of the capacitive transient current (33, 34). In each experiment, P-P/4 procedures are used to digitally subtract leakage and capacitive currents (35).

Patch-Clamp Amplifiers

A patch-clamp amplifier circuit designed and built by M. Yoshii in 1984 is used to record the whole-cell currents. Several patch-clamp amplifiers are available commercially. In our laboratory, Axopatch amplifiers (Axopatch

200A, 200, 1-B, and 1-C, Axon Instruments, Foster City, CA) and an EPC-7 amplifier (Adams & List Associates, Great Neck, NY) are also being used. Current flowing through the cell membrane is measured across a 10-MΩ feedback resistor. This current is then converted to a voltage via operational amplifiers and is stored for later analysis via an AD/DA converter onto hard and floppy disk drives using a PDP 11/73 computer.

Pharmacological Agents

Stock solution (1 mM) of TTX is made by dissolving TTX in 2 mM citric acid and kept frozen until use. On the day of experimentation, stock solution is diluted with external solution to give final TTX concentrations of 0.1 nM–100 μM. The 1 mM stock solution of saxitoxin (31) (STX, Calbiochem, La Jolla, CA) is similarly stored and diluted with the external solution. Deltamethrin is dissolved in dimethyl sulfoxide (DMSO) to make 1, 10, and 100 mM stock solutions, which are refrigerated until use. Each stock solution is subsequently dissolved 1000-fold in external solution. This results in a DMSO concentration of 0.1% (v/v) which has no effect on the sodium currents.

Physiology of TTX-S and TTX-R Channels

The sodium current recorded from a neuron consisted of TTX-R and/or TTX-S channel currents in various proportions. Interestingly, there were direct relationships among sodium currents expressed, animal age (Fig. 2), and cell size. Using the acute isolation procedures described above, DRG neurons isolated from younger animals (2–4 days) tended to be of small diameter (<15 μm) and to express a high percentage of TTX-R current. DRG neurons isolated from animals which were 4–8 days of age tended to be of comparable size, but were more likely to express various proportions of both TTX-S and TTX-R currents. DRG neurons obtained from older animals (8–12+ days and adults) tended to be of large diameter cells (>20 μm) and to express a high percentage of TTX-S currents. These current expression patterns as functions of cell size and animals age were effectively used to select cells for biophysical and pharmacological analyses of TTX-S and TTX-R sodium currents.

TTX-S and TTX-R neuronal sodium channels exhibit distinct biophysical properties. This is clearly demonstrated in Fig. 3A, which shows current records from a DRG neuron simultaneously expressing both TTX-S and TTX-R currents. The two types of sodium currents differ in their overall

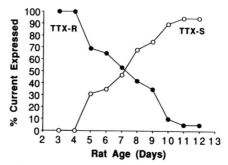

FIG. 2 Developmental expression profiles of TTX-S and TTX-R sodium-channel currents. TTX-R and TTX-S currents are expressed as a function of rat age, as TTX-R current is more likely to be found in neurons isolated from younger animals (3 to 5-days old) and TTX-S currents tend to be expressed by neurons isolated from older animals (8- to 12+-days old). The amplitudes of TTX-S and TTX-R currents plotted as percentages of total current were measured from currents blocked and unaffected by 100 nM TTX, respectively, at each animal age (Day 3, $n = 5$; Day 4, $n = 5$; Day 5, $n = 19$; Day 6, $n = 33$; Day 7, $n = 38$; Day 8, $n = 29$; Day 9, $n = 10$; Day 10, $n = 10$; Day 12+, $n = 4$). Reproduced with permission (6).

waveforms, with the TTX-S-channel activation and inactivation kinetics being much faster than those of the TTX-R channel (Figs. 3B and 3C). Using the steady-state inactivation pulse protocol indicated, the two sodium current components could be separated based on prepulse amplitude (Fig. 3D; $V_{1/2} = -70 \pm 4$ mV for TTX-S and $V_{1/2} = -40 \pm 5$ mV for TTX-R). Thus, at the beginning of each experiment, this steady-state inactivation protocol was used to identify the populations of channels present.

Pharmacology of TTX-S and TTX-R Channels

Based on binding (36) and electrophysiological experiments (31, 37), TTX and STX have been thought to interact with a common region on the extracellular surface of the sodium-channel protein, resulting in the occlusion of sodium ion flow through the channel pore. Like numerous neuronal and muscle preparations, the rat DRG TTX-S sodium channel exhibited nanomolar apparent K_d values for block by TTX and STX (Fig. 4A). In striking contrast to this value, the rat TTX-R sodium channel exhibited an apparent K_d value of 100 μM for both TTX and STX (Fig. 4B). It should be noted that this value, which causes 50% current reduction, is significantly greater than the

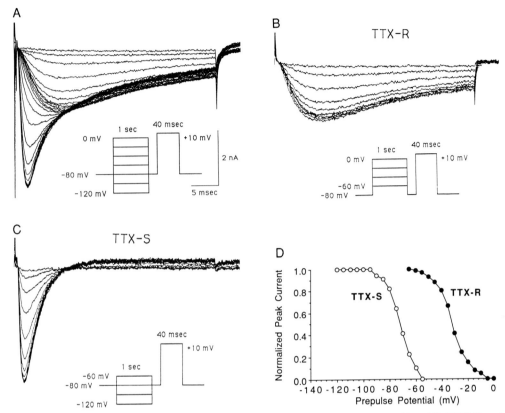

FIG. 3 TTX-S and TTX-R steady-state sodium inactivation profiles. (A) TTX-S and TTX-R currents associated with a 40-msec test pulse to +10 mV following 1-sec prepulses to various potential levels ranging from −120 to 0 mV. Both types of currents are expressed by a single cell. (B) TTX-R currents associated with a 40-msec test pulse to +10 mV following 1-sec prepulses ranging from −60 to 0 mV. (C) TTX-S currents associated with a 40-msec test pulse to +10 mV following 1-sec prepulses ranging from −120 to −60 mV. (D) Peak sodium current plotted as a function of prepulse potential. Current amplitudes were normalized to the maximum current. The potentials at which 50% inactivation occurs are −72.5 and −35.4 mV for TTX-S and TTX-R currents, respectively. Reproduced with permission (6).

1–10 μM TTX K_d values in denervated skeletal and cardiac muscle sodium channels, both of which are completely blocked by higher concentrations of TTX and STX (38, 39). This difference may indicate that the TTX-R channel in DRG neurons differs from that of skeletal and cardiac muscles with respect to the neurotoxin binding site.

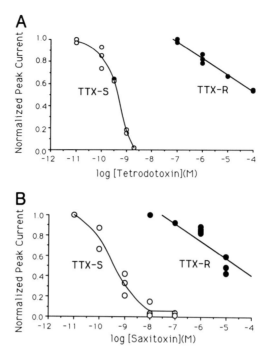

FIG. 4 Dose–response relationships for TTX and STX block of sodium currents. Cells expressing only TTX-S or TTX-R currents were pulsed once per minute to +10 mV to determine peak current amplitude in the presence of increasing concentrations of TTX and STX. The steady-state peak current amplitudes reached at each concentration were normalized to control current amplitudes in toxin-free media and are plotted against toxin concentration. (A) TTX dose–response curves, with K_d values of 0.3 nM (TTX-S) and 100 μM (TTX-R). (B) STX dose–response curves, with K_d values of 0.5 nM (TTX-S) and 10 μM (TTX-R). Reproduced with permission (6).

Pyrethrins, natural compounds derived from the flowers of *Chrysanthemum cinerariaefolium,* and pyrethroids, their synthetic derivatives, have been used as potent insecticides. They prolong the open time of the sodium channel through slowing of the kinetics of activation and inactivation gates, thereby causing hyperactivity in animals (40–43). Deltamethrin, a Type II pyrethroid, exerted differential effects on TTX-S and TTX-R currents (44) (Fig. 5). Although the peak amplitude of TTX-S current was slightly reduced following treatment with 10 μM deltamethrin, the tail current was only negligibly affected. TTX-R current also exhibited a reduction in peak current amplitude, but the tail current was greatly prolonged, such that inward current was still observed over 10 sec following the termination of the pulse.

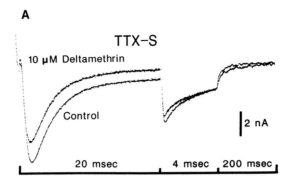

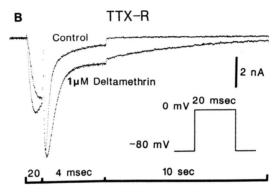

FIG. 5 Effects of deltamethrin on TTX-S and TTX-R sodium currents. (A) A cell expressing TTX-S current was pulsed using the protocol in the inset of (B) in the absence and presence of 10 μM deltamethrin. The amplitudes of peak and tail currents are slightly reduced by deltamethrin. (B) A cell expressing TTX-R current was similarly analyzed in the absence and presence of 1 μM deltamethrin. The TTX-R peak current amplitude is reduced and the tail current is markedly modified. Note the different time scales. Reproduced with permission (44).

The mechanism underlying this profound difference in deltamethrin effects on TTX-S and TTX-R sodium channel is presently unknown. It likely reflects differences in the molecular structure of the channels.

Summary and Conclusions

The procedures described above have proved effective in isolating DRG neuronal cultures for electrophysiological analyses. The above example of experiments shows that the acute dissociation technique can be used for the study of the developmental, biophysical, and pharmacological characteristics of neuronal TTX-S and TTX-R sodium channels. The primary cultured DRG neurons have also been used for the study of ligand-gated channels (17, 21–24). As DRG neurons express a wide range of ion-channel types, this preparation represents one of the most convenient materials for a variety of studies.

Appendix

Preparation of Poly-L-Lysine-Coated Coverslips

1. Glass coverslips (12 mm) are stored in 95% ethanol until time of use.
2. On day of poly-L-lysine coating, two tissue culture clusters are placed in the laminar flow hood.
3. Individual coverslips are removed from the alcohol with forceps, flamed briefly, and placed into cluster wells.
4. Once each of the 48 wells has one sterile coverslip, poly-L-lysine (hydrobromide, Cat. No. P5899, Sigma Chemical Co., St. Louis, MO) solution is prepared. Deionized and distilled water (50 ml) is pipetted directly into the poly-L-lysine bottle, and the bottle is shaken to ensure good mixing.
5. Using a prepackaged sterile graduated pipette, 1 ml of trypsin solution is added to each well.
6. The clusters are then placed in an incubator (35°C, 5–10% CO_2, 100% humidity) for 40 min. Poly-L-lysine will settle and coat the coverslips during this time.
7. Clusters are returned to the hood and the poly-L-lysine solution is aspirated out of the wells.
8. Next, cluster wells are washed three times with filtered deionized and distilled water, by adding 1 ml water to each well and aspirating.
9. The clusters are then dried overnight in the laminar hood, with both blower and ultraviolet light kept on, and the cluster lids are slightly opened to allow exposure to air.
10. The next morning, clusters can be wrapped in aluminum foil and stored in the hood until use.

Acknowledgments

The authors thank Dr. Harry Sontheimer, Yale University Neurology Department, for his assistance in the photography of various stages of the DRG preparation and Vicky James-Houff of Northwestern University Medical School for her unfailing secretarial assistance. This work was supported by NIH Grants F31 MH09839 (M.L.R.) and R01 NS14143 (T.N.).

References

1. P. G. Kostyuk, N. S. Veselovsky, and A. Y. Tsyndrenko, *Neuroscience* **6,** 2423–2430 (1981).
2. M. J. McLean, P. B. Bennett, and R. M. Thomas, *Mol. Cell. Biochem.* **80,** 95–107 (1988).
3. N. Ogata and H. Tatebayashi, *Dev. Brain Res.* **65,** 93–100 (1992).
4. N. Ogata and H. Tatebayashi, *Pfluegers Arch.* **420,** 590–594 (1992).
5. B. R. Ransom and R. W. Holz, *Brain Res.* **136,** 445–453 (1977).
6. M. L. Roy and T. Narahashi, *J. Neurosci.* **12,** 2104–2111 (1992).
7. G. Omri and H. Meiri, *J. Membr. Biol.* **115,** 13–29 (1990).
8. A. Schwartz, Y. Paltie, and H. Meiri, *J. Membr. Biol.* **116,** 117–128 (1990).
9. J. Valmier, M. Simonneau, and S. Boisseau, *Pfluegers Arch.* **414,** 360–368 (1989).
10. S. Yoshida, Y. Matsuda, and A. Samejima, *J. Neurophysiol.* **41,** 1096–1106 (1978).
11. A. A. Elliott and J. R. Elliott, *J. Physiol.* **463,** 39–56 (1993).
12. J. M. Caffrey, D. I. Eng, J. A. Black, S. G. Waxman, and J. D. Kocsis. *Brain Res.* **592,** 283–297 (1992).
13. R. S. Scroggs and A. P. Fox, *J. Neurosci.* **11,** 1334–1346 (1991).
14. R. S. Scroggs, and A. P. Fox, *J. Neurosci.* **12,** 1789–1801 (1992).
15. R. S. Scroggs and A. P. Fox, *J. Physiol.* **445,** 639–658 (1992).
16. C. Yu, P. X. Lin, S. Fitzgerald, and P. Nelson, *J. Neurophysiol.* **67,** 561–575 (1992).
17. O. Arakawa, M. Nakahiro, and T. Narahashi, *Brain Res.* **578,** 275–281 (1992).
18. X. Rabasseda, J. Valmier, Y. Larmet, and M. Simonneau, *Dev. Brain Res.* **51,** 283–286 (1990).
19. T. Amédée, E. Ellie, B. Dupouy, and J. D. Vincent, *J. Physiol.* **441,** 35–56 (1991).
20. S. F. Fan, K. F. Shen, and S. M. Crain, *Brain Res.* **558,** 166–170 (1991).
21. O. Arakawa, M. Nakahiro, and T. Narahashi, *Brain Res.* **551,** 58–63 (1991).
22. M. Nakahiro, J. Z. Yeh, E. Brunner, and T. Narahashi, *FASEB J.* **3,** 1850–1854 (1989).
23. M. Nakahiro, O. Arakawa, and T. Narahashi, *J. Pharmacol. Exp. Therap.* **259,** 235–240 (1991).
24. M. Nishio and T. Narahashi, *Brain Res.* **518,** 283–286 (1990).
25. B. Robertson, *J. Physiol.* **411,** 285–300 (1989).
26. G. White, *J. Neurophysiol.* **64,** 57–63 (1990).
27. J. Y. Ma and T. Narahashi, *Brain Res.* **607,** 222–232 (1993).
28. T. Furuyama *et al., Mol. Brain Res.* **12,** 335–358 (1992).
29. B. P. Bean, C. A. Williams, and P. W. Ceelen, *J. Neurosci.* **10,** 11–19 (1990).
30. R. Giuffrida and A. Rustioni, *J. Comp. Neurol.* **316,** 206–220 (1992).
31. T. Narahashi, *in* "Handbook of Natural Toxins: Marine Toxins and Venoms" (A. T. Tu, ed.), Vol. 3, pp. 185–210. Dekker, New York, 1988.

32. O. P. Hamill, A. Marty, E. Neher, B. Sakmann, and F. J. Sigworth, *Pfluegers Arch.* **391,** 85–100 (1981).
33. A. Marty and E. Neher, *in* "Single Channel Recording" (B. Sakmann and E. Neher, eds.), pp. 107–122. Plenum, New York, 1983.
34. D. R. Matteson and C. Armstrong, *J. Gen. Physiol.* **83,** 371–394 (1984).
35. F. Bezanilla and C. M. Armstrong, *J. Gen. Physiol.* **70,** 549–566 (1977).
36. J. M. Ritchie and R. B. Rogart, *Rev. Physiol. Biochem. Pharmacol.* **79,** 1–50 (1977).
37. T. Narahashi, *Physiol. Rev.* **54,** 813–889 (1974).
38. P. A. Pappone, *J. Physiol.* **306,** 377–410 (1980).
39. C. J. Cohen, B. P. Bean, T. J. Colatsky, and R. W. Tsien, *J. Gen. Physiol.* **78,** 383–411 (1981).
40. G. S. F. Ruigt, *in* "Comprehensive Insect Physiology, Biochemistry and Pharmacology" (G. A. Kerkut and L. I. Gilbert, eds.), pp. 183–263. Pergamon, Oxford, 1984.
41. T. Narahashi, *in* "From Molecule to Organism" (T. Narahashi and J. E. Chambers, eds.), pp. 55–84. Plenum, New York, 1989.
42. H. P. M. Vijverberg, J. van den Bercken, *Crit. Rev. Toxicol.* **21,** 105–126 (1990).
43. T. Narahashi, *Trends Pharmacol. Sci.* **13,** 236–241 (1992).
44. T. Narahashi, J. M. Frey, K. S. Ginsburg, M. L. Roy, *Toxicol. Lett.* **64/65:**429–436 (1992).

[3] Patch-Clamp Techniques for Studying Potassium Currents in Mammalian Brain Neurons

D. James Surmeier, C. J. Wilson, and James Eberwine

Introduction

The study of potassium currents in adult mammalian brain neurons has undergone a profound change in the past few years. With the advent of patch-clamp (Hamill *et al.*, 1981) and acute-dissociation techniques (Kay and Wong, 1986), neurons from virtually every brain region have become accessible to powerful biophysical techniques. This has led to a rapid growth in our understanding of potassium currents in brain neurons. Most of the studies in this burgeoning literature have used the whole-cell variant of the patch-clamp technique. The stated (or unstated) goal of these studies has been to dissect whole-cell currents into unitary components that could be identified with a particular class of channels. Some of the biophysical and pharmacological tools necessary to accomplish this goal are outlined here along with a discussion of some commonly encountered problems, like poor space clamp.

Unfortunately for the physiologist, a rigorous approach is often not enough to ensure that the contribution of a particular channel type will be discriminated at the whole-cell level. Frequently, single channel study is necessary to accomplish this aim. A good example of the problem is found in a series of papers by Hoshi and Aldrich (1989a,b). With whole-cell recording, no clear evidence was found for the existence of multiple potassium channel types. Yet, with single-channel recordings, clear evidence for four different channels was found.

In many situations, there will be no reasonable substituted for single-channel work if an understanding of the channel types contributing to whole-cell behavior is to be the goal. However, there are other tacks on the horizon that will provide a complementary level of molecular information while allowing experimenters the functional perspective afforded by cellular methods. One of these techniques, which is described briefly below, involves combining whole-cell recording with molecular techniques for amplifying cytoplasmic polyadenylated mRNA. This amplified mRNA can be used to generate a "channel profile" for cells that can be reconciled with the concurrently obtained physiological measurements. With the rapid growth in clones of mammalian potassium channels (as well as other channel types), this approach promises to provide an extremely rich source of information about how cells orchestrate ionic mechanisms.

In what follows, we first discuss whole-cell patch-clamp techniques that can be used to characterize potassium currents and then the single-cell mRNA amplification technique. Since the advent of patch-clamp techniques over a decade ago, there have been a number of excellent descriptions of the methodology, starting with Sakmann and Neher's book (1983). There have been several recent descriptions of techniques peculiar to study of potassium currents with whole-cell techniques (e.g., Lancaster, 1992). We do not try to recapitulate these reviews. Our presentation is idiosyncratic in that it revolves on our experience in studying voltage-dependent potassium currents in cultured and acutely isolated neurons from the rat brain (Surmeier *et al.*, 1988, 1989, 1991).

Whole-Cell Patch-Clamp Methodologies

Recording Solutions

The goal in designing internal and external recording solutions is to isolate the conductances of interest without altering their "native" properties. Often, this is a difficult task. Nevertheless, there are some general guidelines that can be followed.

Internal Solutions

Because a very nice review of recording solutions and their properties has been published by Kay (1992), only a brief description is given here. To record outward K^+ currents, the major internal monovalent cation should be potassium, of course. This is not the case if inwardly rectifying currents are of interest. Assuming outward currents are to be characterized, there are situations in which partial replacement of the normal potassium complement (130–150 mM) is desirable to reduce current amplitudes and series resistance errors. Substitutions can be made with impermeant cations such as Tris or N-methyl-D-glucamine but caution should be exercised as these ions may interact with gating processes at the cytoplasmic face of the channel. The role of both extracellular and intracellular potassium in gating is of ongoing research interest (Pongs, 1992).

Choice of the major anion is dictated by the type of preparation. For example, cultured neurons are typically tolerant of anions with high lyotrophic series numbers like Cl^- but acutely dissociated brain neurons will not tolerate concentrations above 50 mM for more than a few minutes. In these neurons, there is a rough correlation between the stability of the recording and lyotrophic number with the longest recordings being obtained with F^-, MeSO4$^-$, and gluconate. One of the complications associated with F^- is that it complexes with trace Al^{3+} to activate G proteins, which could potentially alter current properties.

To minimize "rundown" and enable neuromodulator mechanisms, intracellular solutions should contain adequate levels of ATP (1–2 mM), GTP (0.2 mM), and Mg^{2+} (2–4 mM). In addition, we have found that adding phosphocreatine (10–20 mM), phosphocreatine kinase (50 U/ml), and leupeptin (a neutral protease inhibitor, 0.1–0.2 mM) improves recording longevity and quality (although it is not entirely clear why an ATP-regenerating system is necessary in the presence of 1–2 mM ATP). Although the conventional wisdom is that free Ca^{2+} levels should be buffered with EGTA or BAPTA to less than 50 nM, we have found that extrinsic Ca^{2+} buffers are not always necessary (although it is crucial when using little or no buffer to have HPLC grade water). Moreover, some signaling pathways that depend on fluctuations in internal Ca^{2+} levels are disrupted by buffering strongly to low levels. The role of intracellular Ca^{2+} can be determined by buffering to predetermined levels between nominally zero and about 150 nM; the concentrations of buffer (EGTA or BAPTA), Ca^{2+}, Mg^{2+}, and H^+ necessary to achieve a particular free Ca^{2+} concentration can be determined using programs such as those of Schoenmakers *et al.* (1992).

Recording longevity is enhanced by making the internal solution 10–15% hypoosmolar (265–270 mOsm/liter assuming a 300 mOsm/liter external solution). The pH should be buffered to between 7.2 and 7.4 with 10–15 mM HEPES (PIPES or MOPS buffers also work well but do not block Cl^- currents).

External Solutions

The concentration of K^+ in the extracellular solution will be dictated by several factors. If the characteristics of currents during depolarization are to be the primary focus, the equilibrium potential should be positioned such that driving force for the current is substantial throughout its activation range. This seems obvious but sometimes the reversal potential is pushed toward more depolarized potentials (ca. -70 mV) by recording in relatively high concentrations of K^+ (e.g., 5 mM) without considering that this will significantly reduce the resolvability of currents active near the resting membrane potential of many cells (ca. -70 to -60 mV). If tail currents are to be studied, then the driving force at potentials where deactivation occurs should be the primary consideration. For example, the tails of depolarization-activated currents, like *A* current, are most easily studied in 10–40 m*M* K$^+$. The drawback to recording in elevated K$^+$ is that the kinetics of the tail current may be altered.

Of greater practical importance, perhaps, is the elimination of other current types. The three major classes of conductance that contaminate K$^+$ recordings are carried by Na$^+$, Ca^{2+}, and Cl$^-$. The elimination of Na$^+$ currents is readily accomplished in most brain neurons with tetrodotoxin (0.2–0.4 μM); calcium currents are more problematic. Obviously, the study of Ca-dependent currents requires Ca^{2+} flux; the most compelling characterizations of these potassium currents have used subtraction procedures with channel-specific toxins (e.g., apamin and charybdotoxin). If these currents are not of interest, Ca^{2+} currents can readily be blocked with Cd^{2+} (200–400 μM). At this concentration, Ca^{2+} currents are effectively blocked and charge shielding effects are minimal. Ca^{2+} cannot be completely removed without altering the properties of voltage-dependent channels (Armstrong and Miller, 1990). However, Cd^{2+} may have allosteric effects on the gating of some types of transient current; to determine the extent of this effect, subtraction procedures akin to those used for the Ca-dependent currents should be employed (e.g., with 4-aminopyridine for A currents). Another potential source of contamination comes from Cl$^-$ currents. Although these currents are not widely studied, they appear to be significant in a number of brain neurons (e.g., Stefani *et al.*, 1990; Blatz, 1991). At present, the best strategy for eliminating these currents is to replace extracellular Cl$^-$ with relatively impermeant anions such as MeSO4$^-$, isethionate, or gluconate (intracellular Cl$^-$ should also be eliminated). With the substitution of less-mobile ions for Cl$^-$ in these experiments, care must be taken to compensate for alterations in junctional potentials (see Neher, 1992).

The external solution should also (i) have an osmolarity near 300 mOsm/liter, (ii) have pH buffered to 7.3–7.4 with HEPES (or PIPES or MOPS), (iii) have adequate levels of Mg^{2+} (1–2 m*M*) and glucose (5–10 m*M*).

Biophysical Measurements

Activation

The steady-state voltage dependence of activation for a particular K$^+$ current can be studied in a number of ways. In all of these methods, the most important first step is to determine a condition under which the current can be studied in isolation. Often this can be accomplished by blocking other K$^+$ currents that are active within the same voltage range with pharmacological agents (see Narahashi and Herman, 1992; Lancaster, 1992). Sometimes the most selective blocker is of the current to be studied rather than the extraneous ones; in this case, records taken before and after block can be used to mathematically isolate the current. It should be noted that the block by some ligands is voltage dependent (e.g., 4-aminopyridine and tetraethylammonium at the cytoplasmic face) and their indiscriminate use in subtraction procedures could lead to erroneous conclusions about current kinetics.

Another approach is to use differences in current kinetics and/or inactivation voltage dependence to bring about a separation. For example, in embryonic neostriatal neurons, depolarization from -90 mV activates a rapidly inactivating A current (A_f) and a delayed rectifier current; depolarization from a more positive holding potential (ca. -40 mV) activates only the delayed rectifier current, as the A_f current is inactivated. Therefore, by subtracting the currents evoked by a step depolarization at a holding potential of -40 mV from those evoked from -90 mV, the A_f current can be isolated.

In adult neurons, this strategy does not work because a slowly inactivating K^+ current (A_s) with a similar voltage dependence is also present. However, the A_s current inactivates and recovers from inactivation more slowly than does the fast A_f current. Therefore, the fast A current can be isolated in these cells, by (i) inactivating both currents by holding the membrane potential at -40 or -50 mV, (ii) delivering a short (100-ms) hyperpolarizing step that removes a substantial portion of the inactivation from the fast A current without deinactivating the slower current. Now the fast A current is the only one of the inactivating currents activated by depolarization and the situation becomes like it was in the embryonic neuron. Usually a reasonable degree of isolation can be achieved with a combination of these strategies but, as noted at the outset, there are times when nothing short of single-channel recording will suffice to isolate a current attributable to a single species of channel.

After having defined a condition under which a current can be isolated, the next step is to measure the peak current evoked by a series voltage steps into the appropriate voltage range. In this way, a current–voltage relationship is obtained. From this data, an estimate of the whole-cell conductance as a function of membrane voltage can be made by computing the reversal potential of the channels giving rise to the measured current. In whole-cell recordings, with well-perfused cells, the ionic gradients for all small monovalent cations should be reasonably well defined shortly after seal rupture (Pusch and Neher, 1988). Thus, either the Nernst equation or the Goldman–Hodgkin–Katz equation (see Hille, 1992), the reversal potential can be determined. In most cases, the assumption has been made that the relative permeability of K^+ ions is sufficiently greater than that of Na^+ ions to ignore the contribution of the latter ($P_K/P_{Na} > 50$). However, this may not be true for all channels and some caution should be exercised. Given an estimate of the reversal potential, whole-cell conductance can be estimated from the equation: $G = I/(V_m - V_{rev})$ where V_m is the membrane potential and V_{rev} is the reversal potential for the current.

The conductance estimates can then be plotted as a function of membrane potential and fit with a Boltzmann function of the form

$$G = \frac{G_{max}}{[1 + \exp(-(V_m - V_h)/V_c)]},$$

where G_{max} is the maximal conductance, V_m is the membrane potential, V_h is the half-activation voltage, and V_c is the slope factor (note that the units of V_h and V_c are mV). Although much is frequently made of the meaning of the fitted parameters (e.g., V_c is used to estimate the number of equivalent gating charges required for channel opening), in our opinion, it is best to view the Boltzmann equation as simply a convenient descriptive tool rather than as a way of revealing something about channel mechanisms.

An example of how a component of whole-cell K^+ currents can be isolated and then subjected to this type of analysis is shown in Fig. 1. In these cells there are two inactivating K currents that differ in inactivation kinetics (Surmeier *et al.*, 1991, 1992b). As discussed above, they can be separated using this property. In Fig. 1A, whole-cell currents that were evoked by depolarizing steps from a positive holding potential (-50 mV) are shown. These currents are essentially noninactivating within the time frame of the depolarizing steps. If a brief (50 ms) hyperpolarizing conditioning step to -100 mV is given before these depolarizing steps, the currents change rather dramatically (Fig. 1B). By subtracting these "conditioned" currents from the "unconditioned" currents, a measure of the rapidly recovering A_f currents can be obtained (Fig. 1C). Peak currents were measured and used to estimate whole-cell conductance assuming that the channels were potassium selective ($V_{rev} = -105$ mV). These estimates were normalized by the maximum conductance and plotted. The plotted points were fit with a least-squares criterion procedure with the Boltzmann function defined above. The solid line is the fit and the fitted parameters are shown in the inset. It is important that the voltages spanning the *entire* activation range of the current be probed; often, it is simply assumed that the current will saturate above 0 mV without stepping into this potential range. This is one way in which the adequacy of the subtraction procedure (or whatever strategy is used) can be assessed.

If a Hodgkin–Huxley formalism is desired, there are some changes in the analysis that are necessary. In this type of description, the conductance estimate is given by a product of the maximal conductance and two time-dependent parameters related to the activation and inactivation process; thus,

$$I = m^n h G_{max}(V_m - V_{rev}),$$

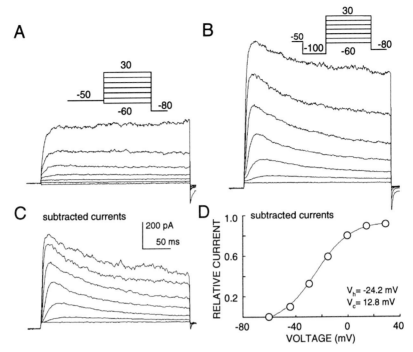

FIG. 1   (A) Whole-cell K^+ currents evoked by depolarizing voltage steps from a holding potential of -50 mV. Inset shows the voltage protocol. (B) The whole-cell currents evoked by the same series of step depolarizations when preceded by a 100-msec conditioning step to -100 mV. (C) Difference currents estimated by subtracting the records in A from those in B. (D) Plot of the peak conductance estimates derived from the currents in C as a function of membrane potential. The data were normalized by the maximum conductance and fit with a Boltzmann function (as described in the text). V_h and V_c are shown in the inset. Recordings were obtained from an acutely dissociated, P32 rat striatal neuron; the extracellular K^+ concentration was 2 mM with Cl^- replaced with isethionate; the principal salt in the internal solution was KF (120 mM) supplemented as described in the text.

where m is the activation parameter, n is an integer (usually 2–3), h is the inactivation parameter. If it is assumed that activation kinetics are much faster than inactivation kinetics (which is not always safe), then we can assume that $h = 1$ near the peak of the current following a step depolarization. Then,

$$m^3 G_{max} = \frac{I}{(V_m - V_{rev})} = G,$$

where G is as defined above. An estimate of the steady-state value of $m[m(\infty, V_m)]$ can be obtained by fitting the conductance estimates with a modified Boltzmann equation of the form

$$m(\infty, V_m) = \left[\frac{G_{max}}{[1 + \exp(-(V_m - V_h)/V_c)]}\right]^{1/n}$$

where m is shown as a function of time and voltage. The value of n can be estimated independently by modeling the activation kinetics (see Hille, 1992). Assuming a square power rule for the currents in Fig. 1, the V_h and V_c were estimated to be -14.1 and 10.9 mV; note how these values differ from those obtained from a model where $n = 1$ (as above). A full discussion of the pros and cons of this procedure is beyond our scope. A more complete treatment of HH models, along with a discussion of alternatives, can be found in Hille's book (1992).

Tail current measurements can also be used to estimate the voltage dependence of activation. Rather than measuring the amplitude of the current evoked by the depolarizing step, the amplitude of the current tail is measured after repolarizing to some fixed membrane potential outside of the activation range of the relevant currents. The traditional conceptual model of these currents is that they reflect ''deactivation'' of channels opened by the preceding step depolarization. Because channel closure from the open state in the original Hodgkin–Huxley model required movement of a single gating particle, deactivation has been thought to be monoexponential for each channel type. As a consequence, tail currents with multiple exponential components have been taken as evidence that more than a single species of channel was contributing to the current.

Although this interpretation may still be accurate for noninactivating currents, it does not appear to be an accurate model for inactivating currents, like the A current (Ruppesburg *et al.*, 1991; Pongs, 1992). Depolarization of these channels produces a sequence of state transitions from closed to open to inactivated or from closed to inactivated (bypassing the open state). Hyperpolarization following a strong depolarizing step changes the transition probabilities such that open channels move to the closed state (deactivation) and inactivated channels move to the closed state (deinactivation). It had been assumed that the inactivated to closed state transitions was direct and did not involve transit through the open state (thus, this transition was *not* manifested in the tail current). Ruppesburg *et al.* (1991) have shown this assumption to be incorrect for Shaker A-type channels expressed in *Xenopus* oocytes. What this means is that if a significant proportion of channels have reached the inactivated state during a conditioning depolarization, tail currents arising from a single-channel type can have multiple exponential components: one reflecting deactivation (movement from the open state), the other de*in*activation (movement from the inactivated state).

In principle, one way of dissecting these different tail current components is to vary the duration of the conditioning depolarization that precedes the tail-eliciting step. As the conditioning depolarization is lengthened and inactivation progresses, the tail current component attributable to deinactivation should grow (as the current evoked by the condition step *declines*). If the inactivation kinetics of the currents of interest are sufficiently slower than their activation kinetics then *short* voltage steps that activate each current component without producing a significant degree of inactivation should be used to elicit currents in the tail protocol. In this situation, tail currents can be fitted with a single exponential function or a sum of exponentials and the amplitude of the current at the time of the voltage step determined by extrapolation. Because this extrapolated amplitude is proportional to conductance (the driving force is constant), the tail currents can be plotted and fit with a Boltzmann function, as described above. The beauty of this approach is that no assumptions need to be made about reversal potentials or relative permeabilities. The voltage to which the membrane is stepped should be chosen to maximize current amplitude and the resolvability of the tails. An example of the application of this strategy is shown in Fig. 2. The test steps are short—not allowing inactivation to develop significantly—and the repolarizing step is negative enough to produce significant inward currents but not so hyperpolarized so as to make the tails so fast that they are difficult to measure accurately. Because it is the tail and not the step preceding it that is to be analyzed, it is best to maximize the driving forces on the tails by recording in elevated external K^+ (5–40 mM). In Fig. 2B, the extrapolated tail current (from a two-exponential model) is plotted as a function of membrane potential during the preceding step; these points are fitted reasonably well with a single Boltzmann function. Obviously, this goodness of fit is not a good criterion to use to determine whether the currents are attributable to a single-channel type. The extrapolated amplitudes of each exponential component peeled from the decay (see inset) are also plotted. The amplitudes of both the slow (squares) and the fast (closed circles) components could be fit with Boltzmann functions that differed only a few millivolts in half-activation voltage from each other (and that of the summed tail currents).

One commonly encountered problem in the analysis of tail currents is that they are inadequately resolved from the membrane-charging transient. Assuming that the membrane resistance is much larger than the electrode across resistance, the charging time constant is approximately equal to R_sC_m, where R_s is the lumped access resistance to the cell and C_m is the membrane capacitance (see Armstrong and Gilly, 1992). The access resistance is not

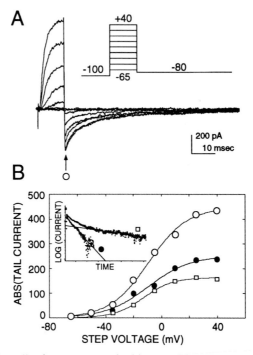

FIG. 2   (A) Whole-cell K^+ currents evoked by a 10-msec step depolarization to -10 mV following conditioning steps to -40 mV of increasing duration. The voltage protocol is shown in the inset. (B) Plot of extrapolated total tail current (open circles) as a function of membrane voltage during the preceding step; the data were fit with a Boltzmann function (as described in the text); $V_h = -9.5$ mV, $V_c = 13.8$ mV. Also shown are the extrapolated tail current amplitudes for the fast (closed circles) and slow (open squares) components of the tail. Boltzmann fits yielded parameter estimates (V_h, V_c) of (-12.6 mV, 10.4 mV) and (-9.5 mV, 14.8 mV), respectively. Inset shows the two components for the tail evoked by the step to $+40$ mV. Recordings were obtained from an acutely dissociated, P28 rat striatal neuron; the extracellular K^+ concentration was 20 mM with Cl^- replaced with isethionate; the principal salt in the internal solution was KF (120 mM) supplemented as described in the text.

equal to the electrode resistance measured in the bath; it normally is two or three times this value. The time constant can be estimated by applying a small voltage step in a passive potential range and measuring the decay time constant. Although for small cells with small C_m values (ca. 5–8 pF) and reasonable access resistances (ca. 10 MΩ), this time constant is small (<100 μsec), for larger cells (50–100 pF) the potassium current deactivation time constants and charging time constants can be of the same order of magnitude.

Adequate series resistance compensation and minimization of extraneous capacitance associated with the electrode (through minimizing bath depth and Sylgarding electrodes) is absolutely necessary for measurements to be interpretable. The adequacy of compensation should be monitored routinely throughout the experiment as the access resistance commonly increases spontaneously. This is more of a problem with smaller cells and smaller electrode tips than with larger cells, in our hands. Small negative pressure at the back of the electrode can be used to reduce the chances that such an increase will occur spontaneously.

Inactivation

An adequate characterization of the inactivation of a current or collection of currents at the whole-cell level requires both kinetic and steady-state information. The first step in the analysis should be a description of the rate at which inactivation develops or is removed (deinactivation) as a function of transmembrane voltage. In the past, these processes have been assumed to be monoexponential for individual currents; although there is convincing evidence that this is an oversimplification (Kasai *et al.*, 1986; MacFarlane and Cooper, 1991), it is a reasonable starting point for our discussion.

The most commonly employed strategy for measuring the development of inactivation is to fit a sum of exponential functions to the current decay following a step depolarization. Most commercially available exponential fitting routines (e.g., those of pCLAMP) use a genearalized exponential model (e.g., $I = A_0 + A_1 e^{-t/\tau 1} + A_1 e^{-t/\tau 2}$ ...). A more traditional approach is to "peel" exponentials by (i) plotting the decay semilogarithmically, (ii) fitting the slow component with a straight line (whose slope is the time constant of the exponential), (iii) subtracting or peeling this component from the original trace, and (iv) fitting the residual with a straight line (or repeating steps 2 and 3 until a single exponential remains). Although the generalized model is more objective, its usual reliance on minimization of the deviation of the fit from the data makes the fit very sensitive to current noise. This can lead to random fluctuations in the fitted time constants. In these situations, it is important for the experimenter to constrain the range over which the algorithm can search for the best fit; although some programs allow this (e.g., Clampfit ver. 5.5, Kaleidagraph) other programs (e.g., Axograph ver. 1.1) do not.

For conductances that exhibit inactivation at membrane potentials where there is not a substantial evoked current (e.g., A-type currents), this procedure provides only a small part of the kinetic information desired. To measure the development of inactivation in these sub- or near-threshold regions, a three-pulse protocol can be used. First, a long hyperpolarizing prepulse is given to remove inactivation (it is not necessary to *completely* remove

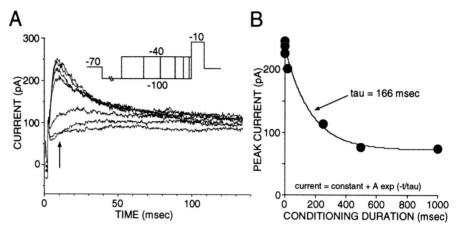

FIG. 3   (A) Whole-cell K^+ currents evoked by a step depolarization to -10 mV following conditioning steps to -40 mV of increasing duration. The voltage protocol is shown in the inset. (B) A plot of peak current as a function of the duration of the conditioning step. The data were fit with a monoexponential decay having a time constant of 166 msec. The cell was cultured from an E17 rat striatum and maintained in culture for 14 days. Recording solutions were as in Fig. 1 except that extracellular K^+ was 5 mM.

inactivation). Next, conditioning steps of variable duration are given prior to a test step that evokes the current of interest. As the conditioning steps increase in duration and inactivation develops, the current(s) evoked by the test step decreases in amplitude. If currents are being separated by fitting exponentials to a decay phase (remember, this makes the assumption that currents can be separated in this way), then the relative amplitudes of each can be plotted as a function of prepulse duration. These plots then can be fit with one or a sum of exponentials to determine the time constant of inactivation development at the conditioning potential. Alternatively, the peak current (or some other point) can be measured, plotted, and fitted to determine whether a single or multi-exponential process is governing decay. An example of this latter variant of the procedure is shown in Fig. 3; here, peak currents were plotted because other results suggested that only the fast A_f current was being evoked.

The recovery from inactivation can be measured with an analogous technique. Instead of delivering a prepulse that removes inactivation, one that inactivates the current of interest is given prior to conditioning pulses of variable duration that remove the inactivation. Figure 4 provides examples of a simple and more complicated recovery time course. In Fig. 4A, the currents evoked by a recovery protocol in an embryonic striatal neuron are shown. Only the A_f current is present and the plot of peak current as a function

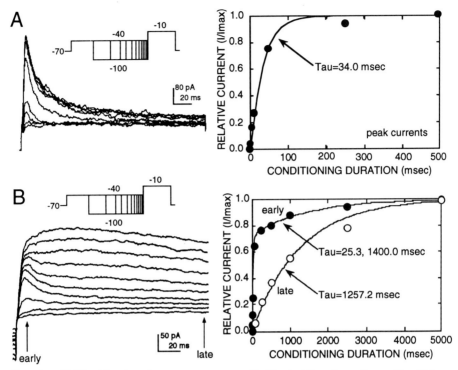

FIG. 4   (A) The kinetics of recovery were studied by holding the cell at -40 mV for 4 sec to inactivate all voltage-sensitive conductances and then stepping to -100 mV for increasing periods of time to remove inactivation prior to the test pulse to -10 mV. The currents evoked by the test pulse are shown at the left. At the right, the normalized peak currents (I'/I'_{max}) are plotted as a function of conditioning duration; $I' = I - I_0$, where I_0 is the current evoked without a hyperpolarizing prepulse. The data points were fit with a recovery function ($I'/I'_{max} = 1 - \exp(-t/\tau)$; $t =$ conditioning duration; $\tau = 34.0$ msec. The cell was cultured from a P1 rat and maintained in culture for 14 days. (B) The kinetics of recovery were studied as in A. The currents evoked by the test pulse are shown at the left. At the right, the normalized peak currents (I'/I'_{max}) (closed circles) and currents at the end of the test pulse (open circles) are plotted as a function of conditioning duration. The peak data were well fit only with a sum of recovery functions ($I'/I'_{max} = \beta(1 - \exp(-t/\tau_1) + (1 - \beta)(1 - \exp(-t/\tau_2))$, $t =$ conditioning duration). $\beta = 0.73$, $\tau_1 = 25.3$ msec, $\tau_2 = 1400.0$ msec; the sustained currents were fit with a single exponential with $\tau = 1257.2$ msec. The cell was dissociated from the neostriatum of a P28 rat.

of prepulse duration is monoexponential (shown at the right). Application of the same protocol to an adult neuron produced a different pattern (Fig. 4B). A plot of the current amplitude early in the step (near the peak of the currents shown in Fig. 4A) as a function of prepulse duration could be fit well only with a sum of exponentials, indicating that two currents with different recovery rates were contributing. The recovery of current at the end of pulse had only a single exponential component that was similar to the slower component of that measured earlier. This finding suggested that the slowly recovering current also inactivated more slowly at the test potential.

It should be remembered that other situations may be considerably more complicated. Individual channels may inactivate with multiple time constants (which would be apparent in the development of inactivation) or have multiple inactivated states (which would be apparent in the recovery of inactivation).

Once the kinetics of inactivation development have been determined, steady-state inactivation measurements can be obtained. The data for these plots are obtained by delivering conditioning pulses of sufficient length (several time constants) prior to a test pulse. Plots of current amplitude as a function of conditioning voltage can then be constructed and fit with Boltzmann functions of the form

$$I = \frac{I_{max}}{[1 + \exp((V_{pp} - V_h)/V_c)]} + C = h(\infty, V) + C,$$

where I_{max} is the maximal current amplitude, V_{pp} is the prepulse or conditioning voltage, C is a constant (measuring the noninactivating current that has been subtracted in these plots), and $h(\infty, V)$ is the inactivation parameter referred to above. An example of such a plot constructed from peak current measurements is shown in Fig. 5A. Here, only the A_f current was present (Surmeier *et al.*, 1988). As we have shown previously, in adult neurons, both A_f and the slower A_s currents are present. The data from the adult cell (Fig. 5B) were well-fit only with a sum of Boltzmann functions of the format

$$I = \frac{\alpha I_{max}}{[1 + \exp((V_{pp} - V_{h1})/V_{c1})]} + \frac{(1 - \alpha)I_{max}}{[1 + \exp((V_{pp} - V_{h2})/V_{c2})]} + C,$$

indicating that there were two inactivating currents contributing to the peaks.

If the conditioning pulses are not of sufficient duration to achieve steady state, the estimates of half-inactivation voltage will be more depolarized than the true value and the contribution of "noninactivating" current components will be overestimated. The extent of the error will increase as the conditioning pulses are shortened.

Another potential complication in the interpretation of inactivation data, as well as that for activation, is poor space clamp.

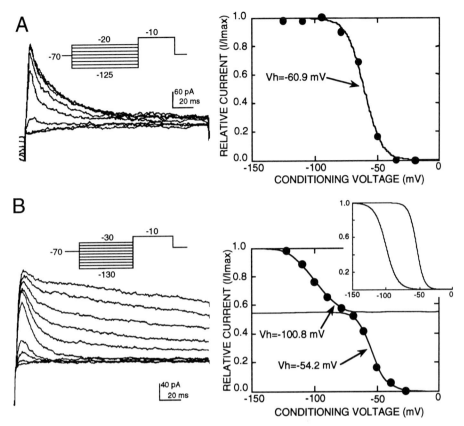

FIG. 5   (A) Steady-state inactivation was studied by holding the cell for 4 sec at potentials between -125 and -20 mV prior to delivery of a test pulse to -10 mV. The currents evoked by this test pulse are shown at the left. At the right, the normalized (I'/I'_{max}) peak currents are plotted as a function of conditioning voltage; $I' = I - I - 20$, where $I - 20$ is the current measured at the time of the peak following a pre-pulse to -20 mV. Although approximate, this procedure eliminated leak currents and most of the delayed rectifier currents from the measurements. The data were then fit with a Boltzmann function $(I'/I'_{max} = 1/(1 + \exp((V - V_h)/V_c)))$. In this case, $V_h = -60.9$ mV, $V_c = 6.9$ mV. (B) Steady-state inactivation was studied as in A; conditioning pulses were from -130 to -30 mV. The currents evoked by the test pulse (to -10 mV) are shown at the left. At the right, the normalized (I'/I'_{max}) peak currents are plotted as a function of conditioning voltage; $I' = I - I - 30$, this gave a reasonable isolation of the inactivating currents. The data were well fit only with a sum of Boltzmann functions (each function is plotted as a solid line) $(I'/I'_{max} = \beta(1/(1 + \exp((V - V_{h1})/V_{c1}))) + (1 - \beta)(1/(1 + \exp((V - V_{h2}/V_{c2})))); \beta = 0.452, V_{h1} = -100.8$ mV, $V_{c1} = 18.5$ mV, $V_{h2} = -54.2$ mV, $V_{c2} = 5.6$ mV. Inset shows the two Boltzmann functions on the same scale to allow easier comparison of their differences.

The Consequences of Poor Space Clamp

The hallmarks of poor voltage control when studying inward, regenerative currents have been described by several authors (see Armstrong and Gilly, 1992). Diagnostics for poor control when studying nonregenerative currents are less commonly discussed. If we assume that series resistance compensation is not a factor (which is *not* always true), then the most frequent source of difficulty stems from voltage-dependent channels in regions of membrane that are electrotonically distant from the access point of the patch electrode. In this situation, the time course of changes in the transmembrane voltage produced by current ejection from the patch electrode are not the same in different parts of the cell. That is, the membrane potential is not clamped at every point in space, hence not "space clamped." This sort of problem is certainly significant in whole-cell recordings from cultured cells or cells in slices. The type and magnitude of the errors that bad space clamp introduces will depend on several factors including the electrotonic dimensions of the cell, the spatial distribution of channels, and the gating properties of the channels (e.g., voltage dependence of inactivation) (see below).

There are two situations that we discuss. The first is when potassium channels giving rise to transmembrane current are electrotonically remote from the soma or access point of the patch pipette. In this case, current kinetics and steady-state voltage dependence will be distorted. There are several simple tests for this eventuality. One is to record from a cell-attached patch or outside-out patch to determine whether the channels are in the membrane at the access point. If so, then the biophysical properties of the currents could be studied in this reduced preparation with the assurance of voltage control. A similar "reduced cell" strategy is to acutely dissociate neurons, this procedure typically denudes cells of distal processes.

The remoteness of channels giving rise to currents could also be tested by determining the reversal potential of tail currents predicted by the Nernst equation (or the Goldman–Hodgkin–Katz current equation for less-selective currents (Surmeier *et al.*, 1989); at K concentrations above 5–10 mM the differences in the predicted reversal potentials for these two models are small even for promiscuous channels). It is reasonably safe to assume that small cations like K^+ equilibrate within a minute or two, even within large cells with substantive dendrites (Pusch and Neher, 1988). Given that the extracellular concentration of K^+ (and Na^+) is known, tail currents largely derived from somatic currents should reverse near the equilibrium potential. On the other hand, currents derived from remote locations should not. It should be noted that this test will only demonstrate that a substantial portion of the current is due to electrotonically proximal channels, not that all of the channels contributing to the current are under good voltage control.

In the event that channels are uniformly distributed, what is a reasonable diagnostic for the existence of channels that lie in regions of the membrane that are not adequately controlled? To answer this question, computer simulations were used. The model neuron was assumed to have inactivating K^+ channels like those we have used in examples above; their properties were derived from biophysical measurements in neostriatal neurons. Currents were assumed to have a reversal potential of -90 mV. Channels were placed uniformly on the surface of a neuron represented by a 10-μm soma and a single equivalent cylinder dendrite (Rall, 1977) adjusted to have 10 times the surface area of the soma. The only other conductance included in the model was a voltage-insensitive leak with an equilibrium potential of 0 mV. The leak conductance was uniform and 10% of the maximal conductance of the transient potassium conductance. The surface area, and therefore the total number of channels, was kept constant as the electronic structure of the model was altered. The diameter and length of the equivalent cylinder were changed together to maintain the surface area and to adjust the electrotonic length of the cylinder (calculated on the basis of the leak conductance) from effectively zero (0.01 λ) to 1.0 length constant. This simple model neuron allowed the effect of dendritic electrotonic structure to be reduced to a single dimension without significant loss of generality. For most simulations, the time constant of activation was voltage sensitive, peaking at 1.0 msec at about the half-activation voltage. The time constant of inactivation was adjusted to increase gradually to a maximum at about the half-inactivation voltage and to stay relatively constant at about 50 msec throughout the activation range. These characteristics were selected to best match the properties of the A-current of mammalian cells. Computer simulations were performed and analyzed with Saber (e.g., Carnevale *et al.*, 1990), using a time step of 50 μs.

Steady-State Activation

Steady-state activation curves were constructed from the peak currents attained during 25-msec voltage steps applied from a holding potential of -90 mV. Examples showing currents generated with dendritic lengths of 0.01 and 0.9λ are shown in Fig. 6. Aside from the increased capacitative transient that arises from redistribution of charge along the dendrites, there were no qualitative differences in the time course of current that could be used to distinguish the two. Several quantitative differences were observed in the voltage dependence of the peak conductance, however. Conductance estimates were obtained by dividing the peak current by driving force ($V_m - V_{rev}$). These conductance estimates were normalized and fit with a Boltzmann function as described above. It should be noted that the Hodgkin–Huxley formalism does not give rise to Boltzmann activation curves but this data can

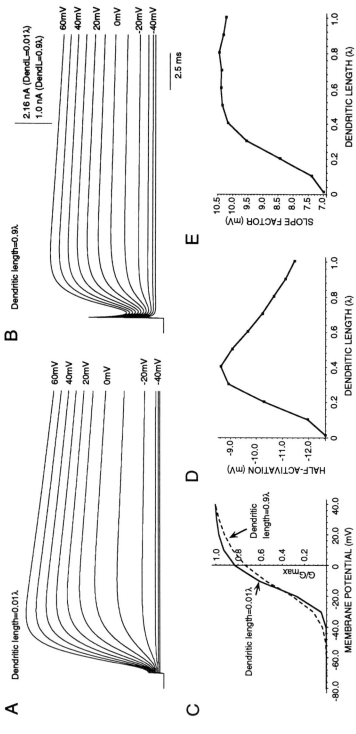

FIG. 6 Effect of dendritic length on the apparent voltage dependence of activation of a transient potassium current distributed uniformly on a model neuron. (A) Activation of the current in an isopotential neuron. Holding potential was −90 mV prior to voltage command steps ranging from −50 to +60 mV (indicated at the right). (B) The same protocol applied to a neuron with the same somatic and dendritic surface area, but more extended dendritic tree. Note the change in scale, corresponding to the overall attenuation of whole-cell currents in the dendritic neuron, and the greatly increased charge redistribution transient. Despite a severe loss of voltage control in the dendrites in B, the time course of the evoked currents is mostly unchanged. (C) Steady-state activation curve obtained from peak currents in A and B. The main effects of increased dendritic length are a reduction in the slope and a slight rightward shift. (D) The shift in half-activation voltage obtained for various dendritic lengths. (E) The change in slope factor for the same model neurons.

be approximated reasonably well using that function (Hodgkin and Huxley, 1952). Fits for the data shown in Figs. 6A and 6B are plotted in Fig. 6C. Two factors were extracted from these fits: the half-activation voltage and the slope factor (which is inversely related to the steepness of the relation). As can be seen in the fits shown in Fig. 6C, the half-activation voltage and the slope factor for activation were not strongly affected by the addition of an electrotonically significant dendrite. There was a very small biphasic effect of dendritic length on the measured voltage at half-activation (Fig. 6D). The biphasic nature of the shift was caused by changes in the shape of the activation curve, rather than a simple shift of the curve to the right. This can be seen in the curves shown in Fig. 6C. The most obvious (but not the only) feature of the shape change is the increase in slope factor (Fig. 6E). This change in slope factor occurred with even relatively short dendrites, being effectively complete by 0.5 λ, because these dendrites become effectively infinite in length when the potassium current is activated. Increasing the density of voltage-gated channels in the membrane shifts this transition of the slope factor to even shorter dendrites.

These relatively small differences probably do not represent any serious problem for the analysis of steady-state activation characteristics in dendritic neurons (unless very accurate values for the slope factor are needed). Neither do they offer any clear diagnostic criteria for identifying the distribution of potassium channels on a dendritic neuron. One reason for this is the relatively good voltage control achieved during the early part of a depolarizing voltage step from −90 mV. Voltage control over a dendritic neuron in this experiment becomes poor as the current turns on. Because the time constant of activation depends on voltage, this good voltage control during the early part of the response to a step change from a hyperpolarized holding potential is responsible for the absence of large changes in activation kinetics. This is not the case for deactivation. In experiments involving deactivation of the current, voltage control is at its worst at the beginning of the test. The conductance in different parts of the dendritic tree will deactivate at different rates and from different starting voltages. Deactivation will be slowest in the most distal parts of the dendritic tree, resulting in the existence of artifactually slow components of deactivation tail currents. In addition to this difficulty, the state of activation of the conductance contributes to the capacitance transient, because it is responsible for most of the membrane conductance at the moment of the voltage step. Thus efforts to compensate for the capacitative transient based on its behavior at different voltages (especially membrane potentials at which the potassium current is not active) will produce unpredictable tail current artifacts.

The largest effect of dendritic length was on apparent maximal conductance. This was due to attenuation of the voltage step in the dendrites, resulting in a decreasing activation of the current with distance from the soma. Conductance measurements dropped off rapidly with dendrites up to about 0.5 λ (ca., 50% attenuation) and the more slowly out to 1.0 λ, where the apparent conductance was about 25% of that which would be measured if the cell was isopotential. This attentuation is much greater than would be expected on the basis of the passive properties of the dendrite. The reason for this exaggerated effect of the dendrite was the alteration of the cable properties of the dendrite that occurred when the voltage-gated conductance was activated (Armstrong and Gilly, 1992). The activation of potassium channels decreased the membrane resistance enormously (by up to 10-fold where activation was greatest), causing voltage control to worsen during the course of activation of the current. Unfortunately, there is no obvious way in which this consequence of dendritic length can be used diagnostically.

Steady-State Inactivation

Steady-state inactivation curves were obtained from peak currents triggered by a voltage step to +40 mV after complete equilibration at holding potentials ranging from −90 to +30 mV. Examples showing the results with a practically spherical neuron (dendritic length = 0.01λ) and a neuron with a long dendrite (dendritic length = 0.9 λ) are shown in Figs. 7A and 7B. As with steady-state activation, the presence of a significant dendrite produced little change in the conductance time course, although the maximal amplitude of the apparent conductance was greatly decreased. In addition to the change in total conductance, a shift in the slope factor (Fig. 7E) and in voltage at half-inactivation (Fig. 7D) was similar to that seen for steady-state activation. Like the steady-state activation curve, the steady-state inactivation curve was relatively unaffected by changes in dendritic length, despite the presence relatively large voltage gradients along the dendritic tree (Fig. 7C). While the voltage clamp at the soma was unable to maintain the desired holding potential in the dendrites, and so a larger than expected proportion of dendritic channels was available for activation, the depolarizing test pulse was also less effectively conducted into the dendrites. This absence of an effect on inactivation was highly sensitive to the degree of voltage-dependence of the time constant of inactivation. In keeping with the experimentally determined characteristics of transient potassium currents, the inactivation time constant in these simulations was made relatively constant over the voltage range in which there is significant activation of the current. Thus the lack of voltage control in the dendrites did not create dramatic differences in the rate of inactivation at various parts of the cell. If the inactivation time constant decreased substantially with depolarization beyond the half-inactivation voltage, the steady-state activation and inactivation curves would both be greatly distorted. Because of the less-depolarized membrane

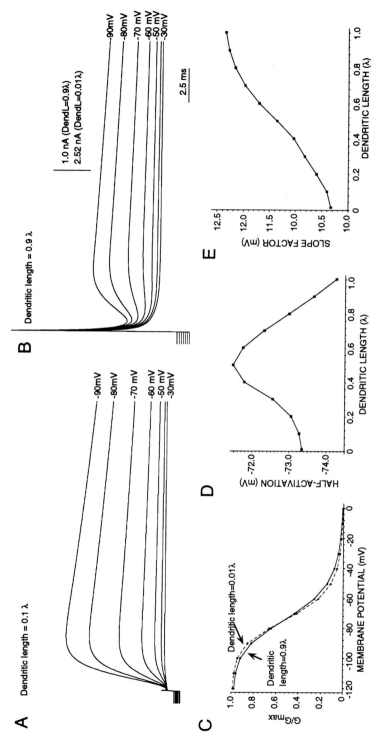

FIG. 7 Effect of dendritic length on the apparent voltage dependence of inactivation of a transient potassium current as in Fig. 6. (A) Inactivation of the current in an isopotential cell. Values at the right of each curve are holding potentials from which a test step to 40 mV was applied. (B) The same protocol applied to a neuron with an extended dendritic tree. (C) Steady-state inactivation curves obtained from the data in B. (D and E) Effect of increasing dendritic length on half-inactivation and slope factor of best-fitting inactivation curves.

potential in the dendrites, inactivation would be slowed there relative to the dendrites, and the locus of current would tend to move out the dendrites during activation of the current. As the current is inactivated, the electrotonic length of the dendritic tree in that region will be reduced, raising the effectiveness of voltage clamp in more distal regions and thus changing voltage there. In simulations performed using greater voltage dependence of the inactivation time constant, this process had large, unpredictable effects on the steady-state properties of the current measured with the patch electrode.

Pharmacological Tools

One of the most reliable ways of dissecting potassium currents is through the use of selective pharmacological blockers (Rudy, 1988). A general review of these agents has been given by Narahashi and Herman (1992). We do not try to reproduce their excellent summary but we discuss some issues relevant to gathering and analysis of pharmacological data.

Because of the heterogeneity in the potassium channels expressed by most brain neurons and overlaping sensitivity to the available set of blockers, quantitative pharmacological approaches need to be employed whenever possible. The first and most important step in generating reliable data is the ability to rapidly change the bathing medium. A number of approaches have been employed successfully. We use a "sewer pipe" system in which a linear array of thin-walled glass capillary tubing (id 500 μm) is placed within a few hundred micrometers of a patched neuron. The gravity flow through each capillary is governed by a solenoid valve that is remotely controlled (Lee Valve Co.). The solution bathing the cell can be switched by advancing the array. In our case, the array is mounted on Newport 461 series XYZ manipulator with the array axis position controlled by an 850 series actuator. The actuator is controlled by a programmable controller (PMC 200, Newport Corp.); the programmable feature is important to allow rapid, well-defined step sizes corresponding to the center-to-center distance between barrels. With this system, complete solution changes can be made in a second or less. The number of different solutions to which a cell can be exposed is limited only by the physical dimensions of the recording chamber and the linear capillary array. With this system, dose–response curves with six or more points can be generated rapidly enough that rundown of current is not a problem.

An example of how this system can be used to study the pharmacology of slowly and rapidly adapting components of depolarization-activated potassium currents is shown in Fig. 8A. Previous experiments had shown that inactivation of the fast *A* current was essential complete by the end of this 400-

ms step to -10 mV and the slowly inactivating A current did not inactivate significantly within this period. Therefore, an estimate of the fast A current could be obtained by measuring the transient portion of the current (see the inset in Fig. 8B) and an estimate of the slowly inactivating and delayed rectifier components from the current at the end of the pulse. The application of low concentrations of 4-aminopyridine (4AP) produced a block of the current at the end of the pulse but not the transient component; higher concentrations blocked both components. A plot of the transient current amplitude was well fit with an isotherm of the form

$$\frac{I}{I_{max}} = \frac{1}{1 + ([4AP]/IC_{50})^n} + C$$

where $n = 1$ and C is a constant. The current at the end of the pulse was a mixture of sensitive and relatively insensitive currents. The insensitive component was estimated from the constant in the initial fitting and subtracted and the data were renormalized. A plot of this component of the current is shown by the closed circles in Fig. 8B. Hill plots of the same data (excluding the extreme points) could readily be fit with a straight line of slope very near one (Fig. 8C), confirming the presence of a single binding site.

Commonly, Hill plots of data derived from experiments like this one do not yield fits with slopes close to one. The most common situation is a slope that is less than one; this indicates that either there are multiple binding sites (e.g., several channel types) or negative cooperativity. In this event, there are several approaches that can be used to tease the affinities of the different sites out of the dose–response data (Limbird, 1986).

Single Cell mRNA Amplification

Despite the power of whole-cell recording techniques, there are limitations in their ability to dissect the contribution of individual channel populations to whole-cell currents. As we mentioned at the outset, single-channel recordings are capable of providing the sort of molecular biophysical information that is necessary to make a complete list of channel types shaping whole-cell currents. The problem with this approach is that it is time consuming and—as with all reductionistic strategies—leads to the loss of more global information about cellular integrative properties.

Another problem faced by both cellular and molecular studies of brain neurons is correlating variation in biophysical properties with variation along other cellular dimensions. In almost any nucleus in the brain, there may be

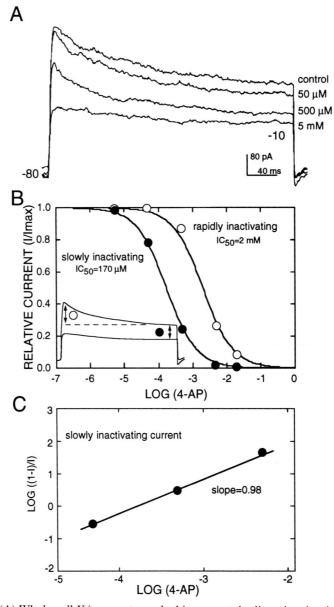

FIG. 8 (A) Whole-cell K^+ currents evoked in an acutely dissociated striatal neuron in increasing concentrations of 4-aminopyridine (4AP). Recording conditions were as in Fig. 1, except that the external K^+ concentration was 5 mM. (B) Plot of the rapidly and slowly inactivating components of the current as a function of 4AP concentration. Each component was operationally defined as in the inset. Currents

from 5 to 500 different cell types based on the types of chemical signals the neurons are capable of transducing, the type of releasable signaling molecules used, intracellular signaling enzymes, etc. Much of the neuroscience community is devoted to characterizing these neuronal properties and how they impact cellular functions. With intact preparations (*in vivo* or *in vitro* slices), cellular staining and subsequent immunocytochemical analysis have allowed for correlations to be drawn between biophysical properties and some of these dimensions. However, with acutely dissociated or even cultured neurons, these approaches have not been profitably applied.

There is a solution to this problem that goes well beyond the traditional methods of phenotyping recorded cells. It is based on a recently developed technique for amplifying polyadenylated mRNA (poly A mRNA) from individual, living cells (Van Gelder *et al.*, 1990; Eberwine *et al.*, 1992). The amplification product is anti-sense-RNA (aRNA) that can be radiolabeled and used to probe Southern blots (or slot blots) of cDNAs. With the burgeoning number of clones for ionic channels, receptors, and other cellular proteins, this technique is capable of providing an unprecedented breadth of information about the molecular constitution of individual cells. Importantly, it makes it possible to correlate functional analysis of channels and signaling systems with the abundance of their mRNAs. We briefly describe the use of the technique to study the potassium channels expressed in striatal neurons. A more complete discussion can be found elsewhere (Van Gelder *et al.*, 1990; Eberwine *et al.*, 1992).

A schematic outline of the technique is shown in Fig. 9. In our application, striatal neurons were dissociated, identified, and recorded from with patch-clamp electrodes that were backfilled with a sterile solution containing dNTPs, reverse transcriptase and a modified oligo(dT) primer which contains a T7 RNA polymerase promoter site just 5' to the oligo(dT) sequence. Positive pressure was maintained on the electrode during the approach to avoid entry of cellular debris. After seal rupture, the electrode and attached cell were lifted into a stream of control salt solution. The cells were subjected to standard recording procedures for 5–30 min, and then the cell was sucked into the pipette.

To generate a first-strand cDNA template of the poly(A) mRNAs, the mixture of cellular mRNA, reverse transcriptase, and oligo(dT) primer was held at 37°C for 1–2 hr. After the first strand cDNA synthesis, double-

---

were normalized by their respective maximum values. (C) Hill plot of the intermediate points of the slowly inactivating component from B. The *y*-axis is plotted as the logarithm of $1 - I/I$, where I is the relative current amplitude.

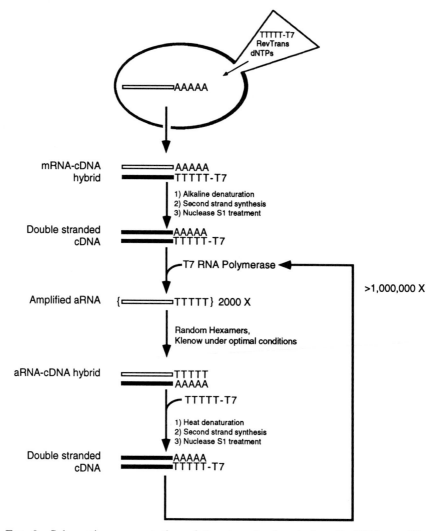

FIG. 9 Schematic representation of the steps in the single-cell aRNA amplification procedure.

stranded cDNAs with the T7 polymerase promoter site were produced as previously described (Van Gelder *et al.*, 1990; Mackler *et al.*, 1992; Surmeier *et al.*, 1992a; Eberwine *et al.*, 1992). T7 RNA polymerase was added to the cDNA mix and the cDNA was copied into aRNA. This step yields about a 2000-fold amplification of the original mRNA population. This amplified

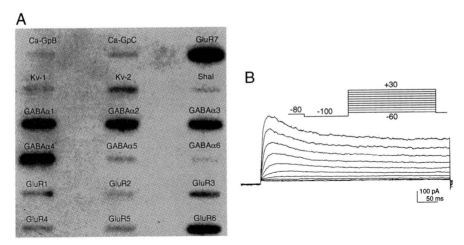

FIG. 10 (A) An autoradiogram developed from a slot blot of cDNAs that had been probed with radiolabeled aRNA amplified from an acutely dissociated striatal neuron (P32). The cDNAs coded for regions of the proteins are indicated by the labels next to the slots. They were potassium channels Kv-1, Kv-2 (Swanson *et al.*, 1990); rat Shal (Baldwin *et al.*, 1991); Group B and C calcium channel α subunit (Ca-GpB, C) (Snutch *et al.*, 1990); GABA$_A$ α1–6 subunits; glutamate receptor subunits GluR1–7. (B) Whole-cell potassium currents recorded from the same neuron. Currents were evoked from a negative conditioning potential (-100 mV) by depolarizing steps from -60 to $+30$ mV (voltage protocol is shown in the inset). Recording solutions as in Fig. 1.

aRNA was then converted back into cDNA and reamplified using published procedures (Eberwine *et al.*, 1992). At the end of this process, the original mRNA population had been amplified more than a millionfold. The final aRNA amplification was performed with high levels of ^{32}P ribonucleotides to radiolabeled the resulting population.

The aRNA population generated in this way is an amplified copy of the mRNA population that was present in the patch-clamped cell. To determine the types and relative abundance of different K$^+$ channel mRNAs, the radiolabeled aRNA was used to screen nitrocellulose filters containing cDNAs for three different K$^+$ channels (Kv-1, Kv-2, and rat Shal); cDNAs for calcium channels (Group A, B), GABA$_A$ receptor subunits (α1–6), and glutamate receptor subunits (GluR1–7) were also included to generate a more complete cellular profile. A autoradiogram developed from a slot blot that had been probed with a single-cell aRNA population is shown in Fig. 10A. This procedure is called a "reverse Northern" because of the symmetry of the protocol when compared to the standard Northern procedure (Mackler *et al.*, 1992; Surmeier *et al.* 1992a). When equimolar amounts of the different cDNAs are blotted to the nitrocellulose, the relative intensity of aRNA hybridization is a reflection of the relative abundance of the corresponding aRNA in the total aRNA population. Since the aRNA population is a copy of the single cell mRNA composition, the results of the reverse Northern are a direct reflection of the corresponding level of each mRNA. Based on the hybridization intensity of the autoradiogram for this cell, the Kv-2 mRNA levels were higher than those of the Kv-1 mRNA and the rat Shal mRNA was present only at very low levels. The strong Kv-2 signal is consistent with the pharmacological (i.e., high sensitivity to TEA, 4AP, and dendrotoxin) and biophysical properties (i.e., slowly inactivating, low-voltage threshold) of the principal K^+ current component activated by depolarization in this (Fig. 10B) and other medium spiny striatal neurons (Surmeier *et al.*, 1991; Swanson *et al.*, 1990).

References

1. C. M. Armstrong, and W. F. Gilly, *in* "Methods in Enzymology" (B. Rudy and L. E. Iversen, eds.), Vol. 207 pp. 100–122. Academic Press, San Diego, CA, 1992.
2. C. M. Armstrong, and C. Miller, *Proc. Natl. Acad. Sci. USA* **87,** 7579–7582 (1990).
3. T. J. Baldwin, M. L. Tsaur, G. A. Lopez, Y. N. Jan, and L. Y. Jan, *Neuron* **7,** 471–483 (1991).
4. A. L. Blatz, *J. Physiol. (London)* **441,** 1–21 (1991).
5. N. T. Carnevale, T. B. Woolf, and G. M. Shepherd, *J. Neurosci. Methods* **33,** 135–148 (1990).
6. J. Eberwine, H. Yeh, Y. Miyashiro, Y. Cao, S. Nair, R. Finnell, M. Zettel, and P. Coleman, *Proc. Natl. Acad. Sci. USA* **89,** 3010–3014 (1992).
7. O. P. Hamill, A. Marty, E. Neher, B. Sakmann, and F. J. Sigworth, *Pfluegers Arch.* **391,** 85–100 (1981).
8. B. Hille, "Ionic Channels of Excitable Membranes." Sinauer, Sunderland, MA, 1994.
9. A. L. Hodgkin, and A. F. Huxley, *J. Physiol. (London)* **117,** 500–544 (1952).
10. T. Hoshi and R. W. Aldrich, *J. Gen. Physiol.* **91,** 107–131 (1988).
11. T. Hoshi and R. W. Aldrich, *J. Gen. Physiol.* **91,** 107–131 (1989).
12. T. Hoshi and R. W. Aldrich, *J. Gen. Physiol.* **91,** 73–106 (1989).
13. H. Kasai, M. Kameyama, and K. F. Yamaguchi, *Biophys. J.* **49,** 1243–1247 (1986).
14. A. R. Kay, *J. Neurosci. Methods* **44,** 91–100 (1992).
15. A. R. Kay and R. K. S. Wong, *J. Neurosci. Methods* **16,** 227–238 (1986).
16. B. Lancaster, *in* "Cellular Neurobiology: A Practical Approach" (J. Chad and H. Wheal, eds.), pp. 97–119. Oxford Univ. Press, New York. 1991.
17. L. E. Limbird, "Cell Surface Receptors: A Short Course on Theory and Methods." Nijhoff, Boston, 1986.

18. S. MacFarlane and E. Cooper, *J. Neurophysiol.* **66,** 1380–1391 (1991).
19. S. A. Mackler, B. P. Brooks, and J. H. Eberwine, *Neuron* **3,** 539–548 (1992).
20. T. Narahashi, M. D. Herman, *in* "Methods in Enzymology" (B. Rudy and L.E. Iverson, eds.). Vol. 207, pp. 620–643. Academic Press, San Diego, CA, 1992.
21. E. Neher, *in* "Methods in Enzymology" (B. Rudy and L. E. Iverson, eds.), Vol. 207, pp 123–131. Academic Press, San Diego, CA, 1992.
22. O. Pongs, *TIPS* **13,** 359–365 (1992).
23. M. Pusch and E. Neher, *Pfluegers Arch.* **411,** 204–211, (1988).
24. W. Rall, *in* "Handbook of Physiology" (J. M. Brookhart, V. B. Mountcastle, and E. R. Kandel, eds.), Sect. 1, Vol. 1. pp. 39–97. American Physiological Society, Bethesda, MD, 1977.
25. B. Rudy, *Neuroscience* **25,** 729–749 (1988).
26. J. P. Ruppersberg, R. Frank, O. Pongs, and M. Stocker, *Nature (London)* **353,** 657–660 (1991).
27. B. Sakmann, and E. Neher, "Single Channel Recording" Plenum Press, New York, 1983.
27a. T. J. Schoenmakers, G. J. Visser, G. Flik, and A. P. Theuvenet, *Biotechniques* **12,** 870–879 (1992).
28. T. P. Snutch, J. P. Leonard, M. M. Gilbert, H. A. Lester, and N. Davidson, *Proc. Natl. Acad. Sci. USA* **87,** 3391–3395 (1990).
29. A. Stefani, D. J. Surmeier, and S. T. Kitai, *Soc. Neurosci. Abstr.* **16,** 418 (1990).
30. D. J. Surmeier, J. Bargas, and S. T. Kitai, *Neurosci. Lett.* **103,** 331–337 (1989).
31. D. J. Surmeier, J. Bargas, and S. T. Kitai, *Brain Res.* **473,** 187–192 (1988).
32. D. J. Surmeier, J. Eberwine, C. J. Wilson, A. Stefani, S. T. Kitai, *Proc. Natl. Acad. Sci USA* **89,** 10178–10182 (1992a).
33. D. J. Surmeier, A. Stefani, R. Foehring, and S. T. Kitai, *Neurosci. Lett.* **122,** 41–46 (1991).
34. D. J. Surmeier, Z. C. Xu, C. J. Wilson, A. Stefani, and S. T. Kitai, *Neuroscience* **48,** 849–856 (1992b).
35. R. Swanson, J. Marshall, J. S. Smith, J. B. Williams, M. B. Boyle, K. Folander, and C. J. Luneau, J. Antanavage, C. Oliva, S. A. Buhrow, C. Bennett, R. B. Stein, L. K. Kaczmarek, *Neuron* **4,** 929–939 (1990).
36. R. N. Van Gelder, M. E. vonZastrow, A. Yool, W. C. Dement, J. D. Barchas, and J. H. Eberwine, *Proc. Natl. Acad. Sci. USA* **87,** 1663–1667 (1990).

[4]   Calcium Channels of Human Retinal Glial Cells

Donald G. Puro

Introduction

The traditional notion that glial cells are permeable only to potassium has been revised. For example, calcium-permeable ion channels also can be expressed by glia (Barres *et al.*, 1990). Pathways for calcium influx are likely to be of importance because an elevation of intracellular calcium levels may be a vital link in the response of glial cells to perturbations in the extracellular environment. Although a calcium current in Müller glial cells of the salamander retina was reported by Newman (1985) in the mid-1980s, studies of calcium-permeable channels in mammalian retinal glial cells were undertaken only recently (Puro *et al.*, 1989; Puro 1991a, b; Puro and Mano, 1991; Uchihori and Puro, 1991).

This chapter details a method for preparing cultures of glial cells from the adult human retina. Applications of various configurations of the patch-clamp technique to the study of the retinal glial cells in culture are highlighted.

Preparation of Cell Cultures

Dissociated cell cultures of human retinal glial cells from postmortem donor eyes are prepared using modifications of methods developed by Oka and co-workers (1985) and Aotaki-Keen *et al.* (1991). Procedures for culturing Muller glial cells from adult rat (Sarthy, 1985; Roberge *et al.*, 1985), rabbit (Trachtenberg and Packey, 1983; Burke and Foster, 1984; Reichenbach and Birkenmeyer, 1984), and cat (Lewis *et al.*, 1988) retinas are also in the literature.

Based on my experience in preparing cultures from approximately 250 pairs of human donor eyes, three parameters appear critical in determining the chances for successful growth of retinal glial cells. These parameters are the age of the donor, the time interval between death and enucleation, and the length of time from death to retinal dissociation. With regard to donor age, the younger the better. Cultures from donors aged less than 40 years are nearly always successful. Although increasing donor age appears to progressively decrease the chances for successful culture, I have had satisfac-

Methods in Neurosciences, Volume 19

tory cultures of retinal glial cells derived from donors as old as 93 years. Thus, in my laboratory , there is no upper age limit for donor eyes that we will use. In addition to donor age, the rapidity of enucleation and refrigeration after death is critical. Less than 18 hr appears to be essential. Less than 4 hr greatly enhances the chances of success. No matter how promptly enucleation is performed, attempts to culture retinal glial cells after 30 hr postmortem have rarely been successful in my laboratory. Nearly all successful cultures are from retinas dissociated within 24 hr of death.

Human eyes are obtained from Eye Banks. Blood from the postmortem donor should be screened for human immunodeficiency virus and hepatitis. Typically, blood results are not available at the time of retinal removal and dissociation, thus adequate safety precautions are imperative. The eyes are disinfected by immersion into ethanol for 2 to 3 sec and a subsequent 5-sec dip into an antibiotic–antifungal solution (2000 U/ml penicillin, 2 mg/ml streptomycin, 5 μg/ml amphotericin B in magnesium-free calcium-free phosphate buffer made by adding 1 ml of Sigma Cat No. A9909 plus 4 ml of GIBCO Cat. No. 310-4190AG). Under sterile conditions, an initial circumferential incision is made near the insertion of one of the rectus muscles, i.e., approximately at the ora serata. Scissors are used to remove the anterior segment which is discarded. This dissection is done in a 60-mm petri dish containing 5 ml calcium-free magnesium-free phosphate buffer (CMF). The posterior segment is then manipulated to allow CMF to cover the retina; this facilitates the separation of the retina from the underlying retinal pigment epithelium. With forceps, the retina is gently pulled away from the retinal pigment epithelium. Curved scissors are used to cut the retina at the optic disk. The retina is then removed and placed in another 60-mm petri dish with 5 ml of CMF. Often the periphery of the removed tissue contains pigment associated with remnants of the pars plana of the ciliary body. Pigmented tissue is excised and discarded. Adhering vitreous is cut away from the retina. The retina is then placed in a 15-ml centrifuge tube containing 1 ml of CMF supplemented with 0.1% trypsin ($\times$3 crystallized, Worthington), 0.2% hyaluronidase (Sigma), and 4% chicken serum (GIBCO) for 45 min at 37°C. Six milliliters of medium A (40% Dulbecco's modified Eagle's medium, 40% Ham's F-12 medium, and 20% fetal bovine serum; all from GIBCO) is then added to the centrifuge tube containing the retinal tissue. The retinal cells are dissociated by pipetting the tissue 10 times with a 10-ml pipette (Costar, Cat No. 4101). The medium containing dissociated cells from one eye is evenly divided into three 35-ml petri dishes (Falcon, Cat. No. 3001; ~2 ml/ petri dish) and kept in a CO_2 incubator (100% humidity; 96.5% air, 3.5% CO_2) at 37°C. The next day the medium with nonadhering cells is removed and replaced with 1.5 ml of medium A. The cells are subsequently fed medium A twice weekly.

Cultures that reach confluency within 2 months are selected for further passages. Each confluent culture is split to three 35-mm petri dishes by washing the plates with CMF, exposing the cells to 0.1% trypsin ($\times$3 crystallized, Worthington) in CMF at 37° for 12 minutes, removing the cells by vigorous pipetting with 2 ml of CMF per petri dish, adding the cell suspension to 2.5 ml of medium A in a 15-ml centrifuge tube, centrifuging the tube for 1 min, decanting the supernatant, redissociating the cells in medium A, and adding 1.5 ml of the cell suspension to the new petri dishes. The newly split cultures usually reach confluency in 10 to 14 days and are again split one to three. Second passage cultures not reaching confluency by 1 month are not used.

At the third passage, virtually all of the cells in the selected cultures are immunoreactive to a monoclonal antibody specific for retinal Müller glial cells (Puro *et al.,* 1990). Also, as illustrated in Fig. 1, the cells in this culture system stain positively with polyclonal antibodies against glutamine synthetase (Uchihori and Puro, 1993). Immunoreactivity for glutamine synthetase is localized in the retina *in situ* to the Müller glial cells and does not occur in astrocytes located on the inner surface of the retina. In addition, the cells typically show some reactivity with antibodies to glial fibrillary acidic protein (Puro *et al.,* 1990), as do Müller cells in the human retina. Thus, immunocytochemical markers indicate that the cells in these cultures are Müller cells. Glial cells that are in culture for three to six passages are used for electrophysiological studies.

Recording Conditions

Various configurations of the patch-clamp technique (Hamill *et al.,* 1981) allow the study of calcium-permeable ion channels in human retinal glial cells in culture. During experiments, the cultures in 35-mm petri dishes are examined at 400$\times$ magnification with an inverted microscope equipped with phase contrast optics. Cells selected for recording have rounded cell bodies with diameters of approximately 50–150 μm. Attempts are made to select cells lacking obvious contacts with other glial cells.

Heat-polished pipettes of borosilicate class (World Precision Instruments, fiber-filled) coated with Sylgard No. 184 (Dow Corning) and having resistances of 5–9 MΩ are used with a Dagan 3900 patch-clamp amplifier. It is my impression that it is quite difficult to obtain gigaseal with these pipettes if the resistance, as tested in the bathing solution, is less than 5 MΩ. Also, it seems that the best time to attempt to form seals with the glial cells is 10 to 15 days after plating. For unknown reasons, pipette–cell seals can be formed more easily in some batches of cultures than in others.

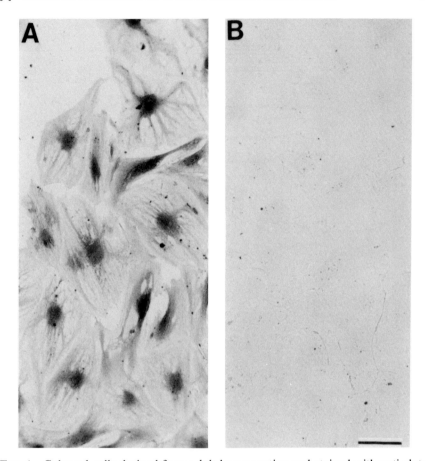

FIG. 1   Cultured cells derived from adult human retina and stained with anti-gluta-
mine synthetase antibodies (A) or control (B). Bar = 100 μm. Reprinted with permission from Uchihori and Puro (1993).

The whole-cell recording configuration has proven useful in the study of
calcium currents in the retinal glial cells (Puro and Mano, 1991). A convenient
bathing solution for these studies contains 125 mM CsCl, 10 mM BaCl$_2$, 10
mM glucose, 5 μM tetrodoxin, and 10 mM barium Hepes (pH 7.3). An often-
used pipette solution consists of 140 mM CsCl, 1 mM MgCl$_2$, 5 mM Cs-
EGTA, and 10 mM cesium-Hepes (pH 7.3). In the designing of solutions,
the osmolarity of the pipette solution is made to be somewhat (5–10%) less
than that of the bathing solution since this difference in osmolarity seems
to enhance the formation of gigaseals.

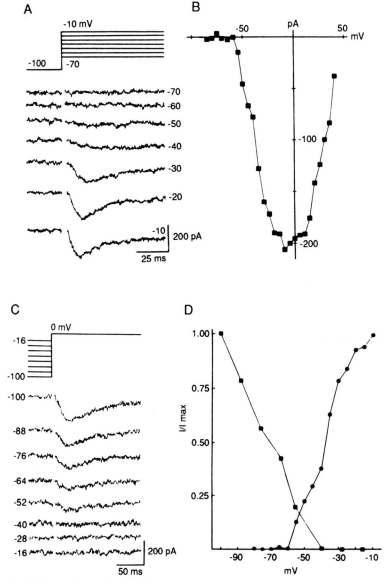

FIG. 2   Voltage dependence of activation and inactivation of the transient inward current, from a human retinal glial cell with predominantly a T-type calcium current. (A) Currents evoked with pulses from a holding potential of −100 mV to test potentials of −70 to −10 mV as indicated. (B) Peak current plotted against test potential for the same cell as in A (holding potential −100 mV). (C) Currents evoked by depolarizations to a test potential of 0 mV from various holding potentials. (D) Squares

Positive pressure (~5 psi with a Biotek pneumatic transducer) is applied to the back end of the pipette as the tip enters the bathing solution. The pipette potential is adjusted to zero current flow before the establishment of a seal with the glial cell. Pipette capacitance is minimized by capacity compensation circuitry on the patch-clamp amplifier. In the whole-cell recording mode, series resistance is compensated by appropriate circuitry. Cell capacitance can be estimated by using circuitry of a Dagan 3910 expander module. Current and signals are usually filtered at 1 kHz. Whole-cell currents are evoked by command pulses that are sampled (250 μsec) by an AT clone minicomputer (Vectra ES/12, Hewlett–Packard) using pCLAMP software (version 5.5, Axon Instruments) and an analog digital converter (TL-1 DMA interface, Axon Instruments). Subtraction of capacity and linear leak currents is achieved using pCLAMP software to take the average of 15 consecutive hyperpolarizing pulses 1/15 of the size of the test potential.

With whole-cell recordings from human retinal glial cells, it is rare to observe rapid rundown of voltage-gated calcium currents as may be seen in other types of cells. Whether the lack of a rapid rundown in the retinal glial cells is due to the relatively small tip diameter of the pipettes used in my studies, the relatively large size of the cells, or other factors is not clear. With the cultured human retinal glial cells, there typically is a progressive increase of 50 to 100% in the size of the peak inward barium or calcium current during the first 5 to 10 min after "going whole-cell." Subsequently, the inward current remains relatively stable for 15 to 30 min.

The use of the perforated-patch technique (Horn and Marty, 1988; Rae *et al.*, 1991) allows the whole-cell calcium currents in human retinal glial cells to be monitored for a relatively long time, i.e., greater than 1 hr. The stability of the amplitudes of the calcium currents is much greater with the perforated patch configuration than with the whole-cell recording technique. Use of Corning glass No. 7052 allows fabrication of pipettes that have resistances of a few Mohm and form gigaseals relatively easily. For perforated-patch recordings of calcium currents, a typical pipette solution consists of 140 mM

show peak current amplitudes (from C), normalized to the maximal current obtained with a holding potential of -100 mV, plotted against the holding potential. Circles show peak current amplitudes (from A), normalized to the maximal current obtained at a test potential of -10 mV, plotted against the test potential. The bath solution consisted of 125 mM CsCl, 10 mM BaCl$_2$, 10 mM glucose, 5 μM tetrodotoxin, and 10 mM barium Hepes (pH 7.3). The pipette solution contained 140 mM CsCl, 1 mM MgCl$_2$, 5 mM Cs-EGTA, and 10 mM cesium Hepes (pH 7.3). Reprinted with permission from Puro and Mano (1991).

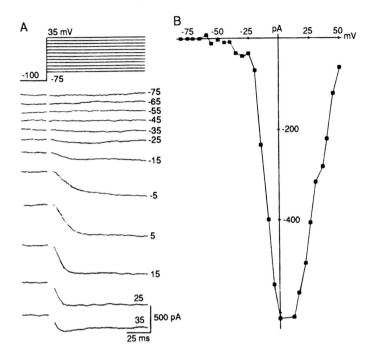

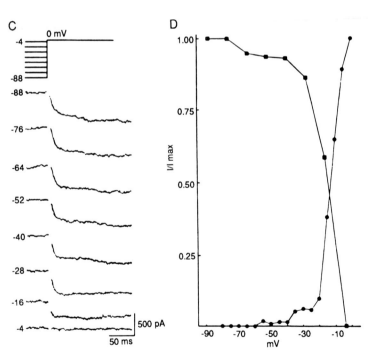

CsCl, 1 mM MgCl$_2$, and 10 mM cesium-hepes, at pH 7.3, plus amphotericin B, which partitions into membrane lipids and forms small channels that allow monovalent ions to permeate the patch (Rae *et al.,* 1991). To fill pipettes to be used for perforated patch recordings, a heat-polished pipette tip is dipped into a drop of the amphotericin-free pipette solution. The ampotericin-free solution is drawn into the pipette tip to a distance of approximately 400 μm by applying suction to the back end of the pipette. An aliquot of the pipette solution supplemented with 120 μg/ml of amphotericin B, is added to the back end of the pipette. This is done using a plastic tuberculin syringe that has been flame-heated and drawn to a fine tip. Bubbles in the pipette are removed by tapping. Microscopic observation of the tip confirms the absence of bubbles. It appears that the chances of forming a gigaseal are enhanced if the seal is made within a couple of minutes after addition of the amphotericin-containing solution to the recording pipette. Subsequent formation of a perforated patch seemed to be maximized if the amphotericin stock solution (7.5 mg amphotericin B in 250 μl DMSO) is mixed within 3 hr and the ampho-tericin-containing pipette solution is mixed within 1 hr of recording.

Standard cell attached and excised patch configurations (Hamill *et al.,* 1981) can also be applied to the study of calcium-permeable ion channels in human retinal glial cells (Puro 1991a,b).

Calcium Channels of Human Retinal Glial Cells

A number of different kinds of calcium-permeable ion channels are present in human retinal glial cells in culture. Approximately 70% of the retinal glial cells sampled with the whole-cell recording configuration of the patch-clamp technique have voltage-gated calcium currents (Puro and Mano, 1991). Of

---

FIG. 3   Voltage dependence of the slowly inactivating inward current, from a retinal glial cell with a predominantly L-type calcium current. (A) Currents evoked with pulses from a holding potential of -100 mV to the test potentials indicated. (B) Peak current plotted against test potential for the same cell as in A (holding potential, -100 mV). (C) Currents evoked by depolarizations to a test potential of 0 mV from various holding potentials. (D) Squares show peak current amplitudes (from C), normalized to the maximal current obtained with a holding potential of -100 mV, plotted against the holding potential. Circles show peak current amplitudes (from A), normalized to the maximal current obtained at a test potential of 0 mV, plotted against the test potential. Solutions were as in Fig. 2. Reprinted with permission from Puro and Mano (1991).

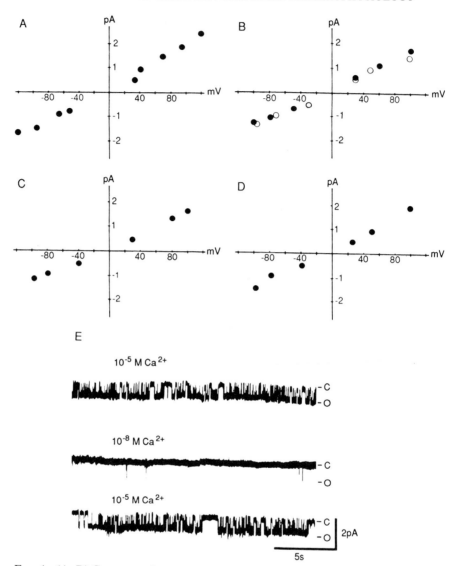

FIG. 4   (A–D) Current–voltage relationships for nonspecific cation channels in cell-attached patches of human retinal glial cells in culture. (A) The pipette solution includes isotonic $BaCl_2$. (B) The closed circles indicate results with 150 mM potassium aspartate in the pipette solution. Open circles show findings with 150 mM KCl. (C) The pipette solution includes isotonic $CaCl_2$. (D) A total of 150 mM NaCl was in the pipette solution. All pipette solutions contained 5 μM tetrodotoxin, 5 mM tetraethyl-ammonium, and 10 mM HEPES. For all conditions tested, the reversal potentials

the cells with detectable inward currents, the mean peak inward current is -223 pA (SE = 22, n = 59). The mean current density is 3.0 pA/pF (SE = 0.4). These values are similar to those found in studies of other types of glial cells that have voltage-gated calcium channels (Barres *et al.,* 1990). In the human retinal glial cells, two types of voltage-gated calcium channels are expressed. One, similar to the T-type current described in various kinds of cells, has a low threshold of activation, a transient response (Fig. 2) and an insensitivity to the dihydropyridine nifedipine. The other type of inward current, which closely resembles the L-type calcium current found in other cells, has a high threshold, has a long-lasting response (Fig. 3), and is inhibited by nifedipine. Of the sampled retinal glial cells with voltage-gated calcium channels, more than 90% have both T- and L-type currents.

In addition to voltage-gated calcium channels, human retinal glial cells in culture have a calcium-permeable, voltage-insensitive, nonspecific cation channel that is activated by cytoplasmic calcium (Puro, 1991a). Approximately 50% of the cell-attached patches in cultured human retinal glial cells sampled by the patch-clamp technique contain channels of this type. With isotonic barium in the pipette solution, the mean unitary conductance is 17 pS. The ionic selectivity of this ion channel was investigated by comparing current–voltage relationships with single channels in cell-attached patches with different ionic species in the pipette solution (Fig. 4). Whether the pipette solution contained 150 mM potassium aspartate or KCl, the reversal potential remained near zero (Fig. 4A), indicating no significant permeability of chloride. Reversal potentials near 0 mV with isotonic potassium, sodium, calcium, or barium in the patch pipettes (Figs. 4B–4D) indicate poor discrimination between monovalent and divalent cations. Excised inside-out patches were used to demonstrate the activity of this cation channel was dependent on calcium at the cytoplasmic surface (Fig. 4E).

Another type of calcium-permeable ion channel detected in human retinal glial cells is one that is activated by applying suction to the cell-attached patch

---

were near 0 mV. (E) The effect of varying the calcium concentration at the cytoplasmic surface of an excised inside-out patch containing a 17-pS cation channel. Immediately prior to patch excision, the bathing solution was changed to one containing 10^{-8} M Ca^{2+} (0.32 mM $CaCl_2$ plus 1 mM EGTA). The top and bottom panels show ion channel activity when a bathing solution with 10^{-5} M Ca^{2+} was perfused in the area of the excised patch. The pipette solution contained isotonic $BaCl_2$. The holding potential was -60 mV. The letters c and o indicate the current level for closed and open channels, respectively. Reprinted with permission from Puro (1991a).

pipette (Puro, 1991b). In nearly 90% of the cell-attached patches sampled, previously silent ion channels could be activated during suction (Fig. 5). With isotonic barium in the patch pipette, the mean channel conductance between -100 to -30 mV for stretch-activated channels is 32 pS (SE = 4, $n = 6$). The ionic selectivity of the stretch-activated channels was investigated using excised, inside-out patches. Permeability to chloride is minimal since replacement of aspartate with chloride in the bathing solution did not significantly change the reversal potential or conductance. As reported (Puro, 1991b), the stretch-activated channels in human retinal glial cells are permeable to both monovalent and divalent cations.

Discussion

The culture system described here appears to be a reasonable system in which to study aspects of the physiology and pathophysiology of human retinal glial cells. For example, this culture system has facilitated the study of the possible role that calcium-permeable channels may play in mediating the responses of retinal glial cells to basic fibroblast growth factor (Puro and Mano, 1991; Puro, 1991a). However, it remains to be demonstrated that the ion channels found in these cultured cells are expressed by retinal glial cells *in vivo*. Although *in vivo* electrophysiological studies of calcium-permeable ion channels in retinal glial cells are lacking, the finding of voltage-gated calcium currents in Müller glia in a slice preparation of the salamander retina (Newmann, 1985) strongly suggests that certain calcium channels are present in these retinal cells *in vivo* at least in some vertebrates.

To more closely approximate the *in vivo* situation, freshly dissociated Müller glial cells from various mammalian retinas can be prepared by methods described by Newman (1987). Despite the availability of these methods, the literature lacks patch-clamp studies of freshly dissociated glial cells from the mammalian retina. Preliminary studies in my laboratory using the patch-clamp technique have demonstrated voltage-gated calcium channels in freshly dissociated Müller cells from the adult human retina.

Methods for the electrophysiological study of a slice preparation of the salamander retina are in use (Werblin *et al.*, 1988; Maguire *et al.*, 1989). Wassle and Boos (1992) recently developed a slice preparation of the rat retina that is satisfactory for whole-cell recording. However, patch-clamp studies of Müller glial cells in a retinal slice preparation have yet to be reported. For whole-cell recordings from astrocytes on the inner surface of the retina, Clark and Mobbs (1992) have used a flat-mount preparation of the rabbit retina (Clark and Mobbs, 1990).

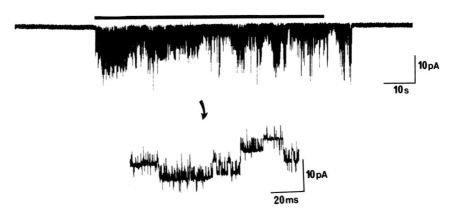

FIG. 5   Stretch-activated ion channels in a human retinal glial cell in culture. The top panel shows current records from a cell-attached patch. The bar above the record marks the period of application of suction (−120 mm Hg) to the recording pipette. The arrow points to a current record displayed at a faster time scale. A downward deflection of the current record indicates an inward current, i.e., positive current passing from the pipette into the cell. A potential of +90 mV was applied to the pipette. Suction was applied to the patch pipette through the sidearm of the pipette holder and was monitored by a pneumatic transducer (BioTek). The pipette was filled with 100 mM $BaCl_2$ and 10 mM barium Hepes (pH 7.3). The bathing solution consisted of 140 mM potassium aspartate, 4 mM $CaCl_2$, 1 mM $MgCl_2$, 10 mM glucose, and 10 mM potassium Hepes (pH 7.3). Reprinted with permission from Puro (1991b). Copyright © 1991, Wiley-Liss, a division of John Wiley & Sons, Inc.

Summary

Methods are available to establish cultures of glial cells derived from the adult human retina. Various configurations of the patch clamp technique can be applied to the study of ion channels expressed by these cells. A variety of types of calcium-permeable ion channels can be found. These include channels that are voltage-gated, mechanosensitive, or activated by cytoplasmic calcium. The specific roles for the various types of calcium-permeable ion channels in regulating retinal glial cell function are unclear though it seems likely that these channels affect the responses of retinal glial cells to physiological and pathobiological changes in the retina. The availability of this culture system and the patch-clamp technique provide the opportunity for changes to be detected in calcium-channel activity with exposure of retinal glial cells to extracellular molecules such as growth factors (Puro, 1991a; Puro and Mano, 1991; Uchihori and Puro, 1991). In addition, the role of putative intracellular messengers in regulating ion-channel activity in human retinal glial cells can be evaluated.

Acknowledgments

This work was supported in part by Grants EY06931 and EY07003 from the National Eye Institute, National Institutes of Health, Bethesda, Maryland. The author is a Research to Prevent Blindness Senior Scientific Investigator.

References

A. E. Aotaki-Keen, A. K. Harvey, E. DeJuan, and L. M. Hjelmeland, *Invest. Ophthalmol. Vis. Sci.* **32**, 1733–1738 (1991).

B. A. Barres, L. L. Y. Chan, and D. P. Corey, *Annu. Rev. Neurosci.* **13**, 441–474 (1990).

J. M. Burke and S. J. Foster, *Curr. Eye Res.* **3**, 1169–1178 (1984).

B. Clark, and P. Mobbs, *J. Physiol. (London)* **426**, 7P (1990).

B. Clark, and P. Mobbs, *J. Neurosci.* **12**, 664–673 (1992).

D. P. Hamill, A. Marty, E. Nehr, B. Sakmann, and F. J. Sigworth, *Pfluegers Arch.* **391**, 85–100 (1981).

R. Horn and A. Marty, *J. Gen. Physiol.* **92**, 145–159 (1988).

G. P. Lewis, D. D. Kaska, D. K. Vaughan, and S. K. Fisher, *Exp. Eye Res.* **47**, 855–868 (1988).

G. W. Maguire, P. D. Lukasiewicz, and F. S. Werblin, *J. Neurosci.* **9**, 726–735 (1989).

T. Mano and D. G. Puro, *Invest. Ophthalmol. Vis. Sci.* **31**, 1047–1055 (1990).

E. A. Newman, *Nature (London)* **317**, 809–816 (1985).

E. A. Newman, *J. Neurosci.* **7**, 2423–2432 (1987).

M. S. Oka, J. M. Frederick, R. A. Landers, and C. D. B. Bridges, *Exp. Cell. Res.* **159**, 127–134 (1985).

D. G. Puro, *Brain Res.* **548**, 329–333 (1991a).

D. G. Puro, *Glia* **4**, 456–460 (1991b).

D. G. Puro and T. Mano, *J. Neurosci.* **11**, 1873–1880 (1991).

D. G. Puro, T. Mano, C.-C. Chan, M. Fukuda, and H. Shimida, *Graefe's Arch. Clin. Exp. Ophthalmol.* **228**, 169–173 (1990).

D. G. Puro, F. Roberge, and C.-C. Chan, *Invest. Ophthalmol. Vis. Res.* **30**, 521–529 (1989).

J. Rae, K. Cooper, P. Gates, and M. Watsky, *J. Neurosci. Methods* **37**, 15–26 (1991).

A. Reichenback and Birkenmeyer, *Z. Mikrosk-anat. Forsch. Leipzig* **98**, 789–792 (1984).

F. G. Roberge, R. R. Caspi, C.-C. Chan, T. Kuwabara, and R. B. Nussenblatt, *Curr. Eye Res.* **4**, 975–982 (1985).

P. V. Sarthy, *Brain Res.* **337**, 138–141 (1985).

M. C. Trachtenberg and D. J. Packey, *Brain Res.* **261,** 43–52 (1983).

Y. Uchihori and D. G. Puro, *Invest. Ophthalmol. Vis. Sci.* **32,** 2689–2695 (1991).

Y. Uchihori and D. G. Puro. *Brain Res.* **613,** 212–220 (1993).

H. Wassle and R. Boos, *Invest. Ophthalmol. Vis. Sci.* **33** (Suppl.), 1172 (1992).

F. Werblin, G. Maguire, P. Lukasiewicz, S. Eliasof, and S. M. Wu, *J. Vis. Neurosci.* **1,** 317–329 (1988).

[5] Axial-Wire Voltage Clamping in Crayfish Medial Giant Axons

John G. Starkus and Martin D. Rayner

Advantages and Disadvantages of the Crayfish Giant Axon Preparation

The crayfish medial giant axon preparation offers substantial advantages for axial-wire voltage-clamp studies on axonal sodium channels in their native membrane. Crayfish are inexpensive and readily available on a year-round basis, either by catching them oneself (if you live in the right climate) or by purchasing them from commercial farms in California and Louisiana. They are easy to maintain in the laboratory and allow one to avoid the bureaucratic problems which arise with use of vertebrate preparations. More significantly, however, crayfish medial giant axons offer two technical advantages which make this a premium preparation for detailed biophysical studies: (i) the axoplasm is fluid and easily washed out of the axon when the internal perfusate is driven by a very low hydrostatic pressure head (see below), to yield a "clean" membrane preparation without requiring either dialysis, or enzymatic or roller techniques for axoplasm removal; (ii) the Schwann cell sheath is permeated by a tubular lattice with openings into the periaxonal space every ~0.2 μm (as opposed to the 5- to 13-μm intervals between Schwann cell clefts in squid axons); see Shrager $et\ al.$ (1). Thus, the crayfish Schwann cells permit very rapid reequilibration of periaxonal ion concentrations (some 25-fold faster than that in squid axons (1)). Furthermore, the close spacing of the tubule lattice openings reduces the possibility that microscopic space clamp errors can arise from the lateral resistance of the Frankenhaeuser–Hodgkin space, as described for squid axons by Stimers $et\ al.$ (2). In almost 15 years of gating current recording we have never seen a gating current "rising phase" which outlasts the clamp rise time in a crayfish axon, an observation which tends to confirm that such rising phases are an artifact arising from microscopic space-clamp errors induced by the geometry of the squid's periaxonal space (2). Finally, the relatively faster kinetics of crayfish axons (even at 6 to 8°C) makes the gating currents in this preparation seem relatively larger and, therefore, easier to record (3).

Against the advantages described above, there is the disadvantage of the smaller diameter of crayfish giant axons. We find that the "best"

Methods in Neurosciences, Volume 19

axons fall between 180 and 250 μm in diameter; axons larger than this (we have seen axons as large as 400 μm) are often of variable diameter and are discarded for this reason. However, the small size of our best axons limits the diameter of the axial wire, makes electrode placement more difficult (our success rate rarely exceeds ~80%), and raises questions with respect to both electrode polarization and potential space-clamp errors which we address below.

Development of the Crayfish Axial-Wire Preparation

The technique for internal perfusion of crayfish giant axons was first developed by Shrager *et al.* (4), who studied the effects of internal perfusion with tannic acid, the insecticide, DDT, and tetraethylammonium ions on action potential propagation. Shrager then explored the use of these axons for axial-wire voltage-clamp studies (5). In this paper, he used a "piggyback" internal electrode assembly (described in more detail below) and built up the thickness of the lacquer (used to cement the capillary voltage electrode to the current-passing wire) so as to provide a mechanical plug, preventing the fluid axoplasm from escaping through the electrode entry hole. By that time, Shrager had developed the experimetnal chamber design which we still use, permitting accurate temperature control of the bathing medium through Peltier cooling devices clamped to a Teflon-coated aluminum block which forms a part of the back wall of the experimental chamber (see below). Shrager's next step was to combine the techniques of internal perfusion and axial-wire voltage clamping, to match the full power of the method which was by then widely in use for squid axons. That step was achieved in his subsequent paper (6), using an internal perfusion pipette inserted into the anterior region of the isolated medial giant axon and permitting the perfusate to escape around the posteriorly located entry hole for the piggy-back electrode. To assist the escape of the perfusion fluid, the lacquer was left thin (on that part of the electrode where the lacquer plug had previously been built up to prevent axoplasm escape). The final step in this developmental sequence was one which has not been specifically mentioned in the Methods sections of the published papers. In Shrager's earlier experiments, problems had occasionally arisen from back diffusion of the high potassium internal perfusate along the outside surface of the axon. A siphon-driven "sewer line" was therefore placed adjacent to the exit hole, to suck up the exiting internal perfusate (see below). This sewer line had become incorporated into the experimental setup by the time that the later Starkus and Shrager study was carried out (7). All subsequent improvements have been merely minor modifications of this basic experimental technique.

Axial-Wire Voltage Clamping of Crayfish Giant Axons

Initial Dissection

As a first step, the entire abdominal cord, from the brain to tail, must be isolated and removed from the crayfish (we use *Procambarus clarkii* as our experimental animal but other regionally available species may have very similar properties).

Before dissection the animal is anesthetized by cooling in ice water. The gross dissection then begins by removing the pinchers/claws and the walking legs. Most of the carapace is removed except for those regions near the brain (i.e., frontal, orbital, and antennal regions). In the tail region, the dorsal (tergal) regions of the six abdominal somites are removed. After removing the carapace, the dorsal tail skeleton, and the lateral gills, the animal is pinned to a cork board with the ventral side down. A push pin in the telson is used to anchor the posterior end, and the anterior end is anchored by attaching a hemostat forceps to the maxilliped (an appendage near the animals mouth). The tail musculature should be removed with caution because the cord is near the ventral surface of this complex muscle mass (8). The cord in the body cavity is exposed by splitting the thoracic endoskeleton with one's index finger and fingernails. After this procedure, the entire cord is exposed from the brain to the tail. With fine Dumont forceps and Castroviejo scissors, the cord can be removed by cutting the segmental nerve roots leading from each ganglion.

The excised nerve cord (see Fig. 1) is laid over two Tackiwax (Cenco Softseal Central Scientific Co., Chicago, IL) supporting "posts" at the edges of the raised Lucite "table" in the center of the experimental chamber (while bathed in circulating extracellular perfusate). The nerve cord is then clamped into place using a pair of stainless steel wire clips (approximately 0.018 inches in diameter) and held at sufficient tension to keep it accurately horizontal throughout the unsupported region between the supporting posts. The height of these posts should be sufficient to ensure both adequate fluid circulation around the isolated axon and appropriate clearances for the "C-shaped" external current electrodes (see below).

Using a fiber optic light source (Dolan Jenner), the preparation is transilluminated by placing the light pipe underneath the chamber. Above the light pipe is a Lucite diffusion panel which provides even illumination of the experimental chamber. With a stereomicroscope, the pair of medial giant axons can be visualized as the largest nerve fibers in the medial part of the

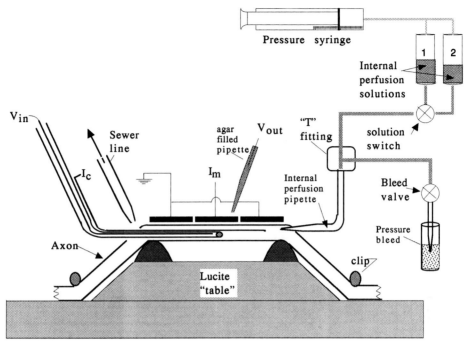

Fig. 1   Schematic diagram of the crayfish giant axon axial-wire voltage clamp. This diagram represents a side view of the experimental setup with the external perfusate level not shown. Abbreviations: V_{in}, intracellular voltage pipette; V_{out}, extracellular voltage pipette; I_m, current collector electrode; I_c, axial-wire electrode. See text for further description.

cord; their average diameter is $\sim$200 μm (range 150–400 μm) by comparison with the two segmented lateral giant fibers (diam $\sim$100 μm) and the remaining nerve fibers all of which are <5 μm.

The nerve cord should be placed such that the portion of the axon which will be voltage-clamped (the region between ganglion 3 and 5) is positioned between the Tackiwax posts. Counting the brain as ganglion 1, the esophageal ganglion (2nd ganglion) should be at the right edge of the table, with ganglion 6 and 7 toward the left edge of the table. Between ganglions 2 and 7, the neural sheath is removed and the selected medial giant is cleaned of smaller nerve fibers by using fire-polished hooked glass needles. Vannas ultrafine micro scissors are used for the fine dissection and the incisions. Two incisions are then made in the cleaned axon: first, for the internal perfusion pipette which enters the axon near ganglion 2 and, second, for the axial-wire electrode which enters near ganglion 6 and 7.

When teaching (or learning) this technique, we find it best to practice inserting only the internal voltage electrode (without the longer piggy-back axial wire) until sufficient expertise has been gained to achieve membrane potentials which remain more negative than -100 mV for at least an hour (when the axon is perfused with the normal potassium internal perfusate and zero potassium external solution).

Electrode Design

The crayfish voltage-clamp electrode designed by Shrager (5) is constructed as a piggy-back electrode (9). This electrode is similar to the piggy-back electrodes used for squid axons, except for the diameter of the current wire and glass capillary which is much smaller in order to fit within the smaller crayfish axon. The crayfish electrode consists of a 20-μm-diameter platinized platinum current electrode glued to a glass capillary with Insl-X lacquer. This L-shaped glass capillary which is approximately 50 μm in diameter also contains an internal floating platinum wire (18 μm) to reduce the AC impedance (10). This glass capillary measures the internal voltage and is filled with a solution composed of KCl, 172 and mM; potassium citrate, 37 mM; adjusted to pH 7.35 with citric acid.

This piggy-back electrode can be used either for the intact axon or for the perfused axon. For the intact axon the lacquer coating on the shank of the glass capillary must be wide enough to seal the electrode entrance hole after the electrode has reached its final insertion distance within the axon. This seal prevents the fluid internal axoplasm from escaping. If axoplasm escapes, then the axon collapses over the electrode which will kill the cell. However, for internal perfusion the shank of the electrode must be smaller in order to permit the internal perfusate to exit the axon. A separate glass capillary called the sewer pipette (see Fig. 1) is placed close to the exit hole to suck up the internal perfusate immediately after it leaves the axon, so as to prevent contamination of the external solution. In addition to external back diffusion of the internal cation, a major concern here is that fluoride from the internal perfusate will precipitate in contact with calcium in the external solution, thus lowering the external Ca concentration. In an uninjured axon, with potassium as the main internal cation and zero potassium in the external medium, resting potential should be within the range of -105 to -115 mV (after junction potential correction). Correction for junction potential is carried out before electrode insertion: the electrode is placed in a small bath containing internal perfusate and connected to external perfusate by an agar bridge. The junction potential (usually about -8 mV) is then backed off to 0 mV difference between the reference solutions. We check the junction potential at the end of each experiment and only rarely note drift greater than 2 to 3 mV.

The external current electrodes consist of three platinized platinum plates, each 2 mm wide, assembled on a common Lucite holder attached to a Prior-type micromanipulator. These electrodes are C-shaped in cross-section with the opening of the "C" large enough that the whole assembly can be moved in (after the internal electrode has been inserted) so as to cover about three-quarters of the axon's surface. The central plate measures the current collected from a 2-mm length of the axon, while the grounded guard electrodes maintain uniform radial current flow at the edges of the collector plate. Near the central plate, a glass capillary is attached which measures the external voltage. This capillary is filled with the external solution and 2% agar.

Internal Perfusion

A short L-shaped glass capillary (see Fig. 1) with a tip opening of 40 μm is inserted into the axon near the esophageal ganglion (2nd ganglion). The pipette is advanced into the opening until it seals the entry hole, preventing back flow and escape of the internal perfusate. The perfusion capillary is attached to a T-shaped Lucite holder mounted on a Prior-type micromanipulator. With PE tubing, one end of the "T" is attached to a control manifold (solution switch) and the other end is attached to an exit bleed valve. The solution switch selects one of two 10-cc syringes which acts as a reservoir for changing the internal perfusates. Hydrostatic pressure (between the syringe fluid level and the tip of the perfusion pipette) of 3–4 cm drives the perfusion. The opening to the syringes is connected to a syringe pump via polyethylene tubing. During solution changes, the exit valve is opened and the syringe pump is turned on. The pump compensates for the decrease in the hydrostatic pressure when the bleed valve is opened and moves the new solution to the entrance of the perfusion capillary (thus greatly reducing the time required for solution changes).

Solutions

The basic external solution for crayfish axons contains (in mM): 210 Na, 5.4 K, 13.5 Ca, 2.6 Mg, 247.6 Cl, and 2 HEPES, adjusted to pH 7.55. However, we normally omit the external K ions and, as discussed below, reduce the sodium concentration so as to keep peak I_{Na} no greater than ~1 mA/cm^2, thus limiting potential artifacts from incomplete series resistance compensation. A more typical solution for sodium current recordings would contain 50 Na, 0 K, and 160 tetramethylammonium (TMA) with the other ions as indicated above. Similarly, the initial internal perfusate contains: 0 Na, 230 K, 60 F, 170 glutamate, and 1 HEPES, adjusted to pH 7.35. For sodium current recordings this would be modified to 0 Na, 0 K, and 230 Cs, with all other ions as shown above. For outward sodium current recordings we reduce external sodium (to 0 Na) by TMA substitution and add 20 to 50 Na internally, with appropriate reduction of internal Cs ions (11). Osmolarity is adjusted to 430 mOsm for both internal and external solutions.

Temperature Control and Experimental Chamber

The experimental chamber consists of two separate portions which are bolted firmly together over a sealing layer of silicon grease: *first*, a Teflon-coated aluminum block which forms its back and sides and, *second,* a Lucite portion which provides the transparent bottom and front. This arrangement permits transillumination from below and viewing of the axon/electrode system from two viewpoints: a lateral view from in front as well as an overhead view. Two Peltier devices are attached to the aluminum block and connected to an electronic feedback circuit. With a thermistor in the bath (located as close as possible to the electrode assembly) the temperature was usually set in the range 6–8°C and maintained at a constant level within ±1°C.

Recording Ionic and Gating Currents

Definitive recording of ionic and gating currents in giant axon preparations requires a constant critical awareness of the artifacts which might distort these data. In this section, therefore, we give considerable attention to the methods we have found particularly useful for evaluation and avoidance of artifacts. Crayfish axons also offer advantages for detailed study of potassium currents, since the relatively fast equilibration of ion concentrations between the periaxonal space and surrounding medium (1) should reduce artifacts resulting from extracellular potassium accumulation. However, it seems appropriate for us to take our examples from those measurements with which we are the most familiar, specifically the ionic and gating currents from crayfish sodium channels.

Separation of Currents

Ionic and Gating Currents
Internal perfusion with cesium or TMA-substituted solutions effectively blocks potassium channels in these axons (11, 12). Conversely, sodium channels are completely blocked by the addition of 100 nM tetrodotoxin (TTX)

to the external medium. Leakage currents are markedly reduced by the use of fluoride (~60 mM) in the internal perfusate. Remember, however, that TTX biases the steady-state inactivation curve to the left by about 10 mV (12). This has little effect when the holding potential (HP) is ~−120 mV but can lead to a 50% suppression of the gating current when holding at −90 mV!

Beyond these simple rules are some interesting permeability properties which can generate artifacts for the unwary. For example, both cesium and potassium will pass through sodium channels albeit with very low permeabilities. Thus, you cannot measure gating currents in the absence of TTX simply by omitting the sodium from both solutions. Fortunately, TMA is impermeant and blocks outward currents through the sodium channels, such that gating currents can be recorded at positive test potentials (in the absence of TTX) when TMA is substituted for monovalent cations in both internal and external perfusates. However, these "gating currents" will be markedly distorted, at negative test potentials, by inward calcium currents (moving through sodium channels) if these distorting currents are not blocked by TTX. Thus, in the absence of TTX, gating currents should be evaluated very critically (both for kinetics and charge movement) for possible distortion by inward calcium currents even at positive test potentials. By contrast, when sodium ions are present in the external solution, calcium ions show up as blocking agents rather than as current carriers. Peak tail currents may become smaller when returning to potentials more negative than ~−80 mV, due to increasing potential-dependent block of inward sodium currents by calcium ions (13). This block can be reduced by lowering the external calcium concentration but cannot be entirely avoided because the crayfish axolemma/Schwann cell system seems to undergo structural changes (which we do not yet understand) when external calcium concentration is reduced below about 1 mM.

Leakage Currents

As mentioned above, leakage currents can be reduced by internal perfusates which contain fluoride and in which the normal potassium has been substituted by an impermeant cation. However, careful consideration of leakage currents remains important, since these currents are distinctly nonlinear in crayfish axons with a disproportionate increase at test potentials more positive than −20 mV. The linear component of I_{leak} is readily subtracted (along with the linear capacity current) by routine P/n protocols (14) which we describe in more detail below. For sodium current recordings, the nonlinear component of leakage current should be removed by subtraction of records obtained using identical protocols in the presence of TTX (100 nM).

Linear Capacity Currents

Additionally, the linear capacity currents are themselves complex, with slow components ($\tau > 100 \mu$s) which presumably arise from field-induced changes in membrane structure. These slow components are small at the start of an

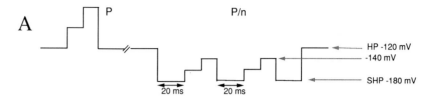

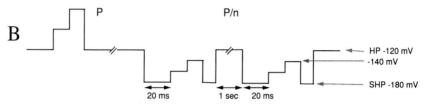

FIG. 2   Two possible *P/n* capacity current subtraction protocols. In both cases, holding potential (HP) is −120 mV; test potential is −60 mV; subtraction holding potential (SHP) is −180 mV, and the subtraction pulse is required to be no greater than 40 mV. Thus, for this test pulse magnitude, the control pulse algorithm permits a *P/2* protocol. Arguments in favor of protocol B and potential artifacts resulting from protocol A are discussed in the text.

experiment but tend to increase with time. (Marked increases in both slow capacitance and leakage current are significant precursors of membrane breakdown.) Fortunately, however, the slow capacitance is accurately linear and cancels out completely in well-chosen *P/n* subtraction protocols. Nevertheless, even at the start of an experiment, the small slow component can produce a significant distortion of gating current kinetics if the test (*P*) and control (*P/n*) pulse protocols are not exactly comparable. This concept is further explained with reference to Fig. 2A, which shows a standard *P/n* subtraction sequence, in which *n* (the number of control pulse repetitions, two in this figure) is adjusted to yield the smallest number of pulses which will fit within the control pulse voltage "window" (normally between −180 and −140 mV). The protocol shown might seem acceptable as long as the intervals between control pulses are long enough for complete decay of the slow capacity component. We would recommend intervals of at least 20 msec and have obtained excellent subtractions even where this duration had to be extended to 100 msec. However, this control protocol would require the axon to remain at the subtraction holding potential (SHP) for times long enough to permit significant recovery from the level of steady-state inactivation induced by the HP. This will cause little problem when the HP is -120 mV, near the top of the steady-state inactivation curve. However, the level of steady-state inactivation would be changed by the control pulse sequence when HP $= -80$ mV. We, therefore, always use the protcol shown in Fig. 2B, which minimizes time at SHP and hence reduces the opportunity for unrecognized changes in steady-state inactivation.

Space-Clamp Evaluation

How can one be sure that all of the membrane area from which current is collected is effectively at the same potential? The axial-wire technique was designed to permit an adequate space clamp, with the low-resistance axial-wire minimizing longitudinal resistances and the guard electrodes minimizing "edge effects" at the boundaries of the current collector region. On the other hand, we use a C-shaped guard and collector electrode configuration (which might fail in producing sufficient radial symmetry of current flow) and a smaller diameter axial wire (which might not adequately prevent longi-tudinal voltage errors). Figure 3 demonstrates a simple method for direct evaluation of space-clamp errors; a prepulse is applied to a saturating positive potential, then a second step is initiated to the previously determined reversal potential (at different times during the rise and decay of the prepulse sodium current). TTX-subtraction records should be obtained to permit subtraction of gating currents and nonlinear leak (see above). If there is no significant space-clamp error, no ionic current will flow at reversal potential and a flat record will be obtained. However, space-clamp deviations greater than about ± 0.5 mV produce sigmoid outward/inward currents (which will fail to cancel completely since they have slightly differing kinetics). The sensitivity of the method is greatest when reversal potential (E_{rev}0 lies within the voltage range of maximum kinetic change. In Fig. 3, E_{rev} is ~ 20 mV but it would be preferable to carry out such a study with E_{rev} adjusted to between 0 and -20 mV.

"Static" space-clamp errors, as shown above, are easy to evaluate and do not seem to be a significant source of error in our system. By contrast, the "dynamic" space-clamp errors caused by invasion artifacts are a much more insidious problem. To look for such errors, use at least 50 mM Na external sodium concentration and examine inward currents recorded from test pulses between about -40 and -20 mV. Invasion artifacts show up as small "bumps" of inward current appearing low down on the falling phase at -40 mV but moving closer to the peak current at more positive test potentials and tending to merge with the I_{Na} peak at potentials >-20 mV. These invasion artifacts result from action potentials occuring beyond the

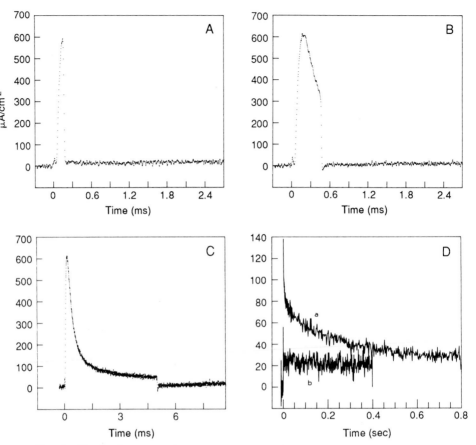

FIG. 3   Testing for space-clamp errors at three different points during a sodium current transient. (A, B, and C) Outward sodium current was initiated by a prepulse to 40 mV followed by a test pulse to E_{rev} (here 21 mV). Prepulse durations were 0.15 msec (A); 0.45 msec (B); 5 msec (C). Note the absence of biphasic currents at E_{rev}. All records were TTX subtracted using identical pulse protocols from the same axon (axon 860617). (D) Additional controls for this pulse protocol. Trace *a* is outward current at 20 mV from a different axon with a more negative reversal potential (note the high gain and slow time base of this record, used here to show continuing slow inactivation of sodium current). Trace *b* shows the absence of slow transients following exposure to TTX at this same potential. (Trace *a*, axon 851213; trace *b*, axon 851118.)

guard electrodes and seem biggest in the best axons, i.e., those in which there is the largest undamaged length between the guards and the regions where the axon has been cut open to permit entry of the perfusion pipette and piggy-back electrodes. Furthermore these errors are insidious since they

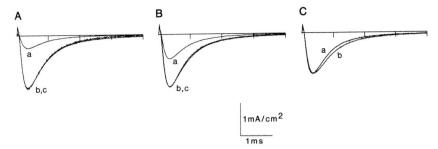

FIG. 4   Testing for series resistance errors. (A and B) Sodium currents were recorded at 0 mV with series resistance compensation set to 10 $\Omega \cdot cm^2$. (A) Reduction in peak sodium current was obtained by shifting holding potential from -100 mV (trace c) to -85 mV (trace a); however, when trace a is scaled to control peak magnitude (see trace b) no significant distortion of sodium kinetics is apparent. (B) Similar data were obtained by lowering the external sodium concentration from 50 to 25 mM. Data in A and B are from axon 830525. (C) Kinetic shifts resulting from a 5-mV change in test potential: trace a, 0 mV; trace b, -5 mV. Data from axon 830425. (Figure 1 of Ref. 11.)

may be almost impossible to pick up at potentials >-20 mV. To avoid this problem, whenever possible we record only *outward* sodium currents using zero sodium external perfusates which will not support action potentials in the surrounding zone. When the experimental logic requires recording *inward* sodium currents, we recommend using the lowest possible external sodium concentrations.

Series Resistance Compensation

Series resistance (R_s) compensation was first introduced to counteract the voltage changes predicted to occur across the axolemma when large ionic currents flow across resistances (such as those caused by the Schwann cell sheath) which are necessarily in series with the axolemma. It was thus a normal practice to use R_s compensation when recording ionic currents but not when recording gating currents, in view of the very small size of the gating currents. Recently, however, it has been pointed out that profound effects on voltage-clamp rise time can result from mismatch between actual and compensated R_s values (2, 15). Thus, clamp settling time may increase by as much as twofold when R_s compensation is turned off (see Fig. 5). Clearly, records to be used either for TTX subtractions or for comparisons of gating and ionic current kinetics should be obtained under identical conditions of R_s compensation.

Many methods have been described for R_s measurement; however, we have been more often concerned with demonstrating the absence of significant R_s error. Presuming that errors in series resistance compensation become significant only if they result in detectable changes in sodium current kinetics, we adjust R_s compensation to achieve constant kinetics in face of >fivefold changes in I_{Na} peak (see Fig. 4). We find this method to be convenient and simple to use, since it can be carried out by changing HP at any stage during a voltage-clamp experiment without unclamping the axon. Furthermore, the accuracy of the method is readily visible at around ±0.5 $\Omega \cdot cm^2$. Having achieved an acceptable initial R_s setting, we then obtain a reference capacity current record for a step from -180 to -140 mV and sampling at 1-μs intervals, in which the magnitude of the initial transient (measured from peak to the beginning of the slow capacity transient) can be used as a measure of reference clamp speed (15). Thereafter, regardless of major changes in solution osmolarity (17), or even introduction of TTX (18), we have a clamp speed reference and can reassess R_s compensation (adjusting when necessary to return to reference clamp speed) after each solution change, drug addition, etc., which might generate errors from changes in uncompensated R_s.

Nevertheless, crayfish axons are capable of generating maximum sodium currents greater than 5 mA/cm^2 at normal physiological external sodium concentrations. Even where R_s has been adjusted to ±0.5 $\Omega \cdot cm^2$, this would suggest a voltage error of ±2.5 mV which is clearly unacceptable. We therefore routinely reduce the external sodium concentration to keep peak currents <1 mA/cm^2.

Electrode Polarization

Although careful electrode preparation should prevent polarization from occurring at normal current magnitudes, it seems appropriate to be always on guard against possible polarization artifacts, particularly when sustained currents are high following pharmacological removal of fast inactivation. Wherever possible we keep the test pulse shorter than the entire record so that both the "initial current" and "tail current" levels are visible on each data trace. If the tail current fails to return to the same level as the initial current then we need to ask what happened. Could this be caused by ion accumulation in the periaxonal space, or is this an indication of electrode polarization?

Special Considerations for Gating Current Recordings

All of the points discussed above are of relevance in gating current recordings, where the accuracy of the system is being pushed to its limits. Thus, Fig. 4A shows the effects of change in R_s compensation on the recorded waveform

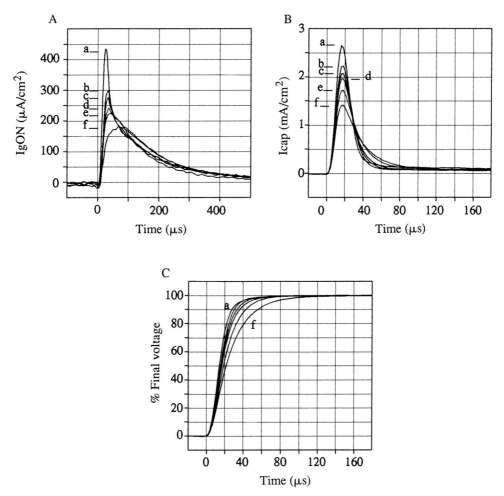

FIG. 5   Changes in gating current (A), capacity current (B), and clamp rise time (C) resulting from changes in series resistance compensation. In each panel traces were recorded at the following compensation levels: *a*, 10; *b*, 9; *c*, 8; *d*, 7; *e*, 5; *f*, 0 Ω · cm^2. (A) Gating currents in a step from -120 mV holding potential to 0 mV. (B) Capacity currents for a step from -180 to -140 mV. (C) Clamp rise times indicated by integration of the fast capacity transients shown in panel B. Membrane capacitance was 1.1 mF/cm^2 in this axon. All data are from axon 871217. (Figure 2 of Ref. 15.)

of ON gating currents (IgON). All waveforms are valid observations, although the fast-relaxing components of IgON can only be observed when clamp speed is high enough to ensure that this component has not relaxed before the clamp settles. It follows that comparable gating currents should be recorded at the same clamp speed and, hence, that great care needs to be taken (see above) to ensure that solution changes, etc., do not change clamp speed without the experimenter noting what has happened (15). Additionally, a digitization frequency of at least 1 MHz is required for adequate delineation of these fast components.

Further problems appear when one attempts to integrate gating current records to measure total charge moved. It may seem reasonable to assume that sodium-related gating charge should cease moving within about two to three times the time to peak I_{Na} at a given test potential, suggesting integration over a standard time period (e.g., 2 msec), even where such a period may fail to correspond to any well-defined "flat" region in the original IgON trace. On the other hand, two recent studies (19, 20) have provided clear evidence indicating continued very slow movements of a sodium gating charge extending over times as long as 100 msec. Additionally, potassium gating currents must occur in these axons and will contribute to the slow charge movements which distort the expected flat at the end of the gating current data trace. We recommend taking frequent control steps (e.g., −120 to 20 mV) during any given experimental sequence where no changes in these control kinetics would be expected. We then integrate each control record and overlay the integrations on the computer screen (scaling, if necessary, to correct for rundown). Unexpected changes in the kinetics of these integrations should lead one to suspect one or other of the artifacts listed below.

Finally, we point out a selection of minor artifacts which can distort gating current records, increasing the difficulty of selecting a good criterion for IgON integrations. Most important is a possible contribution from failure to subtract slow capacity currents, due to use of too short an interval at SHP (see Fig. 2 and associated discussion above). Second, if the SHP is too negative (i.e., <−180 mV) slow "breakdown" of the membrane may occur, giving an increase in leak during the control-pulse sequence leading, following subtraction, to an apparent continuation of test-pulse charge movement. Alternatively, failure of potassium block during the test pulse may yield small outward currents which mask the slower components of the gating currents. We have seen all of these phenomena at different times and there is no simple antidote other than constant vigilance. One approach we have found useful for analysis of suspicious findings is to record the test (*P*) and control (*P/n*) pulses *separately*. They can always be subtracted (or added) later during computer analysis; separate recording of test and control pulses provides an opportunity for close scrutiny of both pulses for their separate crop of possible artifacts.

Summary

Axial-wire voltage clamping in crayfish medial giant axons provides an excellent opportunity to observe the behavior of sodium and potassium channels in their native membrane under accurately controlled conditions. Current magnitudes are sufficient to permit detailed measurements of ionic and gating current kinetics without some of the problems inherent in the squid axon system. Although the axons are smaller than squid giant axons, the fluid axoplasm simplifies internal perfusion to the point that the initial setup time is quite comparable in these two preparations. Given the ready availability of crayfish on a year-round basis, we believe that the crayfish axon is a premium preparation for detailed biophysical studies.

References

1. P. Shrager, J. G. Starkus, M-V. C. Lo, and C. Peracchia, *J. Gen. Physiol.* **82,** 221 (1983).
2. J. R. Stimers, F. Bezanilla, and R. E. Taylor, *J. Gen. Physiol.* **89,** 521 (1987).
3. J. G. Starkus, B. D. Fellmeth, and M. D. Rayner, *Biophys. J.* **35,** 521 (1981).
4. P. Shrager, R. I. Macey, and A. Strickholm, *J. Cell. Physiol.* **74,** 77 (1969).
5. P. Shrager, *J. Gen. Physiol.* **64,** 666 (1974).
6. P. Shrager, *Ann. NY Acad. Sci.* **264,** 293 (1975).
7. J. G. Starkus and P. Shrager, *Am. J. Physiol.* **235,** C238 (1978).
8. M. D. Rayner and C. A. G. Wiersma, *Nature (London)* **213,** 1231 (1967).
9. W. K. Chandler and H. Meves. *J. Physiol. (London)* **180,** 788 (1965).
10. H. M. Fishman, *IEEE Trans. Bio-Med. Eng.* **BME-20,** 380 (1973).
11. J. G. Starkus, S. T. Heggeness, and M. D. Rayner, *Biophys. J.* **46,** 205 (1984).
12. S. T. Heggeness and J. G. Starkus, *Biophys. J.* **49,** 629 (1986).
13. P. C. Ruben, J. G. Starkus, and M. D. Rayner, *Biophys. J.* **58,** 1169 (1990).
14. C. M. Armstrong and F. Bezanilla, *J. Gen. Physiol.* **63,** 533 (1974).
15. D. A. Alicata, M. D. Rayner, and J. G. Starkus, *Biophys. J.* **55,** 347 (1989).
16. J. W. Moore, M. Hines, and E. M. Harris, *Biophys. J.* **46,** 507 (1984).
17. M. D. Rayner, J. G. Starkus, P. C. Ruben, and D. A. Alicata, *Biophys. J.* **61,** 96 (1992).
18. J. G. Starkus and M. D. Rayner, *Biophys. J.* **60,** 1101 (1991).
19. M. D. Rayner and J. G. Starkus, *Biophys. J.* **55,** 1 (1989).
20. P. C. Ruben, J. G. Starkus, and M. D. Rayner, *Biophys. J.* **61,** 941 (1992).

[6]   Biophysical Studies of Ion Channels in Glial Cells and Axons of Myelinated Nerves

S. Y. Chiu and P. Shrager

Introduction

Single myelinated fibers were first successfully dissected for biophysical studies in the early 1950s by Tasaki and Stampfli. In the years since this achievement, single myelinated fibers have been subjected to analysis with increasingly sophisticated biophysical techniques. Previously concealed compartments of the nerve, like the axon region wrapped by the myelin sheath, have been "unlocked" for investigation. Characterization of the currents at the node of Ranvier has been extended from amphibia to mammals. Traditional studies carried out using macroscopic currents are now augmented with single-channel analysis. Schwann cells, the ensheathing cell that once was thought to be a passive partner to the axon, are now found to express voltage-gated channels. In this chapter, we review classical techniques and describe recent methods to study ion channels on single myelinated fibers. We limit the scope to either techniques developed in our laboratories or biophysics methods familiar to us. The emphasis is on voltage-gated ion channels of amphibian and mammalian myelinated fibers.

Morphology of Myelinated Fibers

A single myelinated fiber is composed of a single axon wrapped by the myelin sheath of Schwann cells. The axon is about 10–15 μm in diameter, and each Schwann cell and its associated myelin span about 1000–2000 μm along the axon. The axon covered by the myelin sheath is termed the internodal axon. The axon is exposed at the node of Ranvier, a 0.5- to 2.5-μm gap between two consecutive Schwann cells. Both axons and Schwann cells express voltage-gated ion channels. There are multiple types of channel. Certain channels are strongly segregated. The following describes methods to detect these channels.

Methods in Neurosciences, Volume 19

Schwann Cell

To date, ion-channel studies have been carried out only in mammalian Schwann cells. There have been no reports of ion channels in amphibian Schwann cells. Schwann cells can be obtained in culture, in which sciatic nerves from newborn rats are minced, dissociated into single cells, and placed onto tissue wells. These single cells can then be subjected to standard whole-cell patch clamping. Because Schwann cells undergo dedifferentiation once they are separated from axons, the channel properties of cultured Schwann cells need not represent those found *in vivo*. In this chapter we are only concerned with Schwann cell channels *in vivo*, as studied in freshly isolated cells.

Acute Isolation of Schwann Cells

Schwann cells still retaining a myelin sheath and association with axons can be acutely isolated from newborn and adult rats for patch-clamp analysis (1). Sciatic nerves are taken from these animals. P7 and older sciatic nerves should first be desheathed; those nerves younger than 1 week should be left sheathed to avoid damage to Schwann cells and axons during the process of desheathing. Desheathing can be done using two fine Dumont forceps. While under a saline-type solution, one end of the nerve trunk is gently teased with two pairs of forceps until the sheath at the end comes cleanly off the underlying nerve fibers. The loosened portion of the sheath is tightly grabbed with one pair of forceps, while the nerve trunk at the end where the sheath comes off is held by the other pair. The sheath is then pulled off, in a direction longtudinal to the direction of the nerve trunk, where the rest of the nerve trunk is firmly held down with the forceps.

The deshearthed nerve is then gently teased using two fine sewing needles. The ends of the needle should not be sharp, otherwise the nerves will be severed. To tease, simply press the two needle tips into the middle of the nerve trunk. The distance between the two needle tips is critical; they should be about 10% of the distance across the diameter of the nerve trunk. The needles should be pushed right through the nerve so that the tips touch the surface (using glass or the bottom of a plastic petri dish) on which the nerve is resting. The needle tips are then moved away from one another slowly, in a direction perpendicular to the longitudinal axis of the nerve trunk; the entire teasing process should be monitored carefully under a microscope at 40–60×. A "good" tease is one in which a network of single fibers is produced between the region separated by the two needles. A "bad" tease is one in which the teasing motion simply produces smaller fiber bundles without many single fibers. This teasing is then repeated at intervals of 0.5 cm along the nerve trunk. When teasing is completed, cut the nerve into two to three smaller segments. Typically, one adult animal yields six to eight such segments from the two sciatic nerves. These teased, cut nerve segments are transferred with glass pipettes to a Dulbecco's modified Eagle's medium (DMEM) solution containing 0.3% collagenase and incubated at 37°C for 1–2 hr. The vial containing the enzymes and the nerves should be gently rocked a few times every 15–20 min.

Spreading the Nerves on Glass Coverslips

After collagenase treatment, transfer the nerves with glass pipettes to a petri dish containing enzyme-free saline solution. The volume of this enzyme-free solution should be large (~20 ml) compared with the small amount of enzyme that invariably gets transferred along with the pipette. Once the nerves are ejected into the petri dish, gently blow the solution at the nerves to loosen them. This should be done gently so that the nerve segments do not completely fall apart.

Prepare some clean glass coverslips (No. 1 thickness). We recommend 25 × 25-mm square coverslips cleaned with ethanol and flamed. There should be sufficient area on the coverslip for the nerves to roll over during the spreading process. First put several tiny drops of Vaseline in the middle of the bottom of a 5-cm-diameter petri dish. Then place a glass coverslip over these Vaseline drops and press down on the glass with forceps so that the coverslip is firmly adhered. Next, suck up a small segment of collagense-treated nerves into the tip of a regular pasteur glass pipette. Gently eject the nerve segments onto the coverslip together with a small drop of saline carrying the nerve. Note that the volume of the saline ejected together with the nerve is critical. It should be a small drop only. Then place a second coverslip on top of the nerve and gently let go. This whole procedure should be monitored under a microscope. As soon as the top coverslip presses down on the nerve, the small drop of saline in which the nerve is bathed spreads out evenly across the tiny space sandwiched between the two coverslips. Often, single nerve fibers, already partially loosened from the nerve trunk by the collagenase treatment, are instantly spread out too, being carried along by the gentle fluid current created by the thinning of the original drop of fluid. Now gently pour saline solution into the petri dish so that the "coverslip sandwich" is completely covered by solution. It is critical that the two coverslips not come apart during this step; this can be achieved by pouring the solution directly and gently over the coverslip sandwich. With the coverslips completely immersed in solution, use forceps to gently slide the top coverslip sideways to expose the many single fibers that are already firmly adhered to the bottom coverslip.

Identifying Nonmyelinating and Myelinating Schwann Cells

Two types of fibers can be distinguished on the bottom coverslip. One is the nonmyelinated fiber. These fibers are long and thin, with prominent Schwann cell bodies aligned at regular intervals along the fibers, and completely lacking myelin. Sometimes these fibers appear as small bundles, indicating insufficient enzymatic dissociations. Electron microscopic studies show that fine axons, each ~1 μm in diameter, are still engulfed by the Schwann cells. These axons run longitudinally along the grooves of the Schwann cells, and sometimes individual axons can be seen peeling out from the cell. Gigaseals can be formed on the cell body with patch pipettes having resistances of 4–5 MΩ. In most instances, the seal can be assumed to form on the Schwann cell. It is unlikely, but possible, that some of the seals are formed directly on the 1-μm axons buried in the Schwann cell groove. There are several criteria to distinguish axon from Schwann recordings. First, sodium current amplitude is about 10× larger than a typical Schwann cell recording. Second, lucifer yellow can be introduced into the patch pipette to verify that, in whole-cell mode, the dye fills the cell body and not individual axons. Third, the decay of the whole-cell capacity current is predominantly a single exponential giving an effective total capacity of ~15 pF. Fourth, no action potentials can be elicited. We have seen action potentials in one case in which the seal is deliberately formed on a single 1-μm axon that happened to have been peeled off from Schwann cells during the enzymatic dissociation procedures.

The other fiber type adhered to the bottom coverslip is the myelinated fiber. These fibers are easily distinguished from the nonmyelinated fibers by their larger size (~10–15 μm in diameter) and by the presence of myelin. Both the Schwann cell body and the paranodal processes of the cell can be subjected to patch-clamp analysis.

Cell Body

The cell body usually appears to bulge out prominently at the middle of an internode. We recommend that only Schwann cell bodies associated with intact internodes be used. Thus, use only those internodes with nodes of Ranvier present at both ends of the internode. Avoid using internodes that have been broken during the dissociation procedures; the

Schwann cells associated with these broken internodes may have leaky membranes. Good gigaseals could be formed readily with patch pipettes having resistances ranging from 2 to 5 MΩ. The success of forming good seals depends on how well the collagense treatment cleans off the basal lamina normally covering the Schwann cell. When the membrane under the pipette tip is ruptured to go into whole-cell mode, the capacity current is huge and has multiple exponentials. The multiple exponential nature of the current reflects charging of a distributed myelin sheath. Rigorous biophysical characterization of ion channel kinetics should be done with excised patches (outside-out or inside-out). In an adult myelinating Schwann cell, the cell body does not express detectable voltage-sensitive channels. However, following nerve transecton or in young nerves where myelin is just beginning to form, potassium-channel activities can be detected on the Schwann cell body.

Figure 1 shows (top) a simplified diagram highlighting key features of the dissociation technique and whole-cell recordings at the cell body of (A) nonmyelinated Schwann cells and (B) myelinated Schwann cells (2). Note the absence of voltage-sensitive sodium currents in the cell body of the myelinated cell (B). The differential expression of sodium channels in these two cell bodies can also be demonstrated by excised outside-out patches obtained at the cell body (Figs. 1C and 1D).

Internodal Surface

Gigaseals can also be formed on the internodal surface between the cell body and the paranode. This membrane represents the outermost membrane of the Schwann cell before the spiraling of the membrane toward the internodal axon begins. Channel properties of this outermost membrane have not been examined.

Paranodal Processes

This is the most difficult part of a myelinating Schwann cell to use for patch-clamp analysis (3). The Schwann cell membranes in this region have complex morphology and the surface area is not smooth. Further, it is so close to the nodal axon that is richly endowed with ion channels that care must be exerted to rule out recording from the axonal surface. It is likely that the side of the Schwann cell membrane facing the nodal surface expresses interesting ion channels. To maximize the chance of recording from this side of the Schwann cell membrane, we recommend looking for retracted paranodes. These para-nodes can be recognized by a very wide nodal gap of up to 10 μm flanked on two sides by retracting paranodal Schwann cell bulbs. These retracted paranodal bulbs appear typically swollen, indicating mechanical trauma during the retraction. Gigaseal formations on these swollen paranodal bulbs are not easy to obtain but several steps can increase the success of seal formation. We recommend using very narrow pipette tips (pipette resistance of ~10 MΩ). Apparently, the surface of these swollen paranodal bulbs has many tiny grooves and foldings so that a large pipette orifice, by covering a lot more ruffled surface than a small pipette tip, will run into problems with seal formation.

To exclude recording from axons, use lucifer yellow-filled pipettes. After characterization of the channel activities in cell-attached patches, rupture the membrane and examine the pattern of intracellular dye distribution. If the recording has been from axons, the staining pattern is easily recognized as a long thin fluorescent strand extending into both internodes on both sides flanking the recording site (see dashed lines in Fig. 2C). Note that the width of the staining (i.e., the width of the axon) should be less than the width of the internode. If the recording has been from the paranodal Schwann cell membrane, the lucifer yellow stain should be restricted to one of the two internodes flanking the recording site (see dashed line in Fig. 2B). Further, unlike the axon stain, the Schwann cell stain should be restricted to the outermost thin cytoplasmic layer near the surface of the internode. This literally gives a nice fluorescent outline of the internode, as contrasted with the thin, thread-like axon stain that travels longitudinally through the internode. A telltale sign that you are recording from axons is that axonal channel density is much higher compared with Schwann membranes.

Axon

Three different parts of the myelinated axon can be subjected to ion channel analysis. They are the nodal gap (~1 μm wide), the paranodal axon (extending about 10–50 μm on both sides of the node), and the internode proper (about 1000–2000 μm long). The nodal gap is normally accessible for ion channel studies. In contrast, both the paranodal and the internodal membranes are normally covered by myelin. To expose these axonal membranes, various acute and chronic demyelination methods are applied. Acute exposure reveals channels normally present under the myelin. Chronic exposure (several days or weeks following demyelination) reveals an internodal membrane that may have undergone changes in channel properties. Macroscopic currents can be studied with a gap clamp or loose patch clamp; single channels can be studied with gigaseal patch clamp.

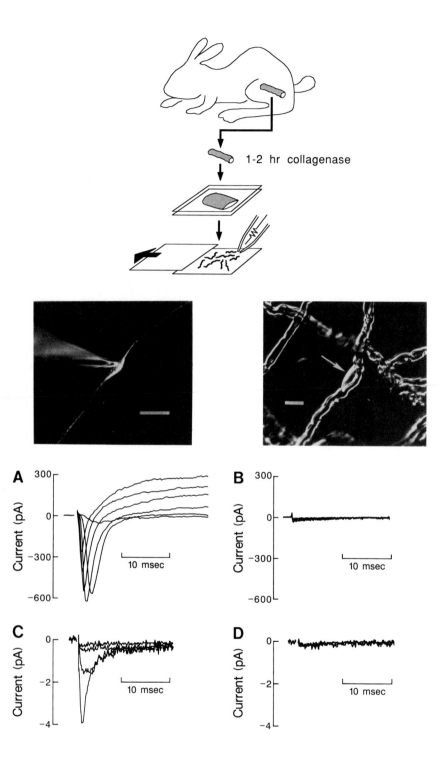

1-2 hr collagenase

A

Current (pA)

300

0

−300

−600

10 msec

B

Current (pA)

300

0

−300

−600

10 msec

C

Current (pA)

0

−2

−4

10 msec

D

Current (pA)

0

−2

−4

10 msec

Gap Clamp

Node of Ranvier

Macroscopic currents from a single node of Ranvier are best studied using the gap voltage clamp. For amphibian fibers (frog), dissection of single axons and mounting of the fibers in the gap-clamp nerve chamber for voltage clamping has been described in detail by Stampfli and Hille (4). Their review should be consulted. Dissection of single mammalian fibers is much more difficult, and a few technical points are pointed out here to ensure a high probability of obtaining a viable fiber. The key point is to avoid longitudinal stretches at all steps, since mammalian fibers are exquisitely sensitive to mechanical trauma. Rabbit sciatic nerves (5) are easier to dissect than those of smaller mammals like rats or mice. The desheathing procedure is identical to that described in the preceding

---

FIG. 1   A schematic diagram illustrating the method for acutely isolating mammalian Schwann cells for patch clamping. Sciatic nerves are obtained from an adult mammal, desheathed, and treated with collagenase (0.3%) for 1–2 hr. After enzymatic treatment, the nerves are placed between two clean glass coverslips and the top coverslip is gently pushed aside under a Locke solution. Single nonmyelinated (left) and myelinated fibers (right) can be found attached to the bottom coverslip, and the associated Schwann cell body is subjected to patch-clamp recordings as early as 2 hr after isolation from the animal. (A, B) Examples of whole-cell recordings of ionic currents at the cell body of a nonmyelinating and a myelinating cell. Note that an inward sodium current and an outward potassium current are detectable in (A) but not in (B). Currents were generated by depolarizations in 20-mV steps (A, from −45 to 55 mV; B, from −35 to 85 mV). The holding potential was −75 mV in both cases; in (B) a prepulse to −120 mV for 200 msec was used to remove sodium channel inactivation. Pipettes contained a 140 mM KCl solution. (Modified from Chiu, 1987, 1988). (C, D) Examples of excised outside-out recordings of ensemble sodium currents from membrane patches obtained from the cell body of a nonmyelinating (C) and a myelinating (D) cell. The patch was depolarized to −60, −40, −20, and 0 mV. The holding potential was −75 mV, and a 500-msec prepulse to −120 mV was used to remove sodium-channel inactivation. Each current trace represents the ensemble average of 50 identical steps at each depolarization. The patch in (C) contained a few single sodium channels which opened stochastically in individual traces (not shown). The patch in (D) contained no observable sodium-channel activities in all single traces. Pipettes contained a 140 mM CsCl solution. [Reproduced with permission from Fig. 1 of Chiu, *Glia* **4**, 541 (1991). Copyright © 1991, Wiley-Liss, a division of John Wiley & Sons, Inc.]

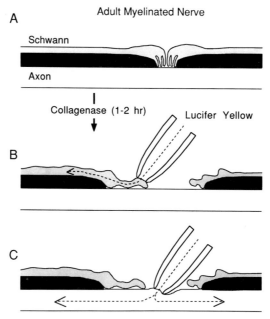

FIG. 2 Schematic diagrams illustrating patch-clamp recordings of Schwann cell membranes at the paranode. Adult myelinated nerves are treated with collagenase to retract the node–paranode. Then a patch pipette, filled with Lucifer yellow, is pressed against the retracted paranode. Gigaseals are formed, and channels, if present, are recorded. Afterward, to identify the membrane (Schwann or axon), the membrane under the pipette tip is ruptured to allow Lucifer yellow to diffuse intracellularly. For Schwann cells, Lucifer yellow (dashed curve) should be restricted to the surface of one of the two internodes. For axons, the pattern of the dye staining (dashed curve) should look like a thin line going through both internodes.

sections on acute Schwann cell isolation. Dissect as much fiber length as is possible toward the sciatic notch or the spinal cord. Pull the sheath off starting from the spinal cord end of the nerve. This should be done under a dissecting microscope. If resistance is encountered, the chances are that tissues have piled up at the junction at which the sheath is separating from the underlying nerve trunk. Should this be the case, slip the sheath back onto the nerve (by reversing the direction of pulling) and reinitiate desheathing. You may have to use fine scissors to cut the blocking tissues at the junction to free the sheath from being caught. Normally, a sheath should come off with no appreciable resistance when no tissue clumps build up along the pathway. In rabbit, there are about two or three small nerve bundles, each about 0.5 mm in diameter, which branch off from the main bundle roughly along the midpoint of the sciatic nerve. We recommend using only these loosened branches for single-fiber dissection. With one pair of forceps, hold the distal end of the main nerve bundle. With another pair of forceps, hold one of the side branches. At this point, the side branch has about 1–2 cm already hanging loose, with the rest still adhering to the main nerve trunk. The objective is to peel the side branch to separate another ~2 cm of length from the main trunk. This extra length of branch nerve is ideal for further teasing to isolate a single fiber because another remnant of the main sheath, left behind after the first desheathing, will be loosened along this ~2-cm length of the side branch.

Teasing is done as described in the preceding sections on Schwann cells. The point here is that a single node should be visually identified as early as possible during the teasing under the microscope, so that its fate (the node) can be followed as the network of single fibers is being created by the two teasing needles. This allows you to monitor the node for any longitudinal stretches, and, if needed, to adjust the teasing needles accordingly to slacken or reduce such stretches. This, in our opinion, is the most critical part in obtaining a viable fiber. It is possible to isolate a node with only mechanical stretches applied perpendicularly to the fiber axis. One contributing factor to the difficulty in the mammalian dissection compared to that in the frog is that it is very difficult to visualize a single mammalian node because it has a much narrower nodal gap than the frog's. Further, never use a mammalian node of Ranvier that is clearly visible; this always means the node has been stretched. Sometimes, a healthy node appears as a slight indentation in the myelin with no obvious gap between successive internodes. That this is a node can be confirmed by noting its distance from the two nodes at the other ends of the two flanking internodes. This distance should be the appropriate length of an internode, namely, ~1000–2000 μm for a large 10- to 20-μm-diameter fiber. Before mounting the single fiber into the nerve chamber, the initial nerve trunk is reduced to two short unteased trunks (each about 1 cm long) bridged by one interrupted single fiber with three nodes exposed.

Internodal Axon Acutely Exposed

Compared to the node of Ranvier, isolation of single internodes for gap-clamp studies is very easy since no care need be taken to avoid stretch. Simply isolate a single fiber as described above, and mount the middle portion of the internode in the recording pool that normally is used for nodal recording (6). An approximately 50- to 100-μm length of the internode is put in the recording pool. To acutely remove the myelin, apply a solution containing 0.2% lysolecithin. Other more potent synthetic analogs of this detergent (lysophosphatidylcholine palmitoyl, 0.1%) are available from Sigma. The membrane capacity of the internodal segment under demyelination should be monitored. We have a program that takes a leakage current trace (hyperpolarzing by 45 mV from a holding potential of -70 mV) and a test trace (depolarizing 100 mV from the holding potential) every 1–30 sec. The leakage trace is used to monitor the increase in capacity, and the test trace is used to monitor the concomittant increase in the outward, delayed potassium current. Figure 3 shows the potassium currents (left column) and capacity currents (right column) at various times after lysolecithin treatment. During the first 30–45 min, the capacity increases slightly, if at all (Fig. 3, 0–45 min). In this same period, no outward potassium current appears. Thereafter, the onset of which is slightly variable, the capacity current (shown as fast downward deflections in the right column of Fig. 3) suddenly enters a phase of extremely rapid increase (Fig. 3, 48–50 min). This signals that the internodal axon is about to be electrically exposed. In parallel, an outward potassium current appears and its size increases rapidly. This moment is critical because the internodal axon can be rapidly destroyed, leaving no time for collecting usable data. Several steps can be taken to prolong the survival of the exposed internodal axon. First, as soon as the capacity undergoes a rapid increase, quickly rinse the recording pool with a detergent-free solution. Sometimes, a low ionic strength solution (like isotonic sucrose) could drastically prolong the survival of the internodal axon. After the internodal axon is stably exposed (as judged by a stable capacity current with a very fast declining phase and by a steady outward current elicited by depolarizations), the low ionic strength solution can be replaced with a normal Ringer's solution, and internodal ionic currents recorded. An internodal axon exposed in this way stays viable for 5–30 min. An experiment usually ends abruptly without warning when the internodal axon breaks.

Patch Clamp

Even though gap-clamp analysis shows that ion channels can be found on the axolemma underneath the myelin. Gap-clamp studies have two disadvantages with regard to further biophysical characterizations of these channels. First, there may be space-clamp problems associated with recording currents from a 100-μm axon segment using a gap clamp. This precludes a rigorous analysis of channel kinetics. Second, the gap clamp gives at best qualitative information concerning ion channel distribution along the axolemma. Both of these disadvantages can be circumvented by using a patch pipette. Single-channel analysis, as well as channel mapping with high spatial resolution, is now possible on axons of myelinated fibers.

Loose Patch-Clamp Recording of Internodal Currents

A central problem in developing therapeutic measures in demyelinating diseases such as multiple sclerosis is the restoration of conduction following myelin destruction. Ionic channels that may have been uncovered in this latter process now take on special importance. In order to characterize the distribution of ionic channels throughout the entire internode, techniques have been developed that allow access by patch pipettes to the axolemma under the myelin. Demyelination is achieved through the use of lysolecithin, a method originated by Hall and Gregson (7). This procedure has been used successfully in both amphibian and mammalian peripheral nerve.

Xenopus are anesthetized by immersion in tricaine methanesulfonate (3.7 g/liter H_2O) for about 15 min. The animal is then placed on wet paper towels and is partially covered by another paper towel that has been soaked in the anesthetic solution. The sciatic nerve in one leg is exposed and is elevated on a glass pipette which serves to provide a firm base. An injection pipette is made by drawing a piece of hematocrit tubing on a microelectrode puller and breaking the tip to a diameter of 18–20 μm. The pipette is backfilled with Ringer's containing lysolecithin (10 mg/ml) that has been sterile-filtered and the meniscus is marked. The injection pipette is held in a micromanipulator and is inserted obliquely into the nerve by tapping gently on the manipulator. One microliter is injected using pressure generated by a glass syringe. Care is taken to avoid leakage of solution at the point of injection. This requires a balance between inserting the pipette too far, which may block it, and too little, which allows leakage. The nerve can be seen to swell slightly during the injection. This procedure is designed to minimize mechanical damage to the axons. The wound is sutured and the frog is allowed to recover in very shallow gently running tap water. Rats are anesthetized, the sciatic nerve in one leg is exposed, and the remainder of the procedure is identical to that for *Xenopus* except that (a) Locke's solution is used in place of Ringer's; (b) the wound is closed with sterile surgical staples, and (c) recovery is aided by warming the cage bottom slightly with a heating pad.

The lysolecithin damages and vesiculates the myelin at several places along an internode. Macrophages are then recruited and all damaged myelin is removed by phagocytosis. Loose patch recordings are possible as soon as the vesiculation is complete, 1 day postsurgery, although it is easier once the bare axolemma has been exposed, a process requiring 1 week (8, 9). On the day of the experiment the nerve is dissected, desheathed as described

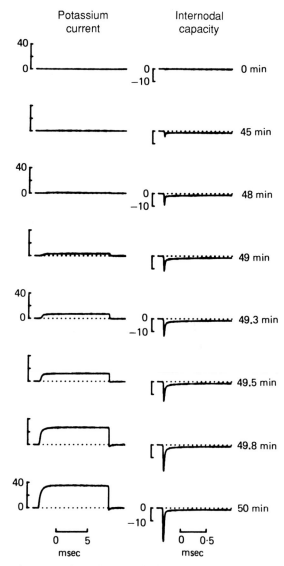

FIG. 3   Uncovering potassium channels in frog internodes by lysolecithin. A single frog internode is mounted in a gap-clamp chamber with 100 μm of internodal segment positioned in the recording pool. Lysolecithin (0.2%) is added to the recording pool to selectively demyelinate the 100-μm segment of internode. Potassium current (left column) and membrane capacity (right column) from the demyelinating segment are monitored at various times (indicated) following lysolecithin addition. Holding potential is −70 mV. The delayed, outward potassium current is monitored with a

earlier, and bathed in collagenase (3.5 mg/ml) for 1.5 hr. Axons can be spread by lowering the bath level and held in place by glass rods (diameter 50 μm) placed horizontally and held by dabs of Vaseline. Axons at the edge of a branch can be pulled gently with sewing needles and tacked in place with Vaseline. The pulled axons are likely to be damaged and are not patch clamped. Only fibers close to the interior are used. The procedures are designed to keep fibers parallel to allow observation over at least two internodes.

Loose patch pipettes are pulled from Kimax-51 glass and fire polished to a tip i.d. of 1.3–1.8 μm. Pipette resistance (R_p) after filling (Ringer's or Locke's) was about 2 MΩ. After mechanical contact of the pipette tip with the fiber, suction is applied and maintained with a Gilson micrometer syringe. The seal resistance rises slowly as the myelin debris enters the pipette tip and then sometimes jumps rapidly to >10 MΩ as contact with the axolemma is established. The minimum seal resistance for stable, low-noise recordings is 2 × R_p, but a value of at least 10 MΩ is preferable. It has been possible to patch successfully internodal sites as soon as the myelin is damaged and before macrophage removal (~1 day postsurgery).

The loose patch-clamp system we employ is based on the bridge circuit of Stuhmer *et al.* (10). In loose patch recording the seal resistance is not stable and tends to drift on a time scale of tens of seconds when the patch is held at its reversal potential. If the clamp uses the pipette resistance (R_p) and seal resistance (R_{seal}) as a voltage divider to determine the potential across the patch, R_{seal} must be measured with every pulse and corrections to both V_m and I_m applied. Even then, errors are inevitable. The bridge circuit eliminates one of these sources of error since, once balanced for R_p, it ensures that V_m at the pipette tip will accurately follow the command potential *independently* of R_{seal}. One important source of error remains since the current calibration also depends on R_{seal}. The original bridge circuit uses a potentiometer to null the seal current in response to a small hyperpolarization. However, we found that again this requires frequent adjustment and is thus cumbersome.

We have introduced two modifications to the basic loose patch circuit. First, the seal potentiometer is replaced by a programmable resistor. This has been constructed as a series combination of 12 elements, ranging from

---

fixed depolarization to +80 mV. The internodal membrane capacity is monitored with a hyperpolarizing pulse to −115 mV. Leakage currents have been subtracted from the potassium current traces. Current scale is in units of nA. [Reproduced from Fig. 3 of Chiu and Ritchie, *J. Physiol.* 322, 485 (1982)].

16 Ω to 32 kΩ, with each successive element increasing by a factor of 2 over the previous value. A set of miniature reed relays (e.g., Magnecraft W107DIP-5) across each element then allows simple control of the total value by a binary 12-bit word provided by a parallel port of the computer interface. Relays are used since their ON resistance is negligible, they provide virtually complete isolation between digital and analog circuitry, and their switching time ($\sim$300 μs) is well below that required. During seal formation hyperpolarizing pulses are applied and the resulting currents are used to calculate the remaining uncompensated seal ($+$leakage) resistance. The programmable resistor is then adjusted to be the parallel combination of the orginal value and the uncompensated value. The correction then converges rapidly. During recording, corrections are made 400 msec prior to each test pulse. Any small residual leakage current can be subtracted by the software in the usual manner from responses to additional hyperpolarizing pulses.

In the second modification we temporarily insert a 1-MΩ variable resistor between the bath reference electrode and ground and read the potential across it with a voltage follower. When the patch pipette is first placed in the bath this resistor is "clamped" and the bridge is adjusted to balance precisely R_p by nulling the difference between the command potential and the voltage across the 1-MΩ resistor. During recording, this resistor can be reintroduced into the circuit when necessary to ensure that the holding current is accurately zero so that the cell is held precisely at its resting potential.

Figure 4 illustrates internodal currents recorded over a wide range of times postsurgery. At 1 day myelin is damaged, but not yet removed. At 7 days macrophage phagocytosis is complete all along an internode. At 31 days remyelination is under way and relatively few internodal sites are still bare. At 54 and 150 days nerves are redemyelinated by an additional exposure to lysolecithin prior to recording. In all cases, K$^+$ currents are reduced by addition of 7.4 mM TEA$^+$. Note that the amplitude of peak Na$^+$ currents is relatively constant over this 5-month period, ranging from the earliest stages of demyelination to virtually complete remyelination with formation of new nodes of Ranvier.

The loose patch clamp can be used to determine whether Na$^+$ channels at a particular site in a cell are activated by signals generated elsewhere. As an example, we have applied patch pipettes to demyelinated sites along an internode and stimulated the axon several millimeters away. This technique rests on the fact that the patch clamp controls the *external* potential, and the internal voltage may still vary if there is a current source. If the patch is held at the resting potential and a depolarzing signal of sufficient strength propagates to the site, all-or-none transient inward currents will be recorded. By separately applying depolarizing pulses to the patch through the clamp in the usual way it is then possible to calculate the fraction of available

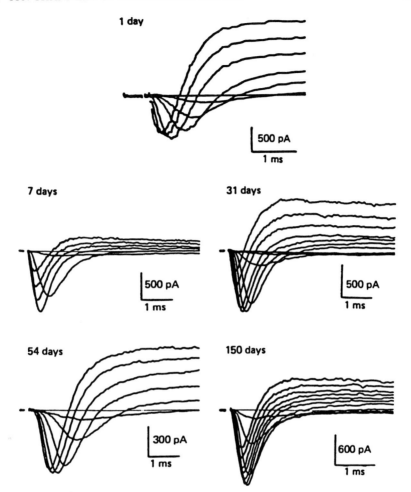

FIG. 4 Internodal macroscopic ionic currents recorded 1 day to 5 months postinjection of lysolecithin. The axons at 54 and 150 days were reinjected with 1.3 μl of 0.2% lysolecithin 1 day prior to the experiment. Holding (V_h) and test (V_t) potentials (mV) relative to rest were 1 day, V_h 0, V_t 45, 55, 65, 85, 105; 7 days, V_h -20, V_t, 40, 60, 80, 100, 120, 140; 31 days, V_h 0, V_t 30, 40, 50, 60, 70, 80, 90, 110, 130, 150; 54 days, V_h -20, V_t 40, 50, 60, 70, 80, 90; 150 days, V_h -40, V_t 70, 80, 90, 110, 120, 130, 140, 150, 160. Pipette and bath solutions were Ringer's with TEA$^+$, 7.4 mM. Reprinted with permission from Shrager, *J. Physiol.* **392**, 587 (1987); **404**, 695 (1988).

channels that is activated by the propagating signal. Figure 5 illustrates one such experiment. The upper family of currents at both sites A and B resulted from depolarizations applied directly to the pipette. The middle traces were elicited by stimulation of propagating action potentials at a proximal site. The sketch shows the recording sites.

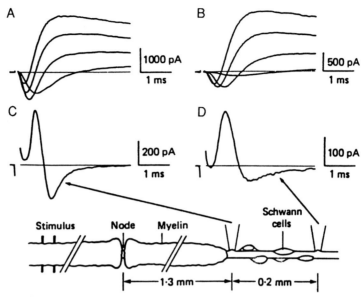

FIG. 5   Ionic channel activation resulting from propagating signals. The axon signal is shown 13 days postinjection. The sketch at the bottom shows the axon and the two recording positions. (A and B) Responses to voltage pulses of 40, 60, 80, and 100 mV from rest delivered through the clamp. (C and D) All-or-none currents recorded in response to external stimulation. Ringer's solution. Reprinted by permission from (8).

This same distinction between external and internal voltage control may lead to unusual currents if there is a site near the patch that has a high density of ionic channels. We have used this idea as a means of detecting the sharp gradient in density of Na$^+$ channels at the paranode. Figure 6 illustrates typical records. A patch has been made at a paranodal site, held at the resting potential, and depolarized by the amount shown next to each trace. The clamp depolarization activates Na$^+$ channels within the patch and an inward current develops. This current spreads along the axon internally, and a fraction of it reaches and depolarizes the node. Below a threshold level the nodal channels are not activated (73 mV depolarization). Above it, the very high density of nodal Na$^+$ channels then results in a large inward current (regenerative since the potential is not controlled at this point), which also spreads, and a fraction of the latter then passes through the patch and into the pipette, appearing as *outward* current through the patch. This pattern, which has been seen only at paranodal sites, persists for several months

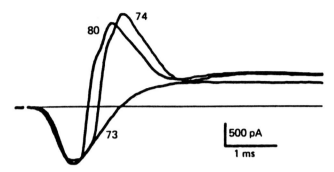

FIG. 6   Ionic currents recorded from a paranodal region, 57 days postsurgery. The patch was held at the resting level and depolarized by the amounts shown in the figure. Ringer's with TEA$^+$, 7.4 mM. Reprinted by permission from (8).

following demyelination, indicating that nodal Na$^+$ channels do not simply break up and diffuse into the internode after myelin disruption.

Single-Channel Recording

The loose patch clamp is useful primarily for fast transient channls, e.g., Na$^+$ and T-type Ca^{2+} channels. With long pulses, artifactual slowly developing currents are encountered that are probably due to an electrostriction of the seal. For this reason, and also because of the wide variety found in various tissues, we study K$^+$ channels primarily with single-channel analysis. It is possile to obtain gigohm seals to internodal sites in lysolecithin-demyelinated fibers. This has allowed recording for the first time of single-channel ionic currents from sites at all locations along the axolemma. Unlike loose patch, however, there is only a narrow window during which this can be done: in *Xenopus*, 6–8 days after injection. Prior to this time removal of myelin debris by macrophages is incomplete. After about 9 days Schwann cells begin to adhere to axons and extend processes preparatory to remyelination. Jonas *et al.* (11) have described an acute method of demyelination that allows gigohm seal recording, but is limited to the nodal/paranodal region. In this technique axons are treated with collagenase and protease and then spread on a grease-covered surface. The myelin retracts from the nodes, exposing sufficient axolemma for pipette access.

The single-channel recording methods used in our laboratory have been developed primarily by Jin Wu and Chaim Rubinstein. Surgery and initial dissection are done as for the loose patch clamp. After spreading mechani-

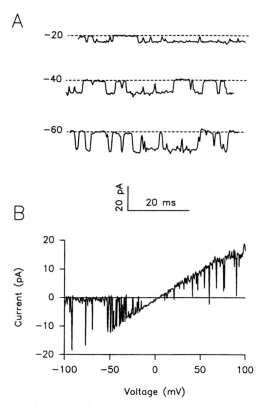

FIG. 7   Single-channel records from an excised patch from an axon 6 days post-surgery. (A) The solutions were symmetrical, 115 mM KCl + 2 mM CaCl$_2$. The patch was held at 0 mV and pulsed to the levels shown. (B) Response to a voltage ramp from −100 to +100 mV with a duration of 600 msec.

cally, a few axons are held by small dots of Cell-Tak (Collaborative Research) that have been placed on a clean coverslip. The remaining axons are then cut away and the chamber is flushed with Ringer's to remove debris. No Vaseline is used. We have been able to characterize several types of K$^+$ channels at internodal sites, including Ca^{2+}-activated, ATP-inhibited, and background channels (12). We have also found several species of Cl$^−$ channels, with differing voltage sensitivities (13).

Figure 7 illustrates records of a Ca^{2+}-activated K$^+$ channel. Traces were recorded from an excised inside-out patch from a demyelinated internode. Symmetrical KCl solutions were used, with 2 mM Ca^{2+}. In Figure 7A sweeps are shown in response to voltage pulses to three different levels and in Fig.

7B the current was elicited in response to a voltage ramp from -100 to $+100$ mV, with a duration of 600 msec. The channel was closed, except for a few very brief openings, until about -50 mV. This channel has a single-channel conductance of 240 pS.

Summary

Combining novel manipulations of axons with modern electrophysiological techniques has allowed a detailed analysis of both axolemma and glial membranes of myelinated nerve fibers. It is clear that the organization of these fibers with respect to ionic channels is far more complex than was originally thought. It now remains to understand more clearly the roles of these channels in development, in normal conduction, and under pathological conditions.

Acknowledgments

Work in the author's laboratories has been supported by Grants NS-23375 (SYC) and NS-17965 (PS) from the National Institutes of Health and RG-1839 (SYC) and RG-1774 (PS) from the U.S. National Multiple Sclerosis Society.

References

1. S. Y. Chiu, *J. Physiol.* **386,** 181 (1987).
2. S. Y. Chiu, *Glia* **4,** 541 (1991).
3. G. F. Wilson and S. Y. Chiu, *J. Neurosci.* **10,** 3263 (1990).
4. Stampfli, R. and B. Hille, *in* "Frog Neurobiology" (R. Llinas and W. Precht, eds.), pp. 3–32. Springer-Verlag, Berlin, 1976.
5. S. Y. Chiu, J. M. Ritchie, R. B. Rogart, and D. Stagg, *J. Physiol.* **292,** 149 (1979).
6. S. Y. Chiu and J. M. Ritchie, *J. Physiol.* **322,** 485 (1982).
7. S. M. Hall and N. A. Gregson, *J. Cell Sci.* **9,** 769 (1971).
8. P. Shrager, *J.Physiol.* **404,** 695 (1988).
9. P. Shrager, *Brain Res.* **483,** 149 (1989).
10. W. Stuhmer, W. M. Roberts, and W. Almers, *in* "Single Channel Recording" (B. Sakmann and E. Neher, eds.), p. 123. Plenum Press, New York, 1983.
11. P. Jonas, M. E. Brau, M. Hermsteiner, and W. Vogel, *Proc. Natl. Acad. Sci. USA* **86,** 7238 (1989).
12. J. Wu, C. T. Rubinstein, and P. Shrager, *J. Neurosci.,* in press.
13. J. Wu and P. Shrager, *Biophys. J.* **64,** A100 (1993).

Section II

Ligand-Gated Ion Channels: Electrophysiology

[7] Electrophysiological Methods for the Study of Neuronal Nicotinic Acetylcholine Receptor Ion Channels

J. G. Montes, M. Alkondon, E. F. R. Pereira, N. G. Castro, and E. X. Albuquerque

Introduction

Advances in electrophysiological techniques introduced over the past 50 years have contributed fundamentally to the characterization of the ionic basis of the function of ligand-gated ion channels and to the identification of the receptors of which these channels are a part. The most consequential of the new developments are the ability to produce intracellular recordings of electrical events, to clamp the voltage of individual cells, and to make recordings of single-channel currents from isolated patches of membranes.

In the mid-1930s, the first intracellular recordings were obtained by Young (1), who used a fine-wire electrode threaded through the length of an isolated segment of squid axon to measure the potential between that electrode and another electrode in the extracellular fluid. This allowed direct measurements of the resting and action potentials of individual giant cells. Soon thereafter, intracellular capillary electrodes were introduced almost simultaneously by the laboratories of Hodgkin and Huxley (2) and of Curtis and Cole (3).

By the late 1940s, Ling and Gerard (4) had developed a technique of recording transmembrane potentials with glass microelectrodes, and within a few years this technique had been used to study the ionic basis of synaptic transmission and of the resting and action potentials observed in a variety of cells. Shortly thereafter the voltage-clamp technique, a method that allows the membrane potential to be held constant, was introduced by Marmont (5), Cole (6), and Hodgkin, Huxley, and Katz (7, 8). This technique, which relies on maintenance of a constant membrane potential by means of a feedback amplifier, has important advantages. In a voltage-clamped cell, any observed changes in transmembrane ionic currents can be attributed to membrane conductance changes, rather than changes in the membrane potential caused by the opening and closing of channels. Hodgkin and Huxley used this technique to study the properties of the Na and K channels in the squid giant axon (9–12). One of their most important contributions was the

Methods in Neurosciences, Volume 19

identification of separate inward Na^+ and K^+ currents during the generation of an action potential. Their studies led to empirical expressions for the changes in conductance during an action potential. Another development of that period was the introduction of microiontophoresis, which made it feasible to deliver agonist to membrane regions in a very fast and discrete manner (13).

Even though all of the above methods relied on the measurement of the electrical behavior of a large number of membrane ion channels acting in concert, inferences could still be drawn about the properties of single ion channels from the characteristics of an entire population of channels. To this end, noise analysis, an approach founded on the principles of Fourier analysis and probability theory applied to random, discrete events, was introduced in the early 1970s (14–17). It was not until 1976, however, that single-channel currents could be measured directly, as demonstrated for the first time in the work of Neher and Sakmann (18) performed on the nicotinic acetylcholine receptor (nAChR) of frog sartorius muscle. It became possible to record not only whole-cell currents from an entire cell, but also single-channel currents from patches of membrane. The patch-clamp technique has also been used in combination with molecular biological and biochemical techniques. For example, RNA-injected frog (*Xenopus*) oocytes expressing reconstituted and mutated versions of the nAChR, have been used to study the relationships between structurally diverse nAChRs and their function (19–22). The legacy of the earlier physiological and pharmacological studies on the readily accessible end-plate region of skeletal muscles, the availability of highly reactive electric organs rich in nAChRs, and the discovery of specific competitive antagonists extractable from biological sources, secured both the background experience and the biological material necessary for rapid progress in understanding of the nAChR. Thus, it happened that the first single-channel current ever recorded was that passing through the muscle nAChR, which subsequently was the first neurotransmitter-gated ion channel to be isolated, purified, and cloned. The status thus far attained in the electrophysiological and related technologies applicable to the characterization of receptor channels is succinctly described in a recent publication (23).

Although much is owed to the nAChR of muscles and electric organs as the prototypic model of nicotinic receptor function and structure, only recently have functional nAChRs in the central nervous system been characterized. Much of our published work on ligand-gated ion channels, including neural *N*-methyl-D-aspartate (NMDA) channels, has depended on the electrophysiological methods mentioned above. The objective of this chapter is to provide an introduction to some of the electrophysiological techniques that have been used by many laboratories to study the structure and function of certain chemosensitive receptors in various tissues, including muscle and the central nervous system.

Electrophysiological Studies of Muscle nAChR: Models for the Study of Neuronal nAChR

Macroscopic Ionic Currents

An end-plate potential (EPP) is generated when the ACh released from the nerve terminal interacts with the nAChR located postjunctionally in the end-plate region of the muscle, causing the receptor ion channel to change conformation, open, and thus conduct inward ionic currents that depolarize the region surrounding the end plate. This results in the propagation of an action potential and, consequently, in contraction of the muscle. The initial functional characterization of the muscle nAChR relied on studies of the kinetic properties of the end-plate currents (EPCs) recorded from voltage-clamped nerve–muscle preparations. [An excellent elementary account of the EPP, EPC, and related matters is given by Nicholls *et al.* (24).] Evaluation of the rising and falling phases of the EPC led to the conclusion that they reflect the kinetics of interaction of ACh with the postjunctional nAChR, whereas the peak EPC amplitude reflects the number of nAChR channels activated. The EPC does not reach its peak instantaneously; this is due to diffusion of the neurally released ACh across the synaptic cleft, the time necessary for opening of the channel after ACh binds to the receptor, and the enzymatic hydrolysis of ACh molecules by acetylcholinesterase (AChE) [see, however, Parnas *et al.* (25)]. The decay phase of the current can be fit to a single-exponential function. At one time it was thought that the decay phase reflected the fast removal of ACh from the synaptic cleft, i.e., via the enzymatic degradation of the ligand and/or by diffusion of the agonist away from the synaptic cleft. Evidence against this possibility was provided in voltage-clamp studies, which showed that the EPC-decay phase is voltage dependent, the decay being faster at positive holding potentials, and temperature dependent, with a Q_{10} too high (2.8) for a diffusion process (26). Thus, Magleby and Stevens (27) concluded that neither enzymatic degradation by AChE of the agonist nor its diffusion away from the synaptic cleft is rate limiting; instead, they attributed the exponential decay of the EPC to the closing of channels in accordance with the principles of thermal vibrations (28) and the time constant of EPC decay (τ_{EPC}) to the rate of relaxation of the conducting species [see also (29)]. The correctness of this explanation has been borne out repeatedly in numerous kinetic studies of either macroscopic or microscopic currents. This is an important consideration that has made electrophysiological studies of the kind discussed here both practical and easier to interpret.

The mechanistic interpretation of electrophysiological data for the interaction between a drug and a receptor ultimately rests on the selection of an appropriate model, typically a reaction sequence, and certain simplifying assumptions that will serve practical and theoretical considerations; above all, the model must be consistent with all experimental observations. For example, one model frequently used to describe the process of activation of the nAChR at the end plate is based on the reaction scheme (1) (30,31):

$$2A + R \underset{k_{-1}}{\overset{k_1}{\rightleftharpoons}} A + AR \underset{k_{-2}}{\overset{k_2}{\rightleftharpoons}} A_2R \underset{k_{-3} = \alpha}{\overset{k_3 = \beta}{\rightleftharpoons}} A_2R^*.$$

(1)

State No.   1   2   3   4

In this model, it is assumed that one molecule of ACh or other agonist (A) interacts with the receptor (R)(state 1) to generate a nonconducting, agonist-bound, intermediate complex (AR)(state 2), which in turn and almost simultaneously interacts with another agonist molecule to form a second nonconducting complex (A_2R)(state 3). This last complex then spontaneously undergoes conformational changes to the conducting species (A_2R^*)(state 4). Evaluation of rate constants for each step in this model has shown that the binding of the agonist to the receptor is governed by the agonist dissociation constant, which is a function of k_1, k_{-1}, k_2, and k_{-2}. Because the binding steps are thought to be faster than those responsible for channel opening, they are not believed to be rate-limiting in ion-channel activation. The opening of the nAChR channel, which is governed by the constant β, is rate limiting, and the channel closure, which is governed by the constant α, controls the decay phase of the EPCs. Both rate constants α and β are voltage sensitive. Details on the calculation of these constants are given by Stevens and Anderson (17). When ACh is the agonist, once the complex A_2R dissociates the ACh released is hydrolyzed by AChE.

The above model is the starting point for the study of the mechanisms responsible for the action of drugs known to stimulate, antagonize, and otherwise modulate the function of nAChRs. For example, certain drugs are known to block the nAChR ion channel in the open conformation, i.e., they act on the conducting species at the right-hand side of Eq. (1) (state 4), while other agents, the closed-channel blockers, may prevent the channel from opening by acting on the second or third complexes in the above model. Thus, on the basis of the model it is possible to infer the mechanism(s) of action of certain nAChR agonists and antagonists from their effects on the amplitude and kinetics of the decay and rising phases of EPCs. For instance, the muscarinic antagonists atropine, scopolamine, and quinuclidinyl benzilate (QNB) block the ion channel of the muscle nAChR primarily in its open conformation (31–33), as do the hallucinogenic drugs phencyclidine (PCP), its analog ketamine (34, 35), and a number of local anesthetics, including bupivacaine (36) [For more details see (30, 31)]. On the other hand, the antidepressant drugs amitriptyline and nortriptyline (37) and the phenothiazine antipsychotic agents (38; reviewed in 39) act as blockers of the nAChR in the inactive or closed state, as revealed by the ability of these agents to produce a voltage- and/or time-dependent reduction in the EPC peak amplitude in the absence of any effects on τ_{EPC}. In addition to helping elucidate the effects of different drugs on the muscle nAChR, the analysis of EPCs and miniature end-plate currents (mEPCs) laid the groundwork for an understanding of synaptic transmission at the neuromuscular junction.

Despite the contributions made by the foregoing techniques, data based on macroscopic currents have to be submitted to mathematical analysis before they can be made to yield information on the electrophysiological properties of single channels. Noise analysis, the most practical of these approaches, is based on analysis of the baseline-corrected, random fluctuations in currents (''noise'') induced by iontophoretic application of an agonist (17). The amplitude fluctuations of a noise recording are due to the summed ionic currents of a multitude of channels that at any given moment are either in the open or closed state, with the fraction of channels in each conformation constantly changing. The duration (or lifetimes) of the open state for each channel should be distributed exponentially; the open-channel lifetime, τ_o, for each channel should equal $1/\alpha$ in the model described by Eq. (1). The number of channels that are closed or open at any given moment for a whole cell varies about an equilibrium value, and any random fluctuations above or below this value will tend to return to it with a relaxation time constant equal to $1/(\alpha + \beta)$. According to the Fluctuation–Dissipation Theorem (28, 40), the behavior of macroscopic currents parallels the behavior of microscopic currents, so that the time constants applicable to the macroscopic currents would also be applicable to single-channel currents (41). The main objective of the noise analysis, therefore, is to extract the relaxation time constant from macroscopic recordings, as this can be used to derive channel lifetimes.

Two methods are generally used in noise analysis: autocovariance analysis and power spectrum analysis. The first approach relies on a direct calculation of the averaged time dependence of relaxation. The second method, usually easier to implement, depends on a Fourier transform of the recording. We describe briefly here some of the features of power spectrum analysis.

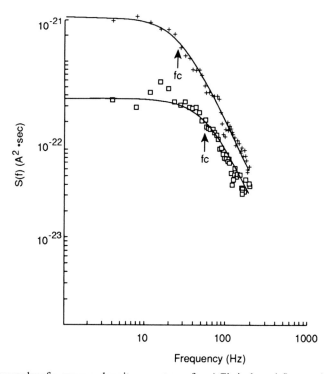

FIG. 1   Example of a power-density spectrum for ACh-induced fluctuations in currents. Recordings were made from frog cutaneous pectoris muscle in the absence (+) or presence (□) of 25 μM bupivacaine. Experiments were conducted at 10°C; holding potential was −80 mV. Arrows indicate the corner frequencies (explained below). The curves indicate the least-squares fit to a single Lorentzian function (see text). Single-channel conductances, estimated once the Lorentzian fit has been determined, were relatively unaffected by the drug (23.7 pS for control vs 21.7 pS in bupivacaine), whereas the mean channel lifetime was drastically reduced from 6.17 msec in the absence of the drug to 2.69 msec in its presence. These results are consistent with bupivacaine as a blocker capable of inducing zero conductance when the nAChR is in its open conformation (i.e., bupivacaine is an open-channel blocker). [Reproduced with permission from S. R. Ikeda, R. S. Aronstam, J. W. Daly, Y. Aracava, and E. X. Albuquerque, *Mol. Pharmacol.* **26,** 293 (1984).]

The Fourier transform process results in a breakdown of the macroscopic currents into their frequency components. These components can then be plotted to produce a graph of the power-density spectrum, which typically is a double-logarithmic plot of current power vs frequency, as in Fig. 1. The power-density spectrum plots of the noise induced by an agonist at the end-plate nAChR can be fitted to the Lorentzian function, shown in Eq. (2), to determine the relaxation time constant, τ,

$$S(f) = S(0)/(1 + (f/f_c)^2) = S(0)/(1 + (2\pi f\tau)^2), \qquad (2)$$

where $S(f)$ is the power of the current component, $S(0)$ is a constant equal to the low-frequency intercept of the function, f is the frequency, and f_c is a constant equal to the frequency ("corner") at which the power equals one-half its maximum. The calculated relaxation time, τ, approximately equals the open-channel lifetime, τ_o, if the Magleby-Stevens model (i.e., Eq. (1)) is applicable and if agonist is applied at low enough concentrations, i.e., well below half-saturation of the receptors. Once τ has been determined, the conductance, γ, of the single channels can be estimated from the area under the spectral density curve, which is approximately equal to the total variance of the agonist-induced noise. However, it should be cautioned that conductance values sometimes have been overestimated (up to 40%), as demonstrated from the direct analysis of microscopic currents (42). Some possible explanations for the underestimated conductance values is the heterogeneity of the ion channels represented in the power spectrum (discussed below), problems with selection of optimum frequency bandwidth, and difficulties that can occur in the event of an unusually high probability of channel openings. See also Silberberg and Magleby (43) for an analysis of errors associated with the use of noise analysis and how to prevent them.

Notwithstanding the limitations inherent in noise analysis, it has been useful, for instance, in demonstrating that open-channel blockers such as bupivicaine can shorten the open time of the nAChR channel in a voltage-dependent manner without modifying its conductance, thus leading directly to the conclusion that this agent blocks the ion channel in its open conformation (36). By the same kind of analysis, another agent, meproadifen, was shown to interact with the nAChR ion channel when it is resting or activated but nonconducting, and only slightly affects the conformation of the channel when it is open, i.e., meproadifen behaves as a closed-channel blocker; similarly, phencyclidine (44) and perhydrohistrionicotoxin (at certain concentrations) (39) have been shown to block the nAChR ion channels in the closed conformation. Table I provides electrophysiological properties determined from power spectra in the presence or absence of four drugs known to block the muscle nAChR noncompetitively. It can be seen in this table that while the conductance of the single channel remains unchanged compared to that for agonist alone, the open-channel lifetime (τ_o) varies with the nature of the blockade, i.e., open-channel blocking drugs cause a decrease in the lifetime, which does not change in the case of drugs that interact with the closed or resting conformations of the ion channel. For a detailed discussion of the

TABLE I   Nicotinic Single-Channel Properties Determined from Power Spectrum
Analysis of ACh-Evoked Noise Currents in the Absence or Presence
of Four Drugs Known to Be Noncompetitive Blockers of the
Muscle nAChR[a]

Drug	Normalized τ_o	Normalized γ	HP (mV)	Temperature °C
Quinuclidinyl benzilate[b] (20 μM)	0.48	1.23[f]	−90	10
Bupivacaine[c] (25 μM)	0.39	0.92	−80	10
Phencyclidine[d] (1 μM)	0.98	1.04	−80	20–22
Meproadifen[e] (1 μM)	1.00	0.94	−90	22

[a] The mean open-channel lifetimes and conductances were normalized with respect to the acetylcholine controls in each study. See text for details.
[b] G. G. Schofield, J. E. Warnick, and E. X. Albuquerque. *Cell. Mol. Neurobiol.* **1,** 209 (1981).
[c] S. R. Ikeda, R. S. Aronstam, J. W. Daly, Y. Aracava, and E. X. Albuquerque. *Mol. Pharmacol.* **26,** 293 (1984)
[d] E. X. Albuquerque, M.-C. Tsai, R. S. Aronstam, B. Witkop, A. T. Eldefrawi, and M. E. Eldefrawi. *Proc. Natl. Acad. Sci. U.S.A.* **77,** 1224 (1980).
[e] M. A. Maleque, C. Souccar, J. B. Cohen, and E. X. Albuquerque, *Mol. Pharmacol.* **22,** 636 (1982).
[f] Statistically not significantly different from unity.

mechanisms underlying the inhibition of ACh-induced single-channel currents, see Papke and Oswald, 1989 (45).

Although the mathematical manipulations of noise analysis were at one time the best that could be hoped for in deciphering the activities and characteristics of individual ion channels, the advent of techniques for the recording of microscopic currents allowed for a much more direct approach, as there are many inadequacies with the indirect methods used for the macroscopic currents. Among the factors responsible for difficulties with noise analysis, including those of a technical nature, is the possible heterogeneity of the channel population being examined. In noise analysis channel currents originating from potentially distinct types or subtypes of receptors are combined to obtain estimates of single-channel parameters that may not in reality apply to a particular channel. On the other hand, in patch-clamp experiments, there is the ability to discriminate among the different receptor subtypes and to ascertain their singular identity. Although it would seem that noise analysis would be supplanted after it was demonstrated that single-channel recordings were feasible, it should be appreciated that it is sometimes technically impractical to measure single-channel currents individually; this may occur, for example, when the cell membrane is inaccessible because of intervening membranous structures, as in myelinated nerve fibers (46) or when whole-cell currents arise from channels of very low conductance (43). In such cases noise analysis is still a viable option.

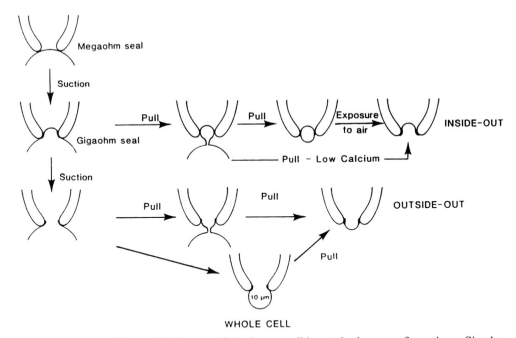

Megaohm seal

Suction

Pull   Pull   Exposure   INSIDE-OUT
to air

Gigaohm seal

Pull – Low Calcium

Suction

Pull

Pull   Pull   OUTSIDE-OUT

Pull

10 μm

WHOLE CELL

FIG. 2   Schematic diagram of the four possible patch-clamp configurations. Simple, mechanical contact of the micropipette with a cell creates a "megaohm seal." If a slight suction is applied, the seal increases considerably in resistance to permit a cell-attached clamp, the simplest of the configurations. Pulling the micropipette away from the membrane leads to the formation of either an inside-out or an outside-out patch, the latter being made if enough suction was previously applied at the cell-attached stage to disrupt the membrane. If, after disrupting the patch, the pipette is left attached to the cell, the whole-cell configuration results. In the case of inside-out patches, manipulations producing the desired configuration include mechanical pulling away of the micropipette from the cell, exposure to air, and/or short pulses of suction or voltage applied inside the micropipette. [Reproduced with permission from E. X. Albuquerque, J. W. Daly, and J. E. Warnick, *In* "Ion Channels" (T. Narahashi, ed.), Vol. I, pp. 95–162, Plenum, New York, 1988.]

Microscopic Ionic Currents

Single-channel currents can be recorded according to standard patch-clamp techniques under four different configurations: cell-attached, outside-out, inside-out, and whole-cell. [For an overview of the patch-clamp technique, see (47).] These four configurations are displayed in Fig. 2. Excised patches are usually preferred over cell-attached patches for a number of reasons. In the cell-attached configuration, each patch of membrane can be exposed to only one drug solution contained in the patch pipette, whereas in the outside-out configuration different drugs or various concentrations of the same agent can be applied to the same patch (but see 48). Furthermore, it is possible to follow the wash-out phase in excised membrane patches, which is not possible in the cell-attached patches.

Recently a patch-recording technique was developed that allows electrophysiological measurements of cells to be made without causing the replacement of the intracellular medium (49, 50). In this method ATP or polyene antibiotics are used to "perforate" the membrane patch sealed against the microelectrode tip, resulting in a reduced access resistance to the cell. Under such conditions the microelectrode remains in contact with an intact cell without disrupting the membrane. This method, which has been used to make measurements of either macroscopic or microscopic ion currents may considerably extend the full lifetime of the recording, leaving cytoplasmic constituents intact and the cell physiologically vital. Its usefulness in making measurements of nAChR ion-channel currents is still to be evaluated.

Molecular Pharmacodynamics

Ion channels are pharmacologically of great interest because they participate in the mechanism of action of many drugs. Some of these channels are activated via binding of drugs to receptors of which these channels are an integral part. The most common way of expressing the mechanism of interaction of such a drug (an agonist) with its receptor is through a reaction scheme, for instance Eq. (1), which serves as a model for the activation of the nAChR channel. In analogy with the theories of chemical kinetics, reaction schemes of the kind in Eq. (1) obey the law of mass action. According to this law, the reaction rates are proportional to the product of the concentrations of the reactants, with the reaction rate constants being the proportionality constants. It is of great interest to determine these constants experimentally, as these constants quantitatively circumscribe the reaction schemes that they subserve.

In the case of the ion channels (and of any macromolecular complex), many different structural conformations can enter in the reaction scheme as if they were distinct chemical compounds or, more appropriately, as if they were distinct states of the macromolecular complex. The number of states incorporated in the scheme and their interconnections can be estimated from experimental observations and theoretical arguments. With respect to the nAChR, biochemical, pharmacological, and structural studies have demonstrated that there are at least two binding sites for acetylcholine on the receptor (for instance, 51 and 52), and this in turn demands two binding steps and the distinction between the second and third states in Eq. (1) above. Using electrophysiological data, the states can be further characterized according to the conductance of the channel. Thus, the receptor in state 4 in Eq. (1) corresponds to a conducting or open state of the channel. The working model presents a simplified scheme, however, insofar as allowance is made for only a small number of states and for sequential transitions without the formation of cycles. In reality, the experimental data can only inform us about the minimum number of states in the scheme. When a signal presents only two current levels, at least two states are necessary to explain it. If besides separable steps there are kinetically distinct states with the same conductance, additional elements have to be included in the scheme to accommodate these observations.

In general, the reaction "constants" may in fact vary with experimental conditions. For instance, the two binding steps of the agonist in the scheme of Eq. (1) are dependent on the concentration of the agonist in the medium. These two states cannot be distinguished from each other on the basis of only the unitary currents; experiments must first be performed with different concentrations of acetylcholine. Many times it is necessary to vary given experimental parameters in order to identify states which otherwise would be kinetically indistinguishable from the others. A particularly important variable is the transmembrane potential, V_m. Perhaps because of their role in the regulation of cellular excitability, ion channels generally have a kinetics of activation dependent on V_m. This is obviously the case with the voltage-gated channels, but chemically activated channels can also manifest these characteristics, albeit in a less marked way. We can incorporate this type of information in the model by making one or more of the constants into functions of the membrane potential, temperature, and other conditional parameters.

At this point, the current trace of an ion channel (e.g., inset in Fig. 3) can be related to the reaction scheme of Eq. (1). During a pulse of current, the channel is in the open conformation, and while closed it is in one of the other three states. The closed intervals inside a burst correspond to a short-lived state, possibly state 3, while the intervals between bursts correspond to closed states of longer duration. The current recording has more information about the kinetics of the system than the obvious distinction between open and closed channels, and the patch-clamp technique is particularly important in pharmacology because it allows the study of the kinetics of ligand–receptor interactions through the direct observation of phenomena occurring at the molecular level.

Before one can interpret the events reflected in single-channel recordings in terms of reaction constants analogous to those derived from principles of chemical kinetics, however, it is important to consider that there is a fundamental difference between the study of reaction kinetics based on the

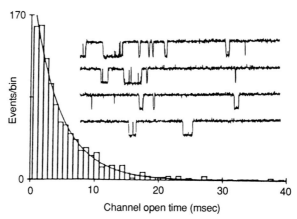

FIG. 3   Histogram of open times of anatoxin-activated single-channel currents recorded from frog hindfoot muscle. A typical histogram of events per bin (open-time intervals) vs open-channel time is displayed for nAChR single-channel currents (sample recordings shown to the right of the histogram) evoked by the strong nicotinic agonist (+)-anatoxin at 80 nM. A mean open time of 8 msec was deduced from a fit of the plot to a single-exponential function; the holding potential was −120 mV. [Adapted from P. Kofuji, Y. Aracava, K. L. Swanson, R. S. Aronstam, H. Rapoport, and E. X. Albuquerque, *J. Pharmacol. Exp. Ther.* **252,** 517 (1990), with permission.]

individual observations of single molecules and that based on the law of mass action when dealing with molar quantities of the molecules. In the latter case, the temporal evolution of the variables of the system reflects mean reaction velocities, which are usually applied in the deterministic modeling through differential equations. On the other hand, unitary currents reflect the essentially stochastic behavior of isolated molecules. The reaction constants have to be interpreted probabilistically, i.e., they are related to the transition probabilities between the states in the molecular system. A postulate analogous to the law of mass action at the molecular level establishes that the present state of a system is independent of its previous states and that the probabilities of transition are independent of the time spent in the current state. In other words, the system has no memory of its past. A consequence of this is that the lifetimes of each state have an exponential distribution, which is a well-known result in probability theory. Based on these theoretical foundations and experimentally observed unitary currents, generally accepted stochastic models treat the process of opening and closing of ion channels as an homogeneous Markov process with discrete states in continuous time. The essence of the Markov model is a matrix with $n \times n$ elements, each element being a transition rate between two of the states in the reaction scheme. Usually several elements will be zero, since some states are thought not to be connected. These transition rates are related to experimental data;

that is, all the intervals that are observed in single-channel data—the distributions of the open times, the different closed times, the bursts, the probability of being open, and so forth—are functions, and in general they will be rather complex functions of those rates in the model. Details of the derivation and application of the foregoing probabilistic concepts and theories are found elsewhere (41, 53–58).

General Analytical Principles

A main objective in the analysis of single-channel currents is to identify a model scheme, as in Eq. (1), which conforms to the data, and then to estimate the parameters of this scheme, in this case the reaction constants. The data at hand will be the sequence of the dwell times at each current level as determined from the single-channel records. In the best of circumstances, the statistics of these measurements can be obtained if the "real model" of the system that generated the currents is known; the analysis is exactly the inverse operation. From the estimates of the probability density function, the covariance, and other statistics from the dwell times, a model can be synthesized that will completely describe the observations. The procedure can be illustrated in a particularly simple case. When a system has only one open state and when there is just one connection between this open state and the other states (as in Eq. (1)), the mean duration of the open times is equal to the reciprocal of the probability rate of closing, or $1/\alpha$ in Eq. (1). The constant, α, is one of the eigenvalues of the transition rate matrix of the Markov model and can be estimated by the measurement of the open lifetimes in the record. In general, the expression that relates the eigenvalues to the constants are more complicated than this simple identity, if one considers all the possible intervening pathways between the states and the chances of correctly identifying the states revealed in the recording (see, for instance, 59).

The simplest reaction scheme frequently incorporates different states of the same conductance. In Eq. (1), states 1, 2, and 3 have zero conductance, and state 4 has the conductance of the open channel, clearly visible as a current pulse in the recording. During periods in which the channel is closed, however, the channel record does not reveal which of the three possible closed states the system is in. If all the states were distinguishable, there would be a better chance of obtaining all of the Markov-model transition rates from the data. The probability density functions describing dwell times of each state would be exponential, and their parameters could be directly estimated from the fit of exponential functions to histogram plots (see below). However, when the closed states cannot be identified separately, the probability density functions will be the sums of exponentials. The parameters of each component obtained from the histograms can be related to the eigenvalues of the transition-rate matrix by a system of linear equations (see 60).

Besides the marginal probability density functions, another set of statistics extracted from the sequence of dwell times has received much attention recently. The joint distribution of successive open times, of successive closed times, and of other combinations of intervals can be very important as an additional source of information for modeling. The aggregate properties of these intervals have been estimated from the correlation between the durations of consecutive intervals [e.g., Jackson *et al.* (61)] as this kind of dependency has been observed in single-channel recordings. Even if the kinetics of the hypothetical system have all the independence properties of the Markov model, the existence of states with the same conductance may generate correlations in the signal depending on the particular structure of the model (55, 57, 62, 63). The number of states of each kind, and the manner in which they are interconnected, are the determinants of this effect. Thus, the analysis of the joint properties of the intervals can help in specifying a particular structure for the model when there are many possible hypotheses.

Practical Aspects of Single-Channel Analysis

The first step in the analysis is to obtain from the recordings the measurements of closed and open times. For the analysis of a large number of events, it is convenient to use computer-based, automated methods to detect channel openings and closings in the recordings. We have used mostly the IPROC-2 program [Axon Instruments (64)], which is a flexible package that incorporates several desirable features, such as user-supplied criteria for acceptance of detected events. The distributions of closed and open intervals can then be estimated by binning the data and fitting multiexponential probability density functions to the resulting histograms. An example of an open-time histogram is shown in Fig. 3; the agonist used was anatoxin-a, a very potent nicotinic agonist (65–69).

Instead of conventional fixed-binwidth histograms, log-binned histograms can be used, which have special advantages when the distributions are complex (70, 71). The fitting can then be done by least-squares or maximum likelihood methods. At this step, a problem arises in the determination of the number of exponential components required to fit the distributions. It is important to combine both statistical measurements of the "goodness of fit" and the working hypotheses on the kinetic scheme to arrive at a minimum number of components to be included. As an alternative to the histograms, the maximum likelihood method can be applied to all the sequences of dwell times as originally described by Horn and Lange (57). The amplitudes (or areas) and time constants of the fitted exponentials can be used to estimate the rate constants in the kinetic scheme. The actual computational methods used to select a kinetic scheme and estimate its parameters involve a higher-level assessment of the goodness of fit of the whole model to the observed data. Different models can be ranked according to their statistical likelihood with respect to a particular set of data [e.g., Ball and Sansom (59)], and the best one(s) picked as working hypotheses.

Although the magnitude of the task of performing a full modeling of the unknown receptor channel can be overwhelming, one is often faced with a much simpler situation. The analysis can be initiated with an already chosen model, the main focus being on the *variations* of the model parameters caused by experimental manipulations. For the muscle nAChR, the scheme in Eq. (1) has served successfully for a long time as a working model for many investigators. Given this model, we can ascribe the changes in the observed interval distributions to changes in the kinetic rate constants. A comparison of the rate constants for different nicotinic agonists and antagonists provides information about how each of them interacts with the nAChR at the molecular level. For instance, reduction of the open time of the single channels activated by nicotinic agonists is expected when a drug interacts with sites inside the ion channel of the nAChR. Charged agents that interact with sites located deep within the ion channel tend to produce a marked voltage-dependent shortening of the open times.

These types of analyses have made fundamental contributions to our understanding of the function of the muscle nAChR under physiological and pathological conditions, as well as the mechanisms of drug interactions with the receptor–ion-channel complex at the neuromuscular junction. Only relatively recently has the neuronal nAChR become the center of focus.

The Limits of the Conventional Analysis

The methodology based on the analysis of dwell times has been widely used but has some limitations. The question of choosing the correct model is of paramount importance, because it cannot be answered with the manipulation of the experimental conditions or the method of analysis. Kienker (72) demonstrated that systems with more than one closed or open state frequently are not uniquely describable by a single Markov model, and when this is true, the possible Markov models are not experimentally distinguishable [see Fredkin *et al.* (58)].

Another limitation stems from the usual assumption that only one channel is active in the patch from which the recordings are made. This is usually not the case in practice. Then, the measurement of the dwell time becomes ambiguous because an open interval or a closed interval can be interrupted by a state transition of a channel different from the one that initiated that interval. Thus, the intervals measured in the record would not necessarily correspond to the dwell times of each of the channels in the various states. The closed intervals are generally shortened by the openings of different channels, and also, there can be superpositioning of openings, with resulting current amplitudes that represent multiples of the unitary current. This complication is usually avoided by not using records in which there is this superposition of events, or ignoring the superimpositions, or restricting analysis of the data to segments of the record in which the chance of having more than one active channel is very small (52). These procedures may introduce a sampling bias of the intervals or may simply represent a waste of information. It is worthwhile noting that spectral analysis is applicable exactly to the currents due to the superpositions of the openings and closings of a large number of channels, but unfortunately, the identification of a model with this kind of data is still more difficult than in the case of the unitary currents.

Finally, the question of the stationarity of the process should be addressed. Analysis based on the histograms of dwell times obviously demands that the statistics of these dwell times be constant during the experiment, that is, the system remain invariant during the observation period. For nonstationary data, e.g., in voltage- or concentration-jump experiments, the analysis protocols that have already been described involve many repetitions of the experiments under the same conditions [e.g., see Bauer *et al.* (73)]. Although analysis of nonstationary data can potentially provide more information about the system than an analysis restricted to the equilibrium (i.e., stationary) state, it may in practice be difficult, or even impossible, to generate the requisite homogeneous ensemble of records. In general, the grouped records must have come from the same membrane patch, which demands a repetition of the experiment at relatively short intervals before the preparation deteriorates. Consequently the slow phenomena of the system kinetics, such as desensitization, cannot be analyzed in this way. These phenomena can actually interfere with the analysis of ensembles, causing the successive samples not to be random, as in studies on sodium channels by Horn *et al.* (74).

Electrophysiological Studies on Neuronal nAChRs

In contrast to the work performed on the muscle nAChR, the study of the neuronal nAChRs is still in its infancy. Patch-clamp techniques have been fundamentally important to the functional characterization of CNS nAChRs, especially those that are sensitive to the snake neurotoxin, α-bungarotoxin

(α-BGT). Although α-BGT binding sites showing characteristics of nicotinic receptors exist in the mammalian brain (75–81), not until recently have more direct techniques been used to demonstrate the existence in the CNS of an α-neurotoxin-sensitive, functional nAChR (68, 82–86).

Neuronal nAChRs have been studied by means of the patch-clamp technique applied to neurons from different brain regions. Studies carried out in our and other laboratories have used this technique to demonstrate the presence of nAChRs on rat hippocampal neurons. The majority of the nicotinic receptors in the hippocampal region can be blocked by three different neurotoxins—α-cobratoxin, α-BGT, and methyllycaconitine (MLA) (67, 82–84)—although a minor population of α-BGT-insensitive nAChRs has also been found in this brain area (68). In contrast, functional nAChRs insensitive to α-BGT in other neurons of the mammalian CNS, such as retinal ganglion cells (87, 88), medial habenular neurons (89, 90), and hypothalamic neurons (91) are reported to be insensitive to blockade by α-BGT.

Given the evidence for the existence of heterogeneous, functional nAChRs in the mammalian CNS, questions have been raised regarding the kinetic and conductance properties of these receptors. Knowledge previously acquired from the studies of muscle nAChRs and the availability of electrophysiological techniques that allow the direct study of the neuronal receptors have made it possible to address such questions.

The following sections deal with the improved techniques that have been used to record and analyze nicotinic ionic currents activated in neurons from different areas of the mammalian CNS. In fact, analysis of these currents has been particularly useful in defining the intrinsic properties of the still obscure nicotinic receptors of the CNS.

Recording and Analysis of Nicotinic Whole-Cell Currents

Whole-cell currents activated by several nicotinic agonists have been studied either in acutely dissociated neurons [described in Kay and Wong (92)] or in cultured neurons. The purpose of most of the experiments carried out in our laboratory has been to characterize the pharmacological and kinetic properties of the nAChRs present in cultured (67) and acutely dissociated hippocampal neurons (work in progress) and to unveil their functional significance. By the end of the 1980s, we were able to demonstrate by means of single-channel recordings the presence of functional nAChRs on hippocampal neurons (86). However, it was not until very recently that we showed that the majority of the nicotinic responses in hippocampal neurons are sensitive to α-BGT (67, 68, 83).

In the experiments carried out in our laboratory, nicotinic whole-cell currents have been induced in cultured hippocampal neurons to which specific agonists had been applied by means of a perforated U-shaped tube ("U" tube) positioned near each neuron to be studied (approximately 50 μm) (42). A U tube with a pore of about 200–400 μm in diameter at the apex can be fabricated from a thin capillary glass (o.d. = 508 μm, i.d. = 380 μm; Garner Glass Co., Claremont, CA). The elements of the fast-perfusion system are depicted in Fig. 4. The solution-exchange time is optimized to avoid differences in current waveforms resulting from a prolongation of the rise time and/or decay time constant of the agonist-induced currents; this can be done by keeping the flow rate constant and by maintaining the position of the U tube in relation to the neuron more or less constant between experiments. In the basic setup the solution-exchange time constant is found to be <15 msec, as measured using sodium ion concentration jumps in the presence of kainic acid (83, 93). Usually, the agonists are delivered via the U tube while the antagonists are applied via bath perfusion. With this protocol, near-equilibrium concentrations of the antagonists can be achieved within 5 min of starting the perfusion. Solution exchanges faster than 1 msec can be achieved for excised patches in optimal circumstances with a piezoelectric translator [see description of technique in Franke et al. (94)]. With this instrument considerable improvement of the time resolution of receptor activation by agonist can be achieved.

In our laboratory, the neurons are perfused continuously with external solution containing atropine (1 μM) and tetrodotoxin (300 nM) (both added in all experiments) and, in some experiments, picrotoxin (50–100 μM). None of these agents at these concentrations alters the peak or decay characteristics of ACh-evoked whole-cell currents. The patch micropipettes used to record whole-cell currents from cultured neurons are pulled from borosilicate capillary glass and have tip diameters ranging from 1.5 to 4 μm. The resistances of the patch micropipettes when filled with the internal solution are between 1.5 and 5 MΩ, and the seal resistances normally range from 10 to 20 GΩ; recordings are typically made with low-pass filtering at 2 kHz (68).

Because responses to the nicotinic agonists display a rundown in most hippocampal neurons, certain precautions have to be taken to minimize or to compensate for the effects of rundown on the evoked currents, as follows: (i) dose–response curves should not be initiated until at least 15 min after patch formation; (ii) different concentrations of agonists should be used in a random manner, and whenever possible more than one set of agonist pulses should be delivered to each neuron; (iii) the test concentrations of the agonists should be bracketed with pulses of saturating concentration (3 mM) of ACh (68).

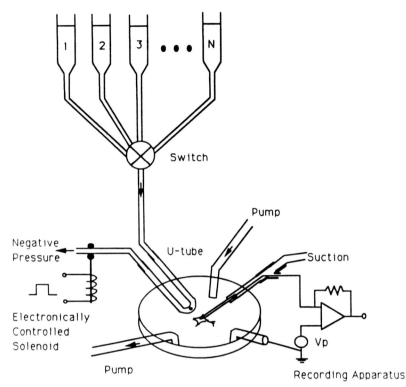

FIG. 4   Fast-perfusion apparatus used in the measurement of whole-cell currents. One end of the U tube is connected to flexible tubing, which is in turn connected to an array of tubes accessed via a common switch (controlled by a solenoid linked to a stimulator) that permits easy switching from one tube to the other, so that up to N different agonist solutions can be tested on the same neuron. The other end of the U tube is connected via a valve to flexible tubing, which allows the agonist solution to drain out by a force of gravity (providing "negative pressure"). The height of the drainage end of the tubing is adjusted so that a small amount of the bath solution from the dish can be continuously removed, thus preventing any leakage of the agonist solution through the pore of the U tube when the valve is not activated. However, when the valve is activated by an electric pulse of known duration (usually 1–2 sec), the agonist rapidly flows out of the pore and completely displaces the solution surrounding the neuronal soma and dendrites. For additional details on setting up a patch-clamp system, see R. A. Levis and J. L. Rae, *in* "Ion Channels" (B. Rudy and L. E. Iverson, eds.), pp. 14–66. Academic Press, San Diego, CA, 1992. [Reproduced with permission from E. X. Albuquerque, A. C. S. Costa, M. Alkondon, K. P. Shaw, A. S. Ramoa, and Y. Aracava, *J. Recept. Res.* **11,** 603 (1991).]

TABLE II   Characteristics of Different Types of Nicotinic Whole-Cell Currents Elicited by 3 mM ACh

Type of whole-cell current	$\%I_1$	$\%I_2$	$\%I_{offset}$	τ_1 (msec)	τ_2 (msec)	Peak current[a] (pA)	Number of neurons
IA	83.2	14.7	2.1	26.7	300	596	43
IB[b]	52.6	33.4	14.0	23.3	311	284	11
II	22.0	78.0	0	108.5	2001	525	2
III	26.0	59.5	14.5	210.5	2272	2250	2

[a] The peak amplitudes were obtained from currents evoked within 10 min after patch formation. No relationship between the peak amplitude and the type of currents evoked was found.
[b] This type of current is probably a mixture of types IA and II.

The decay of nicotinic whole-cell currents is amenable to a mathematical description. The decay phases can be fit either to a single-exponential or to a double-exponential function of the form shown in Eq (3),

$$I(t) = I_{offset} + [I_1 \exp(-t/\tau_1) + I_2 \exp(-t/\tau_2)], \qquad (3)$$

where I is the total peak current, I_{offset} is the offset current during the steady-state condition, and I_1 and I_2 are the amplitudes of current decaying at rates τ_1 and τ_2, respectively.

Based on patterns of nicotinic agonist and antagonist sensitivities seen in cultured hippocampal neurons, and on the kinetics of the decay phase of the nicotinic whole-cell currents evoked in them, it appears that at least three different subtypes of nAChR exist in these neurons (68). The currents evoked from these receptors have been tentatively classified into three major groups based on their sensitivities to α-BGT, κ-BGT, MLA, and dihydro-β-erythroidine (DHβE)—one group (I) would include currents highly sensitive to blockade by α-BGT, κ-BGT, and MLA; the second group (II) would include those very sensitive to DHβE, insensitive to α-BGT, and only moderately sensitive to the other toxins; and the third group (III) is insensitive to the above antagonists. Moreover, group I has whole-cell currents that decay much faster than those in groups II and III. The kinetics of the decay phase and the relative contributions of the whole-cell current components (including offset) elicited in different neurons are in Table II for the three major types of current, including two subdivisions of one of the major groups (i.e., types IA and IB). In this table the first (I_1)- and second (I_2)-component currents, and a residual offset current (I_{offset}), are given as percentage of peak current elicited by 3 mM ACh; the decay-time constants for the first (τ_1) and second (τ_2) phases of decay are also shown.

The categorization of the current types shown in Table II is further justified by the striking differences that were found among the potencies and/or efficacies of other nicotinic agonists that were examined (68). The most obvious divergences among the agonists were seen at their EC_{10}–EC_{25}s applicable to type IA currents. In addition, each current type was associated with a unique hierarchy of sensitivity to the various agonists. For example, ACh, carbamylcholine, (−)nicotine, and suberyldicholine were more effective in activating type II currents than they were in evoking type IA currents. Similarly, cytisine and anatoxin-a were more effective in evoking type III than in eliciting type IA currents. Such a variation in relative sensitivity to agonists supports the notion that neuronal nAChRs exist in a variety of structurally distinct forms in hippocampal neurons. This is consistent with molecular biological evidence for the existence of structurally heterogeneous neuronal nAChRs. Neuronal nAChRs appear to be pentamers made up of two α and three β subunits (95). Unlike the muscle nAChR, which is known to be pentameric (one ε or γ and one δ subunit in addition to two α subunits and one β subunit) but bears only one kind of α and β subunit, the neuronal nAChR molecule thus far appears to be comprised of at least one of seven known types of α subunit ($\alpha2$–$\alpha8$) and probably of at least one of three recognized kinds of β subunit ($\beta2$–$\beta4$), according to the stoichiometry $\alpha_2\beta_3$ (95, 96).

Although the whole-cell currents evoked by different agonists in rat hippocampal neurons have been characterized in detail, the kinetic properties of the single-channel currents, and the subunit structures of the nAChRs subserving them, remain uncertain. Our whole-cell studies, however, have provided some clues of the characteristics of the different nAChRs in the hippocampal region at the molecular level. For example, on the basis of molecular biological and pharmacological information regarding neuronal nAChRs in oocyte expression studies, and our analysis of the sensitivity of the currents to agonists and antagonists, the predominant currents (type IA) evoked in hippocampal neurons appear to be mediated via nAChRs comprised of $\alpha7$ subunits (68).

Recording and Analysis of Neuronal Nicotinic Single-Channel Currents

Neuronal nicotinic single-channel currents have been studied in outside-out patches using the high-resolution standard patch-clamp technique. The agonists can be applied in a stationary condition, i.e., via a perfusion system, to the patches obtained from cultured hippocampal neurons (83, 86, 97). In this protocol the pharmacological agents are dissolved in the external physiological salt solution and applied to the patches via a perfusion system consisting of an array of glass capillaries that are connected to several containers. However, single-channel current recordings of neuronal nAChRs, unlike the whole-cell recordings, may require patch seal resistances of 20–25 GΩ and low-pass filtering at 3–7 kHz, these measures being necessary because the lifetime of the nicotinic single channels of neurons in certain circumstances have been found to be on the order of 200 μsec or less, creating the need for an unusually high signal-to-noise ratio (85). When an outside-out patch is excised from a cultured neuron, the tip of the micropipette bearing the patch is placed inside one of the capillaries through which external solution is delivered. If no "spontaneous" activity is detected, the micropipette is moved to the capillaries that deliver the pharmacological agents to be tested. In rat hippocampal neurons we have observed that either ACh or (+)-anatoxin-a, in the presence of atropine, 2-amino-5-phosphonovaleric acid (APV), and tetrodotoxin, can activate several conductance states. The τ_0 of the 23-pS nicotinic single ion channels is on the order of 0.4 msec. We also observed that in most instances concentrations as low as 1–100 fM of the nicotinic antagonist MLA could block activated nicotinic single channels (83).

The initial characterization of the neuronal nAChR channels in the hippocampal neurons relied on this technique of agonist application. However, our studies have shown that in 83% of the neurons, the whole-cell currents were of type IA, which desensitizes during the time scale of agonist application obtained with the stationary perfusion system. Therefore, it is quite likely that the majority of the events that account for type IA neuronal nicotinic whole-cell currents could have been missed. To determine if a unique single-channel response could account for a particular whole-cell response, a new set-up was devised that allowed both whole-cell and single-channel current recordings to be made from the same neurons using the same fast agonist application technique (85). In this approach, an outside-out patch is excised from neurons that respond to ACh (0.1–1 mM), giving whole-cell currents whose peak amplitudes are greater than 100 pA at −60 mV. Then, the outside-out patch is positioned near the U tube and a 0.1- to 1.0-sec pulse of ACh is applied to the patch. For the whole-cell currents that desensitize fast and that are sensitive to MLA, the single-channel currents present the same characteristics. The frequency of single-channel openings during each 1-sec ACh pulse is high at the onset and decays extremely fast, with the events occurring mostly in isolation or in pairs. Under these experimental conditions, the conductance and the τ_0 of the ACh-activated channels can be determined by pooling the data from several membrane patches. The channel associated with the α-BGT-sensitive type IA currents had a conductance of 73 pS and a lifetime of 0.1 msec at −80 mV, and when activated by ACh, inactivated in <1 msec. It is likely that the kinetic properties of these ACh-activated single-channel currents account for the fast phase of decay of whole-cell currents induced by ACh. Detailed studies of the kinetic properties of the single events underlying this and other nicotinic current types in hippocampal neurons are now being conducted in our laboratory. Certainly, these studies will not only enable the characterization of the neuronal nAChRs at the molecular level but also improve our understanding of the physiological role of these receptors in the mammalian CNS.

Concluding Remarks

The advances that have been made over the past 2 decades in the measurement of single-channel currents have made it possible to rapidly extend our knowledge of the electrophysiology and molecular biology of the nAChR. The methods described here for the study of the nAChR are generally adaptable to the study of other ligand-gated receptor channels, including that of the NMDA receptor. The discovery of considerable heterogeneity of structure and function in the nAChRs of the nervous system, however, has intensified the search for adjunct methods that will enable the pharmacologist, physiologist, and biophysicist to focus on the newer task of identifying, selecting, and delimiting a more diverse set of targets, both cellular and subcellular, present within a more general population of cells constituting a given preparation. Such interdisciplinary approaches may be needed, for example, in further exploring the recently demonstrated role of physostigmine as a nicotinic agonist that acts via a novel site on the nAChR of neurons (98). Nonetheless, the ability to measure currents elicited from single ion channels is essentially complete in its conception. It is difficult to foretell what new quantum leaps will take place in the methods for measuring single-channel currents, but, at least for now, integration of patch-clamp techniques with those borrowed from other areas that have also experienced rapid growth seems to be a worthwhile pursuit with much promise.

The fact that individual hippocampal neurons possess receptors with distinct pharmacological characteristics, for example, underscores the need for refined ways to identify the types of cells involved before electrophysiological measurements are made. This need may be satisfied in the future with the development of certain dyes that may, because of their intrinsic properties, allow the investigator to select, under the microscope, specific cells for study. Perhaps fluorescently labeled antibodies can be created that will specifically target cells of a given receptor type. This would seem imminently feasible, as monoclonal antibodies have already been raised against specific regions of distinct nAChRs (99–101). Such monoclonal antibodies have already been used on nAChRs in our laboratory to selectively block binding sites other than those specific for classical nicotinic agonists (98, 102, 103). It may even be possible to selectively destroy certain subclasses of cells with the help of immunoactive reagents, thus leaving target cells intact. In any event, many of the tools previously used will continue to be necessary [for a recent review of toxins and drugs useful in the study of excitable-membrane ion channels, see Narahashi (104)].

Other adjunct methods that may play an increasingly important role in the future include methods relying on artificially constructed or reconstructed bilayers. These approaches themselves would be aided by molecular biological techniques aimed at the selective extraction and amplification of receptors and their component subunits. Eventually, entire receptors belonging to unique subclasses could be synthesized *de novo* in the laboratory. In this fashion electrophysiological techniques could be applied to well-circumscribed and tightly controlled preparations. The work already done in which nAChRs have been expressed in oocytes can be considered an early forerunner of this type of approach.

Finally, whereas the current emphasis has been on the measurement and elucidation of small-scale events, the ultimate result of a deep understanding of microcosmic principles will be the synthesis of models of phenomena of a much larger scale, such as those applicable to the thinking process and human behavior. Perhaps the heterogeneity of structure and function of central neuronal nAChRs helps to determine the role that each neuron plays within the extremely complex fabric of the CNS.

Acknowledgments

The authors are deeply indebted to Mabel A. Zelle for her helpful comments, suggestions, and criticisms and to Mrs. Barbara J. Marrow for her excellent technical assistance. This work was performed under the sponsorship of USPHS Grants NS 25296 and NS 05730. N.G.C. received a fellowship from CNPq, Brazil. E.F.R.P. holds an appointment in the Department of Basic and Clinic Pharmacology Institute of Biomedical Science, Center of Health Science, Federal University of Rio de Janeiro, Brazil.

References

1. J. Z. Young, *Q. J. Microsc. Sci.* **78,** 367 (1936).
2. A. L. Hodgkin and A. F. Huxley, *Nature (London)* **144,** 710 (1939).
3. H. J. Curtis and K. S. Cole, *J. Cell. Comp. Physiol.* **15,** 147 (1940).
4. G. Ling and R. W. Gerard, *J. Cell Comp. Physiol.* **34,** 383 (1949).
5. G. J. Marmont, *Cell. Comp. Physiol.* **34,** 351 (1949).
6. K. S. Cole, *Arch. Sci. Physiol.* **3,** 253 (1949).
7. A. L. Hodgkin, A. F. Huxley, and B. Katz, *Arch. Sci. Physiol.* **3,** 37 (1949).

8. A. L. Hodgkin, A. F. Huxley, and B. Katz, *J. Physiol.* (*London*) **116,** 424 (1952).
9. A. L. Hodgkin and A. F. Huxley, *J. Physiol.* (*London*) **116,** 449 (1952).
10. A. L. Hodgkin and A. F. Huxley, *J. Physiol* (*London*) **116,** 473 (1952).
11. A. L. Hodgkin and A. F. Huxley, *J. Physiol.* (*London*) **116,** 497 (1952).
12. A. L. Hodgkin and A. F. Huxley, *J. Physiol.* (*London*) **117,** 500 (1952).
13. W. L. Nastuk, *Fed. Proc.* **12,** 102 (1953).
14. B. Katz and R. Miledi, *Nature* (*London*) **226,** 962 (1970).
15. B. Katz and R. Miledi, *Nature* (*London*) **232,** 124 (1971).
16. B. Katz and R. Miledi, *J. Physiol.* (*London*) **230,** 707 (1973).
17. C. R. Anderson and C. F. Stevens, *J. Physiol.* (*London*) **235,** 655 (1973).
18. E. Neher and B. Sakmann, *Nature* **260,** 799 (1976).
19. C. W. Luetje, J. Patrick, and P. Séguéla, *FASEB J.* **4,** 2753 (1990).
20. E. S. Deneris, J. Connolly, S. W. Rogers, and R. Duvoisin, *TIPS* **12,** 34 (1991).
21. R. J. Lucas and M. Bencherif, *Int. Rev. Neurobiol.* **54,** 25 (1992).
22. V. M. Gehle and K. Sumikawa, *Mol. Brain Res.* **11,** 17 (1991).
23. B. Sakmann, *Neuron* **8,** 613 (1992).
24. J. G. Nicholls, A. R. Martin, and B. G. Wallace, *in* "From Neuron to Brain," 3rd Ed., pp. 184–236. Sinauer, Sunderland, MA (1992).
25. H. Parnas, M. Flashner, and M. E. Spira, *Biophys. J.* **55,** 875 (1989).
26. A. Takeuchi and N. Takeuchi, *J. Neurophysiol.* **22,** 395 (1959).
27. K. L. Magleby and C. F. Stevens, *J. Physiol.* **223,** 151 (1972).
28. R. Kubo, *J. Physiol. Soc. Jpn* **12,** 570 (1957).
29. K. Kuba, E. X. Albuquerque, J. Daly, and E. A. Barnard, *J. Pharmacol. Exp. Ther.* **189,** 499 (1974).
30. P. Adams, *J. Membr. Biol.* **58,** 161 (1981).
31. M. Adler, E. X. Albuquerque, and F. J. Lebeda, *Mol. Pharmacol.* **14,** 514 (1978).
32. M. Adler and E. X. Albuquerque, *J. Pharmacol. Exp. Ther.* **196,** 360 (1976).
33. G. G. Schofield, J. E. Warnick, and E. X. Albuquerque. *Cell. Mol. Neurobiol.* **1,** 230 (1981).
34. L. G. Aguayo, B. Pazhenchevsky, J. W. Daly, and E. X. Albuquerque, *Mol. Pharmacol.* **20,** 345 (1981).
35. M. A. Maleque, J. E. Warnick, and E. X. Albuquerque, *J. Pharmacol. Exp. Ther.* **219,** 638 (1981).
36. S. R. Ikeda, R. S. Aronstam, J. W. Daly, Y. Aracava, and E. X. Albuquerque, *Mol. Pharmacol.* **26,** 293 (1984).
37. G. G. Schofield, B. Witkop, J. E. Warnick, and E. X. Albuquerque, *Proc. Natl. Acad. Sci. USA* **78,** 5240 (1981).
38. J. S. Carp, R. S. Aronstam, B. Witkop, and E. X. Albuquerque, *Proc. Natl. Acad. Sci. USA* **80,** 310 (1983).
39. E. X. Albuquerque, J. W. Daly, and J. E. Warnick, *in* "Ion Channels" (T. Narahashi, ed.), Vol. I, pp. 95–162. Plenum, New York, 1988.
40. C. F. Stevens, *Biophys. J.* **12,** 1028 (1972).
41. D. Colquhoun and A. G. Hawkes, *Proc. R. Soc. London* (*Biol.*) **211,** 205 (1981).
42. E. M. Fenwick, A. Marty, and E. Neher, *J. Physiol.* (*London*) **331,** 599 (1982).
43. S. D. Silberberg and K. L. Magleby, *Biophys. J.* **65,** 1570 (1993).
44. E. X. Albuquerque, M.-C. Tsai, R. S. Aronstam, B. Witkop, A. T. Eldefrawi, and M. E. Eldefrawi, *Proc. Nat. Acad. Sci. U.S.A.* **77,** 1224, 1980.

45. F. J. Sigworth, *J. Physiol.* (*London*) **307**, 97 (1980).
46. F. J. Sigworth, *J. Physiol.* (*London*) **307**, 131 (1980).
47. M. Cahalan and E. Neher, *in* "Ion Channels" (B. Rudy and L. E. Iverson, eds.), pp. 3–14. Academic Press, San Diego, CA (1992).
48. A. Auerbach, *Biophys. J.* **60**, 660 (1991).
49. Z. Zhou and E. Neher, *J. Physiol.* **469**, 245 (1993).
50. M. Lindau and J. M. Fernandez, *Nature* (*London*) **319**, 150 (1986).
51. J. L. Rae and J. M. Fernandez, *News Physiol. Sci.* **6**, 273 (1991).
52. N. Unwin, *Neuron* **3**, 665 (1989).
53. D. Colquhoun and D. C. Ogden, *J. Physiol.* (*London*) **395**, 131 (1988).
54. D. Colquhoun and A. G. Hawkes, *Proc. R. Soc. London* (*Biol.*) **199**, 231 (1977).
55. D. Colquhoun and A. G. Hawkes, *Phil. Trans. R. Soc. London* (*Biol.*) **300**, 1 (1982).
56. D. Colquhoun and A. G. Hawkes, *Proc. R. Soc. London* (*Biol.*) **230**, 15 (1987).
57. R. Horn and K. Lange, *Biophys. J.* **43**, 207 (1983).
58. D. R. Fredkin, M. Montal, and J. A. Rice, *in* "Proceedings of the Berkeley Conference in Honor of Jerzy Neyman and Jack Kiefer" (L. M. LeCarn and R. A. Olshen, eds.), pp. 269–285, Wadsworth, Belmont, CA, 1985.
59. F. G. Ball and M. S. P. Sansom, *Proc. R. Soc. London* (*Biol.*) **236**, 385 (1989).
60. D. Colquhoun and A. G. Hawkes, *in* "Single-Channel Recording" (B. Sakmann and E. Neher, eds.), pp. 135–175. Plenum Press, New York, 1983.
61. D. Colquhoun and F. J. Sigworth, *in* "Single-Channel Recording" (B. Sakmann and E. Neher, eds.), pp. 191–263. Plenum Press, New York, 1983.
62. M. B. Jackson, B. S. Wong, C. E. Morris, and H. Lecar, *Biophys. J.* **42**, 109 (1983).
63. I. Z. Steinberg, *Biophys. J.* **52**, 47 (1987).
64. F. Sachs, J. Neil, and N. Barkakati, *Pfluegers Arch.* **395**, 331 (1982).
65. C. E. Spivak, J. Waters, B. Witkop, and E. X. Albuquerque, *Mol. Pharmacol.* **23**, 337 (1983).
66. K. L. Swanson, C. N. Allen, R. S. Aronstam, H. Rapoport, and E. X. Albuquerque, *Mol. Pharmacol.* **29**, 250 (1986).
67. M. Alkondon and E. X. Albuquerque, *J. Recep. Res.* **11**, 1001 (1991).
68. M. Alkondon and E. X. Albuquerque, *J. Pharmacol. Exp. Ther.* **265**, 1455 (1993).
69. P. Thomas, M. W. Stephens, G. Wilkie, M. Amar, G. G. Lunt, P. Whiting, T. C. G. Gallagher, E. F. R. Pereira, M. Alkondon, E. X. Albuquerque, and S. Wonnacott. *J. Neurochem.* **60**, 2308 (1993).
70. O. B. McManus, A. L. Blatz, and K. L. Magleby, *Pfluegers Arch.* **410**, 530 (1987).
71. F. J. Sigworth and S. M. Sine, *Biophys. J.* **52**, 1047 (1987).
72. P. Kienker, *Proc. R. Soc. London* (*Biol.*) **236**, 269 (1989).
73. R. J. Bauer, B. F. Bowman, and J. L. Kenyon, *Biophys. J.* **52**, 961 (1987).
74. R. Horn, C. A. Vandenberg, and K. Lange, *Biophys. J.* **45**, 323 (1984).
75. P. B. S. Clark, R. D. Schwartz, S. M. Paul, C. D. Pert, and A. Pert, *J. Neurosci.* **5**, 1307 (1985).
76. Y. Dudai and M. Segal, *Brain Res.* **154**, 167 (1978).

77. M. J. Marks, J. A. Stitzel, and A. C. Collins, *J. Pharmacol. Exp. Ther.* **235,** 619 (1985).
78. B. J. Morley, J. F. Lorden, G. B. Brown, G. E. Kemp, and R. J. Bradley, *Brain Res.* **134,** 161 (1977).
79. G. Polz-Tejera, J. Schmidt, and H. J. Karten, *Nature (London)* **258,** 349 (1975).
80. P. M. Salvaterra, H. R. Mahler, and W. J. Moore, *J. Biol. Chem.* **250,** 6469 (1975).
81. J. Schmidt, *Mol. Pharmacol.* **13,** 283 (1977).
82. M. Alkondon and E. X. Albuquerque, *Eur. J. Pharmacol.* **191,** 505 (1990).
83. M. Alkondon, E. F. R. Pereira, S. Wonnacott, and E. X. Albuquerque, *Mol. Pharmacol.* **41,** 802 (1992).
84. C. F. Zorumski, L. L. Thio, K. E. Isenberg, and D. B. Clifford, *Mol. Pharmacol.* **41,** 931 (1992).
85. N. G. Castro and E. X. A. Albuquerque, *Neurosci. Lett.* in press.
86. Y. Aracava, S. S. Deshpande, K. L. Swanson, H. Rapoport, S. Wonnacott, G. Lunt, and E. X. Albuquerque, *FEBS Lett.* **222,** 63 (1987).
87. S. A. Lipton, E. Aizenman, and R. H. Loring, *Pfluegers Arch.* **410,** 37 (1987).
88. A. S. Ramôa, M. Alkondon, Y. Aracava, J. Irons, G. G. Lunt, S. S. Deshpande, S. Wonnacott, R. S. Aronstam, and E. X. Albuquerque, *J. Pharmacol. Exp. Ther.* **254,** 71 (1990).
89. C. Mulle and J.-P. Changeux, *J. Neurosci.* **10,** 169 (1990).
90. C. Mulle, C. Vidal, P. Benoit, and J.-P. Changeux, *J. Neurosci.* **11,** 2588 (1991).
91. Z. W. Zhang and P. Feltz, *J. Physiol. (London)* **422,** 83 (1990).
92. A. R. Kay and R. K. S. Wong, *J. Neurosci. Methods* **16,** 277 (1986).
93. L. Vyklický, Jr., M. Benveniste, and M. L. Mayer, *J. Physiol. (London)* **428,** 313 (1990).
94. Ch. Franke, H. Hatt, and J. Dudel, *Neurosci. Lett.* **77,** 199 (1987).
95. L. W. Role, *Curr. Opinions Neurobiol.* **2,** 254 (1992).
96. A. B. Vernalis, W. G. Conroy, and D. Berg, *Neuron.* **10,** 451 (1993).
97. E. F. R. Pereira, M. Alkondon, T. Tano, N. G. Castro, M. M. Fróes-Ferrão, R. Rozental, R. S. Aronstam, A. Schrattenholz, A. Maelicke, and E. X. Albuquerque, *J. Recep. Res.* **13,** 413 (1993).
98. E. F. R. Pereira, S. Reinhardt-Maelicke, A. Schrattenholz, A. Maelicke, and E. X. Albuquerque, *J. Pharmacol. Exp. Ther.,* in press.
99. P. M. Lippiello, K. G. Fernandes, J. J. Langone, and R. J. Bjerck, *J. Pharmacol. Exp. Ther.* **257,** 1216 (1991).
100. P. J. Whiting and J. M. Lindstrom, *J. Neurosci.* **8,** 3395 (1988).
101. G. Fels, R. Plumer-wilk, M. Schereiber, and A. Maelicke, *J. Biol. Chem.* **261,** 15746 (1986).
102. K.-P. Shaw, Y. Aracava, A. Akaike, J. W. Daly, D. L. Rickett, and E. X. Albuquerque, *Mol. Pharmacol.* **28,** 527 (1985).
103. E. X. Albuquerque, Y. Aracava, M. Idriss, B. Schonenberger, A. Brossi, and S. S. Deshpande, *in* "Neurobiology of Acetylcholine" (N. J. Dun and R. L. Perlman, eds.), pp. 301–328. Plenum Press, New York, 1987.
104. T. Narahashi, *in* "Ion Channels" (B. Rudy and L. E. Iverson, eds.), pp. 643–658. Academic Press, San Diego, CA (1992).

[8]   GABA$_A$ Receptor–Chloride Channel Complex

Norio Akaike and Nobutoshi Harata

Introduction

γ-Aminobutyric acid (GABA) is one of the major neurotransmitters in the mammalian central nervous system (CNS). There are three types of GABA receptors: (i) GABA$_A$ receptors, which are bicuculline-sensitive and directly coupled with an integral Cl$^-$ channel, (ii) GABA$_B$ receptors, which are indirectly coupled with K$^+$ and/or Ca^{2+} channels through GTP-binding proteins, and (iii) GABA$_C$ receptors, which operate bicuculline-insensitive Cl$^-$ channels (Drew *et al.*, 1984). Among these receptors, activation of postsynaptic GABA$_A$ receptors is thought to underlie inhibitory postsynaptic potentials ubiquitously in various parts of the CNS. In addition to this inhibitory action, there is another outstanding characteristic of the GABA$_A$ receptor: it contains numerous binding sites for therapeutically significant drugs, including convulsant (picrotoxin and bicuculline), hypnotic-anticonvulsant (barbiturates), anxiolytic (benzodiazepines), and anxiogenic (β-carbolines) agents. Except for picrotoxin and bicuculline, the listed agents bind to distinct but allosterically linked sites on the GABA$_A$ receptor and modulate the GABA-mediated increase in Cl$^-$ conductance. Thus the GABA$_A$ receptor is understood to form a heterogeneous functional conglomerate often termed the GABA$_A$ receptor–Cl$^-$ channel complex. Moreover, endogenous agents such as steroids and intracellular Ca^{2+}, ATP, and cyclic AMP are also known to modulate the GABA$_A$ receptor. The number and mode of actions give rise to a level of complexity, comparable only to that of *N*-methyl-D-aspartate (NMDA) receptors to date. Electrophysiology has been a critical tool for understanding this complexty. In this chapter, we first present the methods that we have employed to study the GABA$_A$ receptor-mediated Cl$^-$ responses and then discuss some aspects of the modulation by extracellular agents.

Preparation of Neurons

Neurons freshly dissociated from the frog dorsal root ganglion (DRG) and the rat CNS are used throughout the experiment. The DRGs are obtained from the adult American bullfrog (*Rana catesbiana*). After stripping away

Methods in Neurosciences, Volume 19

the thick connective tissue by microforceps, the dissected DRGs are treated with 0.3% (w/v) collagenase and 0.05% (w/v) trypsin (pH 7.4) at 37°C for ~18 min. During the enzyme treatment, the preparation is gently shaken by bubbling the bathing medium with 100% O_2. Thereafter, single neurons are isolated mechanically with finely polished pins under binocular microscope. The DRG neurons are maintained in a solution containing equal amounts of Ringer's and Eagle's minimal essential medium at room temperature until used for the experiments. Neurons 30 to 40 μm in diameter are mainly used in the present study.

For the preparation of rat CNS neurons, 300- to 500-μm-thick brain slices are obtained from 1- to 3-week-old Wistar rats with a microslicer (Dosaka, DTK-1000). Following maintenance in an incubation solution well saturated with 95% O_2–5% CO_2 at room temperature for 50 min, the slices are treated either with 1000 unit/ml dispase at 31°C for 50–70 min or with 0.02% (w/v) Pronase for 20 min followed by 0.02% (w/v) thermolysin at 31°C for 20 min. The slices are then washed with Ca^{2+}-free EGTA solution, and the following three areas are micropunched from the slices with an electrolytically polished injection needle: the spinal ventral horn (SVH), the cerebellar Purkinje layer (CPJ), and the cerebral frontal cortex (CTX). The neurons in the punched-out pieces are mechanically dissociated with fire-polished Pasteur pipettes. The neurons adhere to the bottom of the dish within 20–30 min.

Rapid Drug Application and Electrical Recording

The development of fast drug application systems is a hallmark of quantitative studies of GABA$_A$ receptor-mediated Cl^- responses, since the rapidly desensitizing peak component is obliterated in the previous slower systems. Moreover, these methods enable a maintenance of a uniform concentration of an agent in the solution, which is another absolute requisite for quantitative analyses.

Concentration-Clamp Technique

For frog DRG neurons, the "concentration-clamp" technique (Akaike *et al.*, 1986, 1987) is used to record macroscopic currents (Fig. 1A). The concentration-clamp system consists of two parts: (i) a glass suction pipette (Akaike *et al.*, 1978) which holds the neuron at its tip and allows for intracellular perfusion and single-electrode voltage clamp (i.e., the original type of conventional whole-cell configuration), and (ii) a polyethylene solution-exchange tube with a circular hole of ~500 μm diameter on its lateral surface.

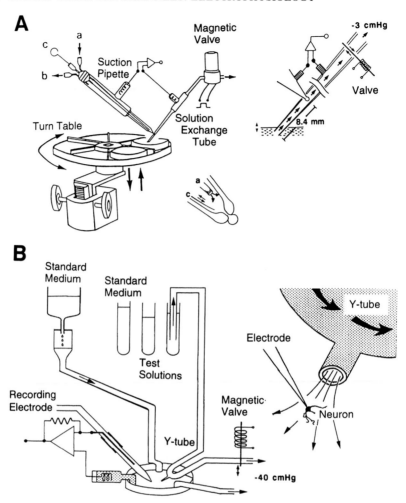

FIG. 1 Two rapid drug application systems used in the present study. (A) The "concentration-clamp" technique used for frog DRG neurons. A suction pipette holds a dissociated neuron at its tip, while the intrapipette solution is perfused by the inlet tube (a) inserted close to the tip and by the outlet (b) connected to negative pressure (−30 cmHg). Pt-Ir puncture wire (c) is used for cleaning the tip of the electrode after each experiment. The suction pipette tip is inserted into a polyethylene tube for solution exchange, and the lumen of the tube is connected to a magnetic valve driven by 12 V DC. Two insets show the details of the electrode tip and the solution-exchange tube. (Modified with permission from Akaike *et al.*, 1987.) (B) The "Y tube" method used for rat CNS neurons. The outlet tip of the Y tube is placed approximately 500 μm from the neuron. When the magnetic valve is opened, the solution in the Y tube is changed by a negative pressure of −40 cmHg. After

Suction Pipette and Solution-Exchange Tube

Suction pipettes are pulled from Pyrex glass with a 3-mm outer diameter to a shank length of 2.5 to 3 mm. The pipette tip is cut at a level with an outer diameter of 40 μm and is then fire polished to give an internal diameter of 5 to 7 μm. A dissociated DRG neuron clearly visible under binocular microscope ($\times$80) is aspirated by the suction pipette wth a negative pressure of 30 cm Hg. The aspirated membrane is ruptured either by application of a negative pressure of 30 cmHg or by a depolarizing square-wave pulse of 5–20 nA in amplitude and 10–50 msec in duration. The tip of the suction pipette is then inserted into the lumen of the solution-exchange tube through the circular hole on the lateral wall. This procedure is performed in the normal external solution. After the arrangement, the suction pipette and the tube are always held stationary, and they work as a concentration-clamp complex.

The suction pipette is equipped with an apparatus for intrapipette perfusion: tube a is an inlet inserted close to the tip, and tube b is an outlet located close to the upper end of the pipette. The suction pipette has an internal volume of 1.5 ml. At a flow rate of 1 ml/min, the ionic environment of the cell interior is completely changed within 10 to 15 min.

Method of Drug Application

Four to six culture dishes containing various kinds of external solutions to be applied to cells are placed on the turntable. By vertically lowering the turntable, rotating it, and elevating it, the lower end of the tube is immersed into another solution. The upper end of the tube is connected to a magnetic valve which was driven by 12 V DC. The opening of the valve leads to a negative pressure of 3 cmHg in the lumen and thereby rapid application of external solutions. The valve is switched on for a desired interval by a stimulator (Nihon Kohden, SEN-7103). With the opening of the valve, the external solution surrounding an isolated DRG neuron can be completely exchanged within 1 to 2 msec. Cells are never exposed to the air during the course of the experiment, due to the small diameter of the insertion hole and the high surface tension of the solution.

Electrical Recording

The resistance of the suction pipette is 200 to 300 kΩ. The electrical signal is measured through an Ag–AgCl plate in a Ringer–agar plug (2%, w/v) mounted on the suction pipette. The reference electrode is also an Ag–AgCl

---

the valve is closed, the solution in the Y tube lumen is applied to the neuron by a hydrostatic pressure. Two other tubes are used as inlet and outlet of the bath perfusion.

plate in a Ringer–agar plug (2%, w/v) located in the middle part of the solution-exchange tube. Both termini are connected to a voltage-clamp circuit, and the membrane potential is controlled under a single-electrode voltage-clamp condition using a laboratory-built sample-and-hold amplifier (Ishizuka *et al.*, 1984). The system allows the recording of 100 nA of membrane current which is measured by a current–voltage (I–V) converter with a 10-$M\Omega$ feedback resistor. The sampling rate is 10 kHz, and 36% of the cycle is used for passing current, thus changes in series resistance do not affect the performance. Both membrane current and voltage are stored on an FM data recorder (Sony PFM-15) and simultaneously monitored on a storage oscilloscope (Rikadenki, R-22) and a pen recorder (National, VP-5730A) after low-pass filtering at 5 kHz. All experiments are performed at room temperature (20 to 23°C).

Y Tube Technique

For rat CNS neurons, the "Y tube" method (Murase *et al.*, 1990) is used (Fig. 1B). In this system, a drug-containing solution is applied through an outlet tip of the Y-shaped tube by gravity.

Y Tube

A polyethylene tube 1 mm in internal diameter, 2.3 mm in outer diameter, and 100 mm in length is bent by heat to a V-shaped form. A small hole, approximately 100 μm in diameter, is made at the curved end, wherein a fine polyethylene tube with 50 μm in internal diameter and 10 mm in length is inserted and fixed by a silicon-based glue to form an outlet tip of the Y tube. One end of the Y tube is connected to a suction bottle (kept at -40 cmHg) through a magnetic valve, the opening of which is driven by 12 V DC. The other end is immersed in a test tube containing a standard external solution, which is placed 20 to 30 cm above the plane of the Y tube. The orifice of the tip is located 500 μm to 1 mm away from a neuron.

Method of Drug Application

Initially, before the start of recording, the valve is opened for 0.5 to 1.5 sec, and the lumen of the Y tube is filled with the standard external solution. After obtaining a stable whole-cell recording, the upper end of the Y tube is transferred to a test tube containing a drug, and then the valve is opened. During the opening for 0.5–1.5 sec, the Y tube lumen is washed with the new solution. When the luminal flow is abruptly halted by closure of the valve, the solution is rapidly applied to the neuron by hydrostatic pressure.

The neuron is continuously perfused with this solution until the standard external solution or another test solution is applied by the same procedure. Thus, the neuron is always exposed to a rapid flow of solutions with a constant concentration of an agent. The exchange speed of the external solution depends mainly on the travel time of the solution from the tip orifice to the cell. By adjusting the position of the Y tube tip so as to obtain the maximal flow without blowing out the cell, the solution surrounding a neuron is exchanged within 10 to 20 msec.

In addition to this rapid application, the experimental setup is equipped with a bath perfusion system that is comprised of (i) an inlet allowing a continuous flow of standard external solution at a rate of 1 ml/min, and (ii) an outlet connected to a bottle aspirated continuously by a negative pressure of 40 cmHg. Bath perfusion is used only for constant washing of the culture dish.

Electrical Recording

Electrical recording is performed with the conventional whole-cell recording. Patch electrodes are pulled from 1.5-mm (outer diameter) capillary glass on a two-stage vertical puller (Narishige, PB-7) and fire polished on a microforge (Narishige, MF-83). The resistance between the recording electrode filled with the internal solution and the reference electrode is 5 to 8 MΩ. The reference electrode is Ag–AgCl wire, making electrical contact with the bath solution through a Ringer–agar plug (2%, w/v). The membrane current and voltage are measured using a patch-clamp amplifier (Nihon Kohden, CEZ 2300 or List Medical, EPC-7), monitored on both a storage oscilloscope (Tektronix, R5113) and a pen recorder (San-ei, Recti-Horiz-8K), and stored on videotapes after digitization with a PCM processor (Nihon Kohden, PCM501ESN). The currents are low-pass filtered at 1 kHz.

Single-Channel Analysis

The patch-clamp recording with the inside-out configuration (Hamill *et al.,* 1981) is used for analyzing GABA-induced unitary currents in frog DRG neurons. Patch pipettes are prepared and electrical signals are processed as in the whole-cell recording from rat CNS neurons. Pipette solution facing the extracellular side of the membrane (i.e., extracellular solution) and bath solution facing the intracellular side (i.e., intracellular solution) are the same as those used for concentration-clamp studies on frog DRG neurons. The patch pipette is internally perfused by a tube inserted deep to the tip of the patch pipette. This technique allows a test solution to be perfused over the patch membrane within a few seconds to approximately 10 min, by proper adjustment of the distance between the tip of the tube and the patch membrane. A distance of 2 mm is adopted in the present study and 10 min is required for complete diffusion.

Composition of Solutions

To isolate the Cl^- current from all other currents in frog DRG neurons, Na^+, K^+, and Ca^{2+} are replaced by equimolar tris(hydroxymethyl)aminomethane $(Tris^+)$, Cs^+, and Mg^{2+}, respectively, in both external and internal solutions. The ionic composition of the external solution is (in mM): Tris–Cl, 90; CsCl, 2; MgCl$_2$, 5; tetraethylammonium (TEA)-Cl, 18; glucose, 5; and N-2-hydroxyethylpiperazine-N'-2-ethanesulfonic acid (HEPES), 10. The pH is adjusted to 7.4 with a Tris base. The ionic composition of the internal solution is (in mM): CsCl, 95; cesium aspartate, 10; TEA-Cl, 25: and ethylene glycol bis(β-aminoethyl)ether N,N,N',N'-tetraacetic acid (EGTA), 2.5; and HEPES, 10. The pH is adjusted to 7.2 with a Tris base. Thus the extracellular and intracellular Cl^- concentrations are 120 mM, giving the Cl^- equilibrium potential (E_{Cl}) of +4 mV, calculated with the Nernst equation considering the Cl^- activities in both solutions.

The standard external solution for rat CNS neurons has the composition (in mM); NaCl, 150; KCl, 5; CaCl$_2$, 2; MgCl$_2$, 1; glucose, 10; and HEPES, 10. The internal solution has the composition (in mM): NaCl, 30; KCl, 50; potassium gluconate, 70; CaCl$_2$, 0.25; MgCl$_2$, 1; Mg-ATP, 5; EGTA, 5; and HEPES, 10. The pH of the external and internal solutions is adjusted to 7.4 and 7.2, respectively, with a Tris base. The incubation solution contains (in mM): NaCl, 124; KCl, 5; KH$_2$PO$_4$; 1.2; NaHCO$_3$, 26; CaCl$_2$, 2.4; MgSO$_4$, 1.3; and glucose, 10. Ca^{2+}-free EGTA solution is prepared from the incubation solution by removal of CaCl$_2$ and addition of 2 mM EGTA.

Neuronal Responses to GABA$_A$ Receptor Activation

Macroscopic Currents

Activation of the GABA$_A$ receptor by agonists such as GABA and muscimol opens a Cl^- channel. Figure 2 shows the responses of frog DRG neurons to rapid application of GABA. At low concentrations of GABA, only steady-state currents were elicited, while at higher concentrations, steady-state currents were preceded by a rapidly desensitizing peak component (Fig. 2A). The half-maximal concentration (EC$_{50}$) and Hill coefficient were 1.3×10^{-5} M and 2.0 for the peak component and 5.2×10^{-6} M and 2.0 for the steady-

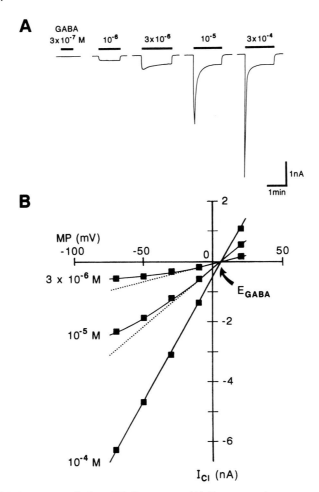

FIG. 2   GABA respones in frog DRG neurons. (A) Concentration–response relationship for GABA at a holding potential (V_H) of -10 mV. Note that at high concentrations, a rapidly desensitizing peak component appears. (Modified with permission
from Akaike *et al.*, 1990.) (B) Current–voltage (I–V) relationship of the peak component of GABA response at three different concentrations. At low concentrations of
GABA, outward rectification is present, which is absent at $10^{-4} M$. E_{GABA} represents
the reversal potential of GABA-induced current, which is close to the theoretical
E_{Cl}. MP represents the membrane potential. (Modified with permission from Akaike
et al., 1986.)

state component, respectively (Figs. 4A and 6B). The reversal potential of
GABA-induced current (E_{GABA}) as estimated from the current–voltage (I–V)
relationship was $+4$ mV, quite close to E_{Cl}, suggesting that GABA activates
Cl$^-$ current (I_{Cl}) (Fig. 2B). The Cl$^-$ channel is also permeable to other anions;

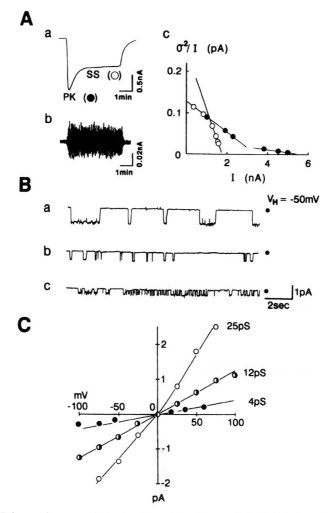

FIG. 3   Noise-variance and single-channel analyses of GABA-induced I_{Cl} in frog DRG neurons. (A) GABA responses obtained by low-pass filtering and high-gain AC recording are shown in (a) and (b), respectively. Open and closed circles in (c) correspond to steady-state (SS) and peak (PK) current components in (a), respectively. The data points fell into three groups, each of which could be fitted by a linear line. The results were obtained from five neurons. (B) Unitary currents recorded in inside-out configuration with a V_H of -50 mV. The patch pipette contained 3×10^{-6} M GABA. The recording was filtered at 1 kHz. Closed circles at the right of traces represent the closed level of single-channel current, and inward current is shown as a downward deflection. The recording was done with equal extracellular and intracellular Cl$^-$ concentrations (120 mM). The openings of large (a), intermediate

internal and external perfusion of the frog DRG neurons suggest a relative permeability ratio of $SCN^- > I^- > NO_3^- > Br^- > ClO_3^- > Cl^- > HCOO^- > BrO_3^- > F^- > CH_3COO^- > CH_3CH_2COO^-$ (Akaike *et al.*, 1989).

Microscopic Currents

In order to understand the basis of macroscopic currents, we analyzed the microscopic currents and reported for the first time that GABA activates at least three separate channels of GABA$_A$ receptor with distinct conductances (Yasui *et al.*, 1985). Figure 3A illustrates the fluctuation analysis of GABA responses in frog DRG neurons. GABA-induced I_{Cl} (Fig. 3Aa) was accompanied by a marked increase in current fluctuation (Fig. 3Ab). The variance (σ^2) of this I_{Cl} fluctuation after subtracting the corresponding value of the background noise is related with the mean I_{Cl} (I) as shown by the equation,

$$\sigma^2/I = i - I/N, \quad \text{with } I = N P i,$$

where i represents single-channel current; N, the total number of channels, and P, the open-state probability of individual channels. Figure 3Ac shows that three linear lines fit the (σ^2/I)–(I) relationship for the peak and steady-state currents. The separate lines represent the three single-channel conductances, <5 pS (small), ~15 pS (intermediate), and ~26 pS (large), readily calculated by dividing the y-intercept (i) extrapolated from the lines, by the Cl^- driving force of 10 mV in this case. The data suggest, moreover, that the steady-state component consists of intermediate conductance channels at lower concentrations of GABA and large conductance channels at higher concentrations. In contrast, the peak component is mediated by either small or intermediate conductance channels, depending on the concentrations used. Single-channel recordings of these channels are shown in Figs. 3B and 3C. The current traces show the opening and closing of large (a), intermediate (b), and small (c) conductance channels, all of which reverse current directions at 0 mV, close to the theoretical E_{Cl}. Although the large conductance channels show periods of burst-like activities in some occasions, subconductance states were never observed. These results unequivocally indicate that

---

(b), and small (c) conductance channels are observed in separate patches. (C) I–V relationship of three types of unitary currents. The data were obtained from the same patches shown in (B). (Modified with permission from Yasui *et al.*, 1985.)

in frog DRG neurons, there are at least three distinct subtypes of $GABA_A$ receptors.

Bormann et al. (1987) applied single-channel recording to cultured spinal neurons of mouse embryo and detected four levels of multiconductance states in $GABA_A$ receptors. The slope conductances measured with equal extracellular and intracellular Cl^- concentrations (145 mM) under outside-out configuration were 12, 19, 30, and 44 pS. Bormann et al. ascribed these different levels to subconductance states of a single channel rather than to multiple channels with distinct conductances, first because the levels were observed with high probability in a single patch, and second because the transition between them was frequently observable. This result deserves noting since the preparations used in the Bormann and the present reports differ in species (mouse and frog), age (embryo and adult), neuron location (spinal cord and DRG), and method of preparation (primary culture and acute dissociation). These differences may suggest the possibility of species specificity, ontogenetic and/or regional alteration in expressed $GABA_A$ receptors, and await further clarification.

Modulation of $GABA_A$ Receptors

Benzodiazepines

One of the well-studied extracellular modulators is benzodiazepines. Diazepam (DZP) is a benzodiazepine derivative that exhibits anxiolytic, anticonvulsant, and sedative actions in humans; its potentiating effect is shown in Fig. 4 (Yakushiji et al., 1989a). DZP treatment shifts the concentration–response curve of the peak component to the left without affecting the maximal current or the Hill coefficient. The EC_{50} value of the peak component was $1.3 \times 10^{-5} M$ and was reduced to $6.0 \times 10^{-6} M$ in the presence of $3 \times 10^{-6} M$ DZP. The EC_{50} value of the steady-state component was also reduced from 5.4×10^{-6} to $9.0 \times 10^{-7} M$, but the maximal level was suppressed. At a single-channel level, DZP increased the open time (Fig. 4B, 1 min) and the number of available channels (Fig. 4B, 16 min). These results, along with the fact that the E_{GABA} of macroscopic and microscopic currents was not affected in the presence of DZP, suggest that DZP increases the affinity of the receptor to GABA and activates GABA receptors at a subthreshold level.

Various benzodiazepine-related compounds have been classified into three categories according to their potencies to enhance GABA responses (Yakushiji et al., 1989b). Full agonists including DZP maximally potentiate GABA-induced I_{Cl}. Partial agonists also potentiate the GABA responses, but the efficiencies are weaker than those of full agonists. Interestingly, partial

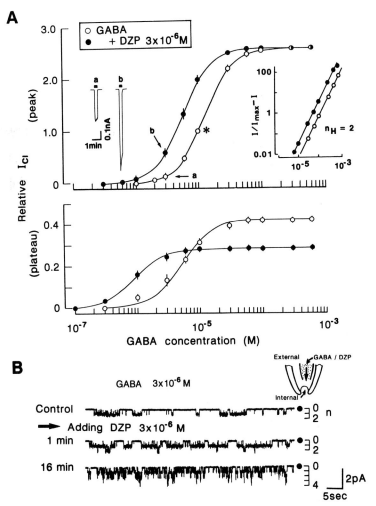

FIG. 4   Potentiation of GABA-induced I_{Cl} by diazepam (DZP) in frog DRG neurons.
(A) Concentration–response curves with (closed circles) and without (open circles)
$3 \times 10^{-6} M$ DZP. The amplitudes of the currents were normalized to that of I_{Cl} induced
by $10^{-5} M$ alone (∗). Inset on the left shows the currents (a and b) corresponding to
the points on the curves. Inset on the right shows the linear-regression analysis
for Hill coefficients (n_H). (B) Inside-out recording of microscopic currents. Gradual
increase in channel-opening events was due to intrapipette perfusion of DZP (inset).
DZP markedly recruited channels to open. (Modified with permission from Yakushiji
et al., 1989a.)

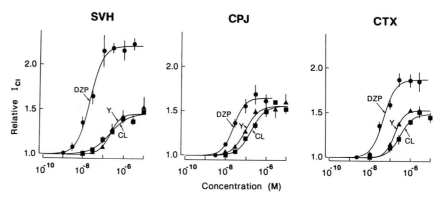

FIG. 5   Concentration–response curves for three benzodiazepine-related compounds in different areas of the rat CNS. CL 218,872 (CL) and Y-23684 (Y) are both classified as partial agonists. The changes in current amplitudes were normalized to the amplitude of I_{Cl} induced by 3×10^{-6} M GABA alone. SVH, spinal ventral horn neurons; CPJ, neurons in the cerebellar Purkinje layer; CTX, neurons of the cerebral frontal cortex. V_H was -50 mV. (Modified with permission from Yakushiji *et al.*, 1993.)

agonists suppress the maximal potentiation by full agonists when applied in combination. The third category is an inverse agonist; it suppresses rather than potentiates the GABA-induced I_{Cl}. The actions of these three categories of agents are blocked by an antagonist, Ro 15-1788.

Based on the binding affinity to a partial agonist, CL 218,872, the pharmacological division of the $GABA_A$ receptor complex into two types has been proposed. Type I denotes the population with a high affinity to CL 218,872 and type II is the population with a lower affinity. We have successfully demonstrated a regional difference in potentiating effects of DZP in the rat CNS (Fig. 5). (Yakushiji *et al.*, 1993). Partial agonists, CL 218,872 and Y-23684, equally potentiated the GABA responses in three areas of the rat brain. DZP exerted a full-agonist activity in SVH neurons, but the effect was as weak as that of partial agonists in neurons of the CPJ. In neurons of the CTX, the effect of DZP was intermediate. Ligand-binding study showed that type II receptors are abundant in SVH and type I receptors are abundant in CPJ, while both types are found in CTX. A recent cloning study (see Conclusion) revealed that α_2 or α_3 subunits coexpressed with β_1 and γ_2 subunits in mammalian cells manifested a property comparable to type II, whereas the α_1 subunit in the same condition exhibited the property of type I receptors (Pritchett *et al.*, 1989). *In situ* hybridization of α subunits demonstrated that α_2 and α_3 are abundantly expressed in SVH neurons, whereas α_1 is predominant in CPJ neurons. These results of binding and cloning studies are consistent with the functional study presented in this paper, and thus our electrophysiological data add more criteria to the classification of GABA$_A$ receptors.

Barbiturates

Ligand-binding studies showed that pentobarbital (PB) increased the binding of [^{3}H]GABA and [^{3}H]muscimol to GABA$_A$ receptors. Electrophysiological studies from our laboratory supported this notion as shown in Fig. 6 (Akaike *et al.*, 1990). The GABA-induced I_{Cl} was augmented by the simultaneous application of 10^{-4} M PB (Fig. 6A). The concentration–response curves in the presence or absence of PB are shown in Fig. 6B. PB markedly shifted the curve of a peak I_{Cl} to the left in a concentration-dependent manner, without affecting the maximum current or the Hill coefficient. The EC$_{50}$ value of peak I_{Cl} was reduced from 1.3×10^{-5} M in the absence of PB to 2.9×10^{-6} M in its presence (10^{-4} M). In contrast, the curve of the steady-state component of I_{Cl} was slightly shifted to the left and the maximal level was markedly increased, indicating an increase in the available receptor–channel complex. The EC$_{50}$ value was reduced from 5.2×10^{-6} M for control to 3.1×10^{-6} M in the presence of 10^{-4} M PB. The Hill coefficients were 2.0 for all the curves shown in the figure. These results indicate that PB augmented the GABA-induced I_{Cl} mainly by enhancing the apparent affinity of the GABA$_A$ receptor to GABA. The potency among several barbiturates was in the order of secobarbital > pentobarbital > hexobarbital > phenobarbital.

At a concentration higher than 10^{-4} M, PB alone elicited I_{Cl}, which was concentration-dependently potentiated by DZP and suppressed by competitive and noncompetitive GABA$_A$ receptor antagonists, bicuculline, and picrotoxin, respectively. This GABA-mimetic action further strengthens the close interaction operating in the GABA$_A$ receptor–Cl$^-$ channel complex.

Steroids

The search for an endogenous modulator of GABA$_A$ receptor has raised steroids as a possible candidate. A synthetic steroidal anesthetic, alfaxalone, was first proposed to exert anesthetic action through enhanced inhibitory transmission in the CNS. The effect of a progesterone metabolite, pregnanolone (5β-pregnan-3α-ol-20-one), on GABA-induced I_{Cl} in frog DRG is shown in Fig. 7 (unpublished data). At 10^{-7} M, pregnanolone induced hardly detectable current, and simultaneous application of pregnanolone potentiated the response to GABA. The augmentation was due to increased apparent affinity of the receptor to GABA, since the

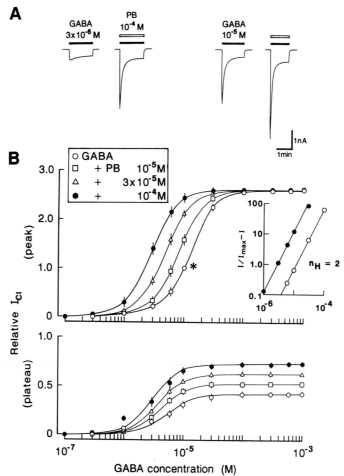

FIG. 6 Potentiation of GABA responses by pentobarbital (PB) in frog DRF neurons. (A) PB ($10^{-4}M$) was simultaneously applied with GABA. (B) Concentration–response relationship of GABA responses in the presence of PB at various concentrations. Both the peak and steady-state currents were normalized to the peak I_{Cl} induced by 10^{-5} M GABA alone (∗). (Modified with permssion from Akaike *et al.*, 1990.)

concentration–response curve shifted to the left without changes in maximal response or Hill coefficients and since the reversal potential was not altered by pregnanolone treatment. Similar action was reported with structurally related compounds such as progesterone metabolites, deoxycorticosterone metabolites, and androsterone (Lambert *et al.*, 1987). Single-

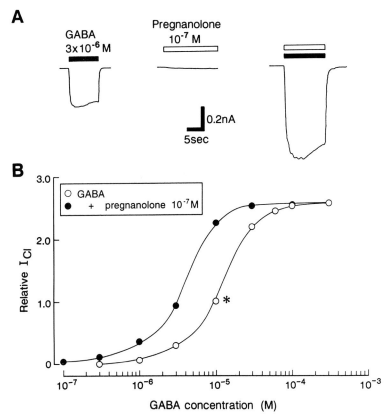

FIG. 7   Potentiation of GABA-induced I_{Cl} by pregnanolone in frog DRG neurons. (A) The membrane currents elicited by GABA alone, pregnanolone alone, or a combination of the two compounds. (B) Concentration–response curves show that the steroid enhances the affinity of the receptor to GABA. The currents were normalized to the peak I_{Cl} induced by 10^{-5} M GABA alone (∗).

channel analysis revealed that pregnanolone concentration-dependently prolonged the burst durations of single GABA channel currents in bovine chromaffin cells (Lambert *et al.*, 1987). At a concentration higher than 3×10^{-7} M, pregnanolone exhibited a GABA-mimetic action.

Based on pharmacological studies, steroidal site of action is thought to be different from benzodiazepine-binding sites, but close to barbiturate-binding sites. However, the exact site of steroidal action and the mechanism of interaction between different binding sites are currently not known and are under extensive investigation.

Conclusion

We have demonstrated that rapid drug application systems are suitable for quantitative analyses of $GABA_A$ responses. Since both the quickly desensitizing peak component and the nondesensitizing steady-state component can be recorded without reduction in amplitude, the techniques are applicable not only to kinetics studies but also to a wide range of functional modulations. Recently we used multi-barelled pipettes, devised by Carbone and Lux (1987), to show the suppressing and rebound-potentiating effects of penicillin on glycine-induced Cl^- current in the continued presence of glycine (Tokutomi *et al.*, 1992). This promising technique will facilitate the assessment of pharmacological differences between the desensitizing and nondesensitizing components of $GABA_A$ response.

In 1987, molecular cloning of the $GABA_A$ receptor revealed the sequence of two subunits, α and β, and thereafter a wide variety of subunits were discovered based on the peculiar characteristic that each subunit in homo-oligomer configuration responded to GABA, if not as completely as in native receptors. The $GABA_A$ receptor belongs to a ligand-gated receptor superfamily with a typical four-transmembrane-spanning motif and is currently thought to be a heteromeric pentamer of at least five types of subunits with numerous isoforms: $6\alpha s$ (α_{1-6}), $4\beta s$ (β_{1-4}), $3\gamma s$ (γ_{1-3}), 1δ, and 1ρ (Olsen and Tobin, 1990). α subunits are thought to affect affinity for GABA and benzodiazepine pharmacology. β subunits might affect GABA affinity and carry a single protein kinase A site in each subunit. γ subunits are essential for binding of benzodiazepines but the detailed functions of γ, δ, and ρ are still unclear. Simple combination of these subunits plus alternative splicing and posttranslational modification theoretically yield an enormous number (3000 to 4000) of possible receptor types. In contrast to the wealth of information on subunits expressed in *in vitro* systems, only limited data are available on the actual composition of native receptors *in situ*. Moreover, the mechanisms of region-specific expression of a given subtype are far from being understood, and ontogenetic changes in the subunit composition, if present, are currently not known in detail.

The characteristic properties of $GABA_A$ receptors are (i) numerous sites of modulation and (ii) structural diversity inferred from molecular biology. Part of the results presented in this chapter indicate nonuniform properties of the $GABA_A$ receptor–Cl^- channel complex, widely distributed in the nervous system. Functional studies are expected to further reveal novel properties of this multifaceted receptor–channel complex, which would provide definitive understanding of physiological and pathological functions of GABAergic neurotransmission.

Acknowledgments

The authors thank Mr. D. Bowman for critical reading of the manuscript. This work was partially supported by a Grant-in-Aid for Scientific Research (No. 04044029) to N. Akaike from the Ministry of Education, Science and Culture, Japan.

References

C. A. Drew, G. A. R. Johnston and R. P. Weatherby, *Neurosci. Lett.* **52,** 317 (1984).

N. Akaike, M. Inoue, and O. A. Krishtal, *J. Physiol.* **379,** 171 (1986).

N. Akaike, T. Yakushiji, N. Tokutomi, and D. O. Carpenter, *Cell. Mol. Neurobiol.* **7,** 97 (1987).

N. Akaike, K. S. Lee, and A. M. Brown, *J. Gen. Physiol.* **71,** 509 (1978).

S. Ishizuka, K. Hattori, and N. Akaike, *J. Membr. Biol.* **78,** 19 (1984).

K. Murase, M. Randic, T. Shirasaki, T. Nakagawa, and N. Akaike, *Brain Res.* **525,** 84 (1990).

O. P. Hamill, A. Marty, E. Neher, B. Sakmann, and F. J. Sigworth, *Pfluegers Arch.* **391,** 85 (1981).

N. Akaike, N. Inomata, and T. Yakushiji, *J. Neurophysiol.* **62,** 6 (1989).

S. Yasui, S. Ishizuka, and N. Akaike, *Brain Res.* **344,** 176 (1985).

J. Bormann, O. P. Hamill, and B. Sakmann, *J. Physiol.* **385,** 243 (1987).

T. Yakushiji, N. Tokutomi, and N. Akaike, *Neurosci. Res.* **6,** 309 (1989a).

T. Yakushiji, T. Fukuda, Y. Oyama, and N. Akaike, *Br. J. Pharmacol.* **98,** 735 (1989b).

T. Yakushiji, T. Shirasaki, M. Munakata, A. Hirata, and N. Akaike, *Br. J. Pharmacol.* **109,** 819 (1993).

D. B. Pritchett, H. Lüddens, and P. H. Seeburg, *Science* **245,** 1389 (1989).

N. Akaike, N. Tokutomi, and Y. Ikemoto, *Am. J. Physiol.* **258,** C452 (1990).

J. J. Lambert, J. A. Peters, and G. A. Cottrell, *TIPS* **8,** 224 (1987).

E. Carbone and H. D. Lux, *J. Physiol.* **386,** 547 (1987).

N. Tokutomi, N. Agopyan, and N. Akaike, *Br. J. Pharmacol.* **106,** 73 (1992).

R. W. Olsen and A. J. Tobin, *FASEB J.* **4,** 1469 (1990).

Section III

Cardiac and Smooth Muscle Voltage-Gated Ion Channels: Electrophysiology

[9]   Whole-Cell Voltage Clamp of Cardiac Sodium Current

Michael F. Sheets and Robert E. Ten Eick

Introduction

The nature of the sodium current in excitable membranes and its role in the generation of action potentials have been the objects of intensive investigation for many years. As a result of techniques developed within the recent past, we now can voltage clamp the Na current from cardiac cells as well as that elicited by the cells of many tissues found in a broad spectrum of animal species and rigorously study the channels responsible for its conductance.

The cardiac cell provides the unusual advantage in that it allows whole-cell Na current (I_{Na}), Na-channel gating current ($I_{g\text{-}Na}$), and unitary Na-channel currents (i_{Na}) to be investigated using the same cell type for all studies. In addition, cardiac Na channels are structurally different than those from nerve (1–3) and consequentially possess a number of special characteristics. Whole-cell cardiac I_{Na} exhibits a slower onset and decay than neuronal I_{Na}; in heart the voltage dependence of the mean Na-channel open time is biphasic, exhibiting a maximum at approximately -30 to -40 mV, whereas in neuronal cells it reaches a sustained plateau (4–7).

The pore properties of cardiac Na channels are also different. Their sensitivity to block by divalent cations such as Cd^{2+} and Zn^{2+} applied extracellularly is greater than that for neuronal Na channels (8–11). In addition, their sensitivity to block by the guanidinium toxins, tetrodotoxin and saxitoxin, is less (12, 13), yet to the anti-arrhythmic drug, lidocaine, it is greater (14).

Whole-cell voltage clamp is but one of several experimental approaches that can be used to characterize the electrophysiological behavior of cardiac Na channels. However, it is the most powerful experimental technique and thus is the most frequently used. For these reasons we review the whole-cell voltage-clamp technique in detail. In addition, we discuss several other important experimental methods currently being applied to the study of cardiac Na channels; some important considerations involving data analysis are also discussed.

Methods in Neurosciences, Volume 19
Copyright © 1994 by Academic Press, Inc. All rights of reproduction in any form reserved.

General Considerations

Experimental Preparations

Presently, most experimental preparations use freshly isolated single cells from fetal, neonatal, or adult hearts. The notable exceptions are vesicle preparations used for lipid bilayer experiments and the multicellular tissue preparations used to record action potentials and measure the maximum rate of rise of the action potential upstroke or the conduction velocity of the action potential as rough indices of the Na current. Although many methods have been developed to prepare isolated single heart cells, most are variations of the method pioneered by Powell and Twist (15) and use enzymes such as crude collagenase or trypsin to digest the connective tissue that holds the myocardial cells in place within the cardiac skeleton. Some variations perfuse the heart with enzyme-containing solutions while others incubate finely minced pieces of heart tissue in similar enzyme-containing solutions (16). A mechanical technique which does not involve exposure of the sarcolemmal membrane to proteolytic enzymes, the cut-slice technique, has been used to expose sarcolemma membranes from cardiac tissue (17). Data obtained from preparations made with this latter technique should help answer the question of whether treating the hearts with an enzyme to isolate cells has altered the electrophysiology of sarcolemmal channels, which after all are protein in nature.

Either atrial or ventricular myocytes can be isolated, with ventricular myocytes being most commonly studied because they can be obtained in relatively large numbers. Although cardiac I_{Na} and i_{Na} have been studied in cells from hearts of many different species, guinea pig cardiac cells have been the most frequently studied. They are relatively small cells and appear to have a relatively low Na-channel density. These attributes permit the quality of the voltage clamp to be enhanced (relative to what it would be were cell size larger) by minimizing both the time constant of the capacity current and the possibility for a transient loss of membrane voltage control during peak I_{Na}. Other species and cell types offer other special advantages. For example, canine cardiac Purkinje cells appear to offer specific advantages for the study of Na-channel gating currents ($I_{g\text{-}Na}$). The density of their Na channels is high compared to that for calcium channels; this permits measurement of gating current which is essentially an uncontaminated Na channel $I_{g\text{-}Na}$ (18). Cat ventricular cells are very tolerant of the isolation procedure; one can harvest a very large proportion of the myocytes present in the heart and most will be Ca^{2+}-tolerant (19). They have proven to be very amenable to forming high-resistance seals and display vigorous Na currents, and most of the other commonly described cardiac current systems appear to be present. In addition, the finding that adult feline ventricular myocyte can be placed in primary culture for up to several weeks appears to make them unique among adult mammalian cardiac cells (20). Recently, hearts obtained from patients undergoing cardiac transplantation have provided a source of isolated myocytes for the study of human cardiac I_{Na}. This may finally allow testing of the assumption that conclusions based on studies of animal heart are applicable to the human heart and the clinical situation (21, 22). However, because explanted hearts are frequently diseased and because a very small fraction of the total cells from the heart (<1–2%) can be typically isolated in a viable, Ca^{2+}-tolerant condition, data obtained from these human cardiocytes must be interpreted with utmost caution, particularly when asociations between the electrophysiology of the isolated myocytes and the clinical state of the patient's explanted heart are made.

Electrophysiological Approaches to the Study of Cardiac Na Channels

Voltage clamp is one, if not the most powerful, approach to the study of the voltage-gated Na channel. In contrast to action potentials where the membrane potential is free to spontaneously change, voltage clamp enables the membrane potential to be controlled (i.e., clamped) and the resultant current response to be measured. Even though an instantaneous change in membrane potential during voltage clamp would require an infinitely large current source, which is theoretically impossible, selection of an appropriate experimental preparation can permit nearly instantaneous changes in membrane potential when compared to the speed of ionic channel kinetics. Voltage clamp of ionic channels (e.g., cardiac Na channels) allows assessment of the voltage and time dependence of the opening and closing behaviors and the channel's conductance. Without control of membrane potential, an analysis of the channel's kinetic and pore properties would be extremely complicated because both are governed by membrane voltage.

There are many different approaches to voltage-clamp cardiac Na channels (see Table I). Each approach can be defined in terms of the specific type of electrode employed and how it is used to obtain recordings during voltage-clamp experiments. Included are (i) the whole-cell patch technique using either small- or large-pore suction pipette electrodes to record whole-cell I_{Na} or $I_{g\text{-}Na}$, (ii) an oil-gap voltage clamp that utilizes a 30- to 40-μm-wide oil gap to electrically isolate the voltage-clamped end of a single cell from the other end that has had its cell membrane physically disrupted to allow for current injection and internal perfusion (23), (iii) the single microelectrode "switch

TABLE I   Electrophysiological Techniques for the
Study of Cardiac Na Channels

---

1. Whole-cell voltage clamp with
   a. Patch electrode
   b. Large-pore suction pipette
   c. Oil-gap voltage clamp
   d. Microelectrode switch clamp
   e. Perforated patch voltage clamp
2. Cell-attached macropatch voltage clamp
3. Cell-attached single-channel patch clamp
4. Lipid bilayer voltage clamp with reconstituted Na channels
5. Two-microelectrode voltage clamp of multicellular tissue

---

clamp'' that achieves voltage clamp by very rapidly switching between current injection and voltage measurement, (iv) a whole-cell I_{Na} recording technique that employs a small-pore suction pipette filled with internal solution containing amphotericin B or nystatin to permeabilize the sarcolemma within the pipette pore and to permit intracellular electrical access, (v) the macropatch technique to record many (i.e., >50, <200) Na channels from a cell-attached patch using a relatively large-pore (i.e., 10–20 μm) patch pipette, (vi) the cell-attached patch technique to record the activity of single Na channels using small patch pipettes, (vii) a technique to record single-channel currents from Na channels derived from membrane vesicles and reconstituted into lipid bilayers, and (viii) the two-microelectrode voltage clamp of small multicellular strands of cardiac tissue (24). In addition to the aforementioned, a number of whole ''cell-style'' patch clamping approaches have been developed to enable the study of cloned Na channels that have been expressed in a variety of expression systems including *Xenopus* oocytes and several types of stable or immortalized mammalian cell lines. This review concentrates on voltage-clamp techniques for recording whole-cell cardiac I_{Na} and cardiac Na-channel gating current from native membranes.

During the past 10 years almost all investigators have used computers to generate voltage-clamping protocols and to acquire and analyze data. IBM-compatible PC computers have been the most popular but Unix-based workstations and Macintosh computers also are used. Each has its advantages and disadvantages. The advantage of Unix-based workstations is their ability to record and manipulate large data sets more easily than a DOS-based personal computer. On the other hand, the Intel 80486-driven family of PC computers is relatively inexpensive and has a large number of useful, relatively inexpensive DOS-based applications software packages available (e.g., pClamp, Axon Instruments, Inc., Forster City, CA, and Sigmaplot, Jandel Scientific, Corte Madera, CA).

Whole-Cell Voltage Clamp

Applications and Advantages

Whole-cell voltage clamp of cardiac myocytes has become one of the most commonly used techniques to study cardiac I_{Na}. This technique employs either a single small-pore (0.3- to 5-MΩ, hereafter referred to as a patch electrode) or a single large-pore (50–80 KΩ) suction pipette to enable single isolated cardiac cells to be voltage clamped. Both types of electrodes allow both the internal and the external solutions to be controlled. The internal solution filling a patch electrode controls the intracellular milieu by allowing diffusion of solutes into and out of the cell through the disrupted sarcolemmal membrane surrounded by the patch electrode pore; ordinarily the internal solution cannot be changed during the course of an experiment (25). The large-pore suction pipette, in contrast, readily permits rapid changes of the internal solution during an experiment (see below). With either small or large electrodes the extracellular solution can be controlled by changing the bath solution; this readily allows I_{Na} obtained before and after a change in extracellular solution to be compared.

Although the patch electrode is suitable to answer many experimental questions involving voltage clamp of I_{Na}, their relatively large series resistance and the associated relatively long duration of the capacitance transient time constant ($\tau = {}>100–200$ μmsec) can be a significant limitation particularly when attempting to measure very fast events such as sodium-channel gating currents. In contrast, the large-pore suction pipette with a series resistance of 50 to 80 KΩ typically exhibits a primary capacitance τ of only 5–15 μsec; this allows for a more rapid voltage clamp and enables more accurate measurements of $I_{g\text{-}Na}$ and I_{Na} when membrane potential is changed rapidly (e.g., during "instantaneous" I_{Na} recordings).

Whole-cell recordings have specific advantages over single-channel recordings. First, the extracellular and intracellular milieu can be more easily modified during an experiment with whole-cell voltage clamp than with single-channel recording. Second, I_{Na} elicited at very positive test potentials near the I_{Na} reversal potential (E_{rev}) can be easily measured using whole-cell voltage clamp but this is difficult to do with cell-attached patch single-channel recordings. This is because the magnitude of the single Na-channel currents are small and become submerged in the background noise. Third, changes in Na-channel kinetics or conductance usually are more easily recognized and characterized from whole-cell Na currents than from single-channel currents. With whole-cell voltage clamp, I_{Na} kinetics can be easily derived from a family of currents elicited by a relatively short series of voltage clamps, typically no more than 15 to 20 pulses that are 5–10 mV apart from each other covering the pertinent range of membrane potential. In contrast, to derive the same information from single-channel recordings, a lengthy series of pulses (typically >250 pulses) at each test potential for both the control state and after an intervention would be required. In addition, the analysis of single-channel data typically requires more time and more model-dependent assumptions than does the analysis of whole-cell I_{Na} recordings. However, if one wishes to examine characteristics such as single-channel conductance, channel open times, and latency to channel openings, etc., the unique advantages of single-channel patch recordings for these purposes are clear.

Patch Electrodes

Voltage clamp of cardiac cells using patch pipettes has been adapted from the method detailed by Hamill et al. (26) and typically employs a patch electrode made of borosilicate glass pulled in two stages so that the taper on the electrode tip is blunt rather than long and gradual. The blunt taper typically permits patch electrodes to have tip resistances ranging from 0.5 to 2 MΩ with the lower resistances enabling better voltage clamp, particularly when large currents flow (i.e., > 1–2 nA). It is usually not necessary to heat polish the electrode tip to form high-resistance seals. Likewise, coating the tips with Sylgard (Dow Corning Corp.) to lower electrode capacitance is unnecessary because the resistances of the electrodes are relatively low and the additional contribution of the electrode to the total capacitance of the equivalent circuit is small when compared to that from the cell membrane.

Large-Pore Suction Pipettes

The large-pore suction pipette is made from a 2–3 mM outside diameter borosilicate double-barreled (θ) glass that is fabricated so that the septum extends to only 500 μm from the pore. This allows for an inflow of internal solution on one side and its outflow on the other side (27). The pore is heat-polished so that its diameter is nearly the same as the cross-sectional diameter of the cell to be voltage clamped. For example, the electrode pore diameter suitable for canine Purkinje cells typically is about 30 μm while that for the smaller guinea pig ventricular cells is about 12 μm.

Extracellular and Intracellular Solutions

Because I_{Na} in cardiac cells normally can be a very large current, extracellular Na$^+$ is usually reduced to decrease I_{Na} amplitude. The reduction of I_{Na} amplitude helps to minimize the voltage drop (e.g., loss of membrane control) resulting from current flowing across the series resistance (that resistance in series with the cell's sarcolemma) in the voltage-clamp circuit. Typically peak sodium current magnitudes greater than 4 to 5 nA will show clear evidence of loss of membrane potential control during peak I_{Na} (see section entitled Special Considerations in Voltage Clamp of Cardiac I_{Na}). Reduction of extracellular (i.e., bath) [Na$^+$] to 5–25 mM is usually necessary to reduce I_{Na} to magnitudes sufficiently small so that adequate voltage clamp can be achieved.

There are several considerations that come into play when selecting the compositions of soluions used for I_{Na} studies. Frequently cation substitutes for Na$^+$ include Cs$^+$ and tetramethylammonium (TMA$^+$). However, neither are completely impermeant to the Na channel with Cs$^+$ and TMA$^+$ having permeabilities of 0.02 and <0.01 (respectively) to that of Na$^+$ (28, 29). Large cation substitutes such as tris(hydroxymethyl)aminomethane (Tris$^+$) and N-methyl-D-glucamine (NMDG$^+$) also can be used as substitutes for Na$^+$ but they appear to alter cardiac whole-cell I_{Na} kinetics (21, 30). Internal F$^-$ (typically >30 mM) appears to "help" form a high-resistance seal between the patch pipette and the myocyte's membrane. It also acts to minimize the "spontaneous" shift in the voltage dependence of Na-channel kinetics (see section entitled Special Considerations in Voltage Clamp of Cardiac I_{Na}). However, F$^-$ may complicate investigation into the metabolic regulation of ion channels because it activates adenylate cyclase (31) and it binds to Ca^{2+} and Mg^{2+}. Although the anion substitutes aspartate or glutamate can be used to minimize current conducted through chloride channels, it is important to note that divalent cations such as Ca^{2+} bind to carboxylic groups of aspartate and glutamate and can lower free divalent cation activity by 50% or more (32). N-Morpholino-ethanesulfonic acid (Mes$^-$) does not appear to chelate Ca^{2+} or to conduct through chloride channels, and therefore it may serve as a suitable anion substitute (21). Also, small cations such as Cs$^+$, Na$^+$, and Cl$^-$ usually have a higher conductivity in the pipette solution than the larger cations, TMA$^+$, Tris$^+$, etc., and will result in a lower series resistance (R_s). Fortunately, neither HEPES buffer nor bicarbonate buffer appear to affect I_{Na} kinetics (21).

Changing Extracellular and Intracellular Solutions

The use of small bath chambers, of short inflow tubing, and of drip chambers that permit an electrical break between the bath chamber and the inflow and exhaust flow will minimize electrical noise and facilitate changing the bath solution quickly. Because the large-pore suction pipette stabilizes the cell within the electrode pore well above the bottom of the bath chamber, relatively rapid extracellular solution changes can be accomplished by physically moving the electrode with the attached cell from the inlet of one bath chamber to the inlet of a second bath chamber, the second being connected to the first by a short, narrow, solution-filled channel in the Plexiglas separating the two chambers. Another method to rapidly change extracellular solutions during whole-cell voltage-clamp experiments using patch electrodes involves moving the patch electrode with an attached cell through the streams jetting from one or more simultaneously flowing large inflow tubes, each with a different solution, which displaces and replaces the control solution immediately around the cell (33). In our experience, only rarely will the patched cell be dislodged from the pipette by the force of the solutions jetting from the inflow tubes.

Junction Potentials

It is important to properly compensate for junction potentials between solid–liquid and liquid–liquid interfaces because they can become large. Additionally, the stability of any given junction potential is important because further accurate adjustment during an experiment cannot be reliably accomplished. $Ag\text{-}AgCl_2$ electroplated on a silver wire or in the form of a $Ag\text{-}AgCl_2$ pellet is typically employed as the solid in contact with a solution (or an agar bridge). One of the most stable solid–liquid junctions consists of $Ag\text{-}AgCl_2$ in contact with 3 M KCl; however, diffusion of the KCl into the bath or the pipette solution may become problematical. Usually solid $Ag\text{-}AgCl_2$ is in contact with the pipette solution, and if the bath solution is continually flowing, one end of a 3 M KCl agar bridge can be placed in the bath outflow and the other end inserted into a $Ag\text{-}AgCl_2$ electrode filled with 3 M KCl solution to complete the circuit. If the bath solution is stagnant, a 3% agar bridge made with the control extracellular solution can be used.

Because the solid–liquid junctions are different between the pipette and bath electrodes, the sum of their junction potentials will not be zero and will require compensation. To correct for the solid–liquid junction potentials, an offset potential is applied such that the current magnitude is zero when the pipette is lowered into a grounded puddle or bath chamber filled with the *same* solution as in the pipette. This is important because if the current is set to zero when the pipette is placed in a bath chamber containing standard extracellular solution, the offset correction will also include the liquid–liquid junction potential arising from the electrolytic differences between the pipette (internal) solution and the extracellular control solution. This should be avoided because the liquid–liquid junction potential between the internal and extracellular solutions will no longer exist after the formation of a high-resistance seal, and the prior correction will now introduce an offset error that can vary from experiment to experiment. This method is similar to that used by Oxford (34) for voltage clamp of squid giant axons. When the large-pore perfused patch pipette is used (see below), both the internal and external solutions are connected to 3 *M* KCl bridges. This results in a nearly equal but opposite liquid–solid junction potentials, so normally no voltage offset is required to set the current to zero.

Temperature Control

Reduced temperatures are commonly employed principally to slow I_{Na} kinetics but also to reduce the magnitude of I_{Na}. Typically used experimental temperatures range from 12 to 18°C and can be obtained with a Peltier cooling device coupled to a current source either with or without feedback control regulated via the temperature error signal.

Formation of a High-Resistance Seal and Initiating Whole-Cell Voltage Clamp

When patch electrodes are used, the methodology is similar to that used for recording I_{Ca} and other currents and has been recently reviewed in *Methods in Neurosciences* (35). Many investigators think it is important to maintain a slight outward flow of pipette solution by positive pressure within the patch pipette in order to keep the tip clear of any debris in the bath solution prior to establishing pipette-to-cell contact. Others do not regard this as a serious problem. However, only after the pipette tip has come into contact with the myocyte's membrane as determined by an increase in the resistance to current flow during short (10 msec) voltage pulses should negative pressure be applied. Application of continued negative pressure or a slight increase in pressure after the formation of a high-resistance seal will rupture the sarcolemma underneath the patch pipette pore and result in a large, formerly absent, capacitance transient that is associated with the addition of the sarcolemma's membrane capacity to the recording circuit (16).

For the large-pore suction pipette voltage clamp, the cell is positioned in the pore by mild negative suction until the desired amount of the cell remains on the outside of the pore in the bath solution. After the cell has sealed to the rim of the suction pipette's pore, the sarcolemmal membrane inside the pipette is ruptured with a manipulator-controlled platinum wire that is positioned in the outflow tube of the pipette (27). This allows for both internal perfusion and membrane voltage control of the portion of the cell remaining outside the electrode pore. Steady-state seal resistance is calculated from the linear, time-independent currents between -100 and -190 mV, and typically should be greater than 30 MΩ before data collection. Although this value represents the resistances of the resting cell membrane and the seal of the membrane to the glass in parallel, it provides a good estimate of the quality of the seal because membrane resistance is very high relative to the seal resistance in the absence of K^+ (both intra- and extracellularly). Typically the seal resistance is greater than 100 MΩ and approximates a gigaohm seal with a conventional patch pipette after accounting for the larger area of glass–cell membrane interface.

Recording Sampling Rates, Filter Frequencies, and Aliasing

A general understanding of the relationship between filter frequencies and sampling rates of digitized signals will help optimize recordings of voltage-clamp recordings. A 4-pole or 8-pole Bessel filter is usually recommended because it results in a fairly constant delay in the filtered signal over a large frequency range. The appropriate sampling rate will depend on the filter frequency and the inherent frequency response of the voltage-clamp electronics. A generally accepted rule (i.e., the Nyquist criteria) is to sample at a rate which is at least twice as fast as the filter frequency to avoid an error resulting from aliasing (a mismatch between the digital sampling rate and the analog signal frequency). It is also important to recognize that these criteria should be met for the smallest desired resolution of the recorded signal. For example, if an analog-to-digital converter with a 12-bit resolution and a range of ± 10 V (i.e., a resolution of 4.88 mV/bit) and an 8-pole Bessel filter with a corner frequency at 10 kHz were used to record to the least significant bit, a sampling frequency of about 60 kHz would be appropriate (4.88 mV/20 V $= -36$ dB) because the filter passes signals with an amplitude of ≈ -40 dB at 30 kHz. If the frequency response of the voltage-clamp electronics is limiting, then the appropriate sampling interval can be determined from a plot of its frequency vs amplitude characteristics (i.e., the Bode plot).

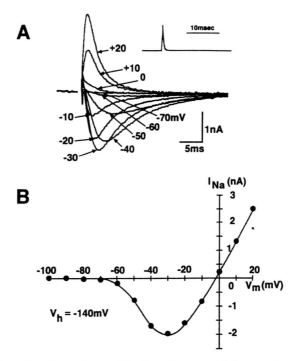

FIG. 1   Whole-cell voltage clamp of I_{Na} from human atrial myocytes recorded with a patch electrode. (A) Typical family of leak- and capacity-corrected current recordings to various test potentials from a holding potential of -140 mV. The inset shows the capacity transient resulting from a 10-mV step depolarization. (B) The peak I_{Na}–V_t relationship from the recordings shown above. Na_o/Na_i, 5 mM/5 mM; C_m, 116 pF; 17°C. Reprinted with permission from (22).

Identification of I_{Na}

Voltage-clamped sodium currents are usually readily identified from other ionic currents based on channel kinetics, channel pore properties, a relatively large Na channel density in cardiac myocytes, and the binding by specific toxins. As already mentioned, I_{Na} kinetics are rapid compared to most other ionic currents and consequently the onset, peak, and decay of I_{Na} occurs in milliseconds even at low temperatures (Fig. 1). By using the internal and external solutions recommended above, I_{Na} should be isolated from other ionic currents because their channels should be blocked or nonconductive. The high selectivity of the Na-channel pore for Na^+ results in the experimental reversal potential of I_{Na} to be well approximated by the equilibrium potential for Na^+ calculated from the Nernst Equation (29). By comparison to many other ionic currents, the magnitude of I_{Na} is usually very large because of the relatively high density of Na channels (18). If the ionic current under study is blocked by the addition of extracellular tetrodotoxin or saxitoxin, both of which specifically bind to Na channels, then it is very strong evidence that Na channels are involved in conduction of the ionic current. It is the combination of these properties that allows for the identification of I_{Na}.

Special Considerations in Voltage Clamp of Cardiac I_{Na}

Voltage Clamp Considerations

Both the membrane capacitance and the series resistance of the voltage clamp requires special attention during voltage clamp of I_{Na} because the current has both a large magnitude and very rapid kinetics. Maximal deviation of the voltage across the cell membrane from the command potential is defined by Ohm's law and is equal to the series resistance (R_s) multiplied by the maximal peak I_{Na} (assuming that the leak current is small.) The series resistance results from all resistances except for the membrane resistance and includes the solid–liquid junctional impedances, the resistance arising from the solution in the electrode, the resistance associated with the pore, the longitudinal resistances of the perfused cell, and the bath solution. The conductivity of each of the involved solutions and their contribution to the overall series resistance depends on both the ionic mobilities of the constituent ions and their concentrations. Reduction of the resistance of any component contributing to R_s will improve the control of membrane potential (V_m) during voltage clamp. In addition, reducing the transmembrane $[Na^+]$ gradient will decrease peak I_{Na}, thereby minimizing the series resistance voltage error. The judicious selection of a cardiac cell preparation that has a relatively low density of Na channels will also help minimize the possibility of losing control of membrane voltage.

Loss of control of membrane potential during peak I_{Na} can be detected from peak I_{Na}–V_m plots in which there is an abrupt increase in peak I_{Na} during step depolarizations to voltages just positive to threshold. The quality of membrane control can be evaluated by determining the slope factor from a fit of a Boltzmann distribution to the peak Na current conductance–voltage relationship. Hanck and Sheets (36) found that a slope steeper than -6 mV (i.e., > -6 mV) indicated loss of membrane control, although a less steep slope (i.e., < -6 mV) did not necessarily confirm voltage control. In 43 single Purkinje cells voltage clamped with a large-pore suction pipette that were judged to be well controlled, they found the mean slope to be -8.6 ± 1.2 mV.

Because the kinetics of the channel can cause I_{Na} to activate within a few milliseconds, the membrane capacity must be fully charged before the onset of ionic current so that voltage across the membrane is constant and none of the capacity current contributes to the current measured as I_{Na}. Therefore, in addition to accounting for R_s errors, voltage control must be attained rapidly to properly voltage clamp I_{Na}. The time constant (τ_{cap}) of the current charging the membrane capacitance (C_m) during a step in membrane potential can be estimated by calculating the time constant of the relaxation of the capacity transient. To optimize the speed of the voltage clamp, reductions in both cell capacitance and R_s are typically accomplished through the selection of smaller cells, blunt electrode tapers, large electrode pores, and solutions having a higher conductivity, particularly the pipette solutions. In addition, reducing the temperature slows Na-channel kinetics and permits more time for the capacity current to settle before the Na channels activate.

Voltage clamp with a large-pore suction pipette typically achieves charging of the capacitance with a primary τ_{cap} of 5–15 μsec. This rapid charging is possible because: (i) the resistance of the large-pore pipette is low and about one-half of the cell is positioned inside the pipette and therefore is outside the circuit, (ii) the amount of cell C_m under voltage clamp is reduced, and (iii) the contribution of the intracellular longitudinal resistance of the cell to the total R_s is also reduced.

Background Shifts in the Kinetics of Cardiac I_{Na}

Cardiac I_{Na} kinetic parameters (such as the half-point of voltage-dependent availability and of peak conductance) shift spontaneously to more negative potentials during whole-cell voltage clamp of many mammalian cardiac preparations (36, 37). This spontaneously occurring change in kinetics does not appear to result solely from an affect of internal cell perfusion because single-channel recordings in cell-attached patches also demonstrate negative shifts in the voltage dependence of kinetic parameters (38–40). Interestingly, there appears to be no correlation between the size of the electrode pore (presumably a large pore would be more effective in exchange of the cytoplasm) and the rapidity of the kinetic shifts. A shift in I_{Na} kinetic indices has been reported to occur at rates ranging from 0.5 to 1.0 mV min^{-1} during whole-cell voltage clamp of I_{Na} using patch pipettes with resistances of 0.3–1 MΩ (33, 37) and 0.5 mV min^{-1} with 10-fold lower resistance pipettes (0.05–0.07 MΩ) (36). In contrast, it has recently been reported that the kinetic indices of cardiac I_{Na} do not change during whole-cell voltage clamp using a perforated patch technique with amphotericin B (41). However, because the perforated patch usually has a relatively high R_s, this technique is expected to suffer the associated limitations imposed by an inability to clamp as quickly as required for some measurements. Although the causes for the spontaneous changes in cardiac I_{Na} kinetics that can occur during an experiment are not understood, it is important to recognize that they do occur and to design experiments which permit an accounting for these potentially confounding effects.

Whole-Cell Voltage Clamp for Recording Na-Channel Gating Currents

Applications and Advantages

Another powerful approach enabling study of voltage-gated Na channels involves measurements of Na-channel I_{g-Na}. Although their existence had been postulated by Hodgkin and Huxley in 1952 (1), gating currents associated with Na channels were not demonstrated until 1973 in the squid giant axon (42). Gating currents occur in response to changing the membrane potential and are postulated to result from a change in the conformation of a charged polypeptide portion(s) of the Na-channel protein. These conformational changes are believed to confer voltage sensitivity to the Na channel's kinetic behavior. I_{g-Na} always exhibits a positive polarity in response to depolarizing steps and is superimposed on the capacity current that is concomitantly produced in response to the depolarizing step that evoked I_{g-Na}.

Figure 2B shows a plot of the total charge (charge from both the capacitance of the sarcolemma and from I_{g-Na}) as a function of step potential from a holding potential of -150 mV. Note that at potentials from -190 to -110 mV the total charge is a linear function of test potential (V_t) reflecting the change in charge held by the membrane capacity. At more positive potentials the total charge is greater than that predicted for a linear capacitor. In Fig. 2C this excess charge is plotted as a function of test potential to obtain the typical gating charge vs V_t relationship. At very negative potentials no gating charge is moved while at positive potentials the total charge plateaus at a maximal value. The utility of I_{g-Na} measurements is that they provide direct information about all kinetic transitions including those that are nonconductive (43).

Measurement of Na-Channel Gating Current

Recording Na-channel I_{g-Na} is similar to that already discussed for whole-cell I_{Na}. However, because I_{g-Na} signals are small and integration of a small signal over time is easily influenced by small background currents the re-

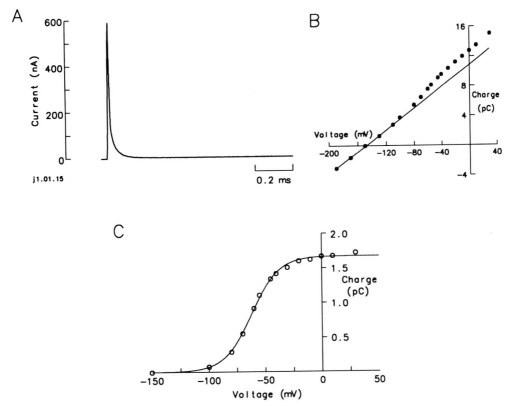

FIG. 2   Charge transfer during depolarizations in an isolated cardiac Purkinje cell. (A) The current in response to a voltage test step (V_t) of 0 mV from a holding potential of -150 mV. Note that most of the capacity transient was complete within 200 μsec. (B) The area under the capacity transients (i.e., total charge) as a function of V_t. At very negative V_t, charge was a linear function of potential. The linear charge corresponded to a cell membrane capacitance of 75 pF. (C) $I_{g\text{-Na}}$ (i.e., total charge $-$ linear charge) as a function of potential. The line is a fit of a Boltzmann distribution to the data with a midpoint of -62 mV and a slope of 12 mV. Cell J1.01.15, R_L 180 MΩ, 15°C. Reprinted with permission from (46).

cording technique requires additional attention to the voltage-clamp methodology. All I_{Na} and time-dependent currents must be eliminated because even a very small ionic current integrated over time will grossly contaminate Na-channel gating charge measurements. Electronic noise, particularly 60-Hz noise, will also severely interfere with integration of $I_{g\text{-Na}}$. Additionally, a 16-bit A–D converter is preferable to achieve adequate signal resolution to record the small $I_{g\text{-Na}}$ signal superimposed on the large linear capacity transient. Because of the high frequencies in $I_{g\text{-Na}}$ and its small magnitude, careful design of the filtering characteristics becomes important to prevent aliasing of the signal. Finally, voltage-clamp protocols for the recording of $I_{g\text{-Na}}$ typically require repetitive pulsing to allow for signal averaging to minimize further the effect of background noise.

Although several investigators primarily interested in Ca^{2+}-channel gating current have recorded Na-channel $I_{g\text{-Na}}$ (44, 45), only Hanck *et al.* (18) have published methods for recording gating currents which are predominately associated with cardiac Na channels. An approach employing the large-bore suction pipette to record $I_{g\text{-Na}}$ elicited in isolated voltage-clamped Purkinje cells is described below.

To isolate gating currents, it is necessary to minimize all ionic currents. We have found the following solutions to be best. The extracellular solution contains (in mM) 150 TMA$^+$, 150 Mes$^-$, 2 Ca^{2+}, 4 Mg^{2+}, 0.010 Cd^{2+}, 0.010 STX, and 10 HEPES (pH 7.2). Mg^{2+} and Cd^{2+} will block most of the calcium current while TMA$^+$ and Mes$^-$ appear to be impermeant to both K$^+$ and Cl$^-$ channels, respectively. The internal perfusate contains (in mM) 150 TMA-F, 10 EGTA, and 10 HEPES (pH 7.2). The temperature of the experimental bath chamber is typically cooled to 12 to 15°C to slow the gating kinetics. Current responses are recorded with a 16-bit A–D converter at 300 kHz with an intrinsic corner frequency of $\approx$125 kHz for the voltage-clamp electronics. To permit the capacity current that results from charging the sarcolemmal membrane to be subtracted from the total current, thus obtaining $I_{g\text{-Na}}$, a linear capacity transient is constructed from eight consecutive voltage steps to voltages too negative to permit movement of Na-channel gating charge (i.e., 40-mV steps from -150 to -190 mV). Typical $I_{g\text{-Na}}$ voltage-clamp recordings are composed of the averaged signal derived from four sequential records each acquired $\frac{1}{4}$ of a 60-Hz cycle out of phase from the preceding one to minimize high frequency and 60-Hz noise. The scaled linear capacity transient is subtracted from the signal-averaged recording and $I_{g\text{-Na}}$ is obtained. A typical $I_{g\text{-Na}}$ recording, its integral, and its corresponding I_{Na} response are shown in Fig. 3.

Correlation of Gating Currents with Na Channels

Because the heart has many voltage-gated ion channels which presumably all have corresponding gating currents, it is necessary to ascertain that the gating charge measured is in fact related to Na channels rather than to some other channel type. One of the most important criteria to satisfy is that the presumptive $I_{g\text{-Na}}$ exhibits kinetics similar to those of I_{Na}. For example, the

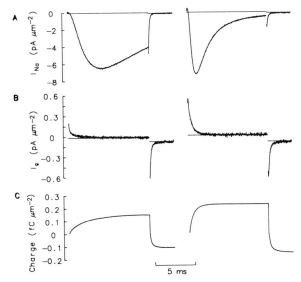

FIG. 3   Capacity-corrected I_{Na}, $I_{g\text{-}Na}$, and integral of $I_{g\text{-}Na}$ (i.e., gating current charge) in an isolated Purkinje cell. Data are signal-averaged traces from four sequential sweeps that were recorded at 300 kHz, digitally filtered at 20 kHz, and decimated by six (i.e., 20-μsec intervals). (A) Linear capacity-corrected but not leak-corrected I_{Na} after depolarizations to -50 mV (*left*) or to 0 mV (*right*) and repolarization back to -150 mV. Horizontal lines indicate the value of the linear leak. (B) $I_{g\text{-}Na}$ recorded in the absence of Na$^+$ and in the presence of 10 μM saxitoxin. Scale is 8$\times$ greater than that in (A). (C) Integral of (B). Cell c1.02, 65 pF, R_L 200 mΩ, 10.5°C. Reprinted with permission from (18).

kinetics of the Na channel are fast compared to most other channels; therefore, it is expected that $I_{g\text{-}Na}$ should also be commensurately rapid. In heart Hanck *et al.* (18) have shown that the prominent τ for $I_{g\text{-}Na}$ relaxations corresponds to the activation time constant, τ_m, of I_{Na} in a Hodgkin–Huxley model (1). This is the association that would be anticipated if most of the measured gating charge is from Na channels and is produced by a transition of the channels from a closed to an open state.

There are other types of evidence that can be used to help demonstrate a relationship between gating charge and I_{Na}. For example, the magnitude of gating charge should be proportional to the magnitude of I_{Na} and estimates of Na-channel density predicted from gating charge measurements should be similar to those obtained from toxin-binding studies, assuming that each channel has about $6e^-$ of gating charge (29, 43). By satisfying a combination of at least several such criteria, the gating currents measured under specific experimental conditions can be determined to represent signals arising pre-dominately from cardiac Na channels.

Potential Limitations of the Internally Perfused Whole-Cell Voltage-Clamped Cardiac Preparation

Although the internally perfused whole-cell voltage clamp is an extremely powerful experimental approach, it is worthwhile to recognize potential limitations of the technique as currently employed. Measurements of I_{Na} are performed at nonphysiological low temperatures which alter both channel kinetics and membrane fluidity. Low temperatures may also affect the results of studies on Na channels intending to characterize drug binding and un-binding. Intracellular perfusion may wash out or alter important intracellular regulatory constituents. It is possible that sarcolemmal membranes or the channels themselves become altered during the enzymatic isolation of single cells. Despite these potential limitations, investigations of cardiac Na chan-nels and their differences with neuronal Na channels have improved and should continue to improve our understanding of this important ionic current responsible for conducting action potentials throughout most of the *in situ* heart as well as our understanding of the pharmacology of agents employed to modify cardiac Na channels and other voltage-gated channels. However, it is important to note that other factors such as intracellular electrical coupling contributing to anisotropic conduction, effective capacitance, and intercellu-lar clefts all influence conduction of the action potential throughout the heart and must be taken into account in the overall understanding of cardiac impulse conduction. For a complete understanding of the Na current in heart all of these types of questions will require answers; however voltage clamp of whole-cell I_{Na} will provide a solid foundation.

Acknowledgments

Supported by NIH Grant HL-R29-44630 to M. F. Sheets and Hl-R01-27026 to R. E. Ten Eick.

References

1. A. L. Hodgkin and A. F. Huxley, *J. Physiol.* **117,** 500 (1952).
2. R. B. Rogart, L. L. Cribbs, K. Muglia, D. D. Kephart, and M. W. Kaiser, *Proc. Natl. Acad. Sci. U.S.A.* **86,** 8170 (1989).

3. J. S. Trimmer, S. S. Cooperman, S. A. Tomiko, J. Zhou, S. M. Crean, M. B. Boyle, R. G. Kallen, Z. Sheng, R. L. Barchi, F. J. Sigworth, R. H. Goodman, W. S. Agnew, and G. Mandel, *Neuron* **3**, 33 (1989).

4. D. T. Yue, J. H. Lawrence, and E. Marban, *Science (Washington, D.C.)* **244**, 349 (1989).

5. M. F. Berman, J. S. Camardo, R. B. Robinson, and S. A. Siegelbaum, *J. Physiol.* **415**, 503 (1989).

6. G. E. Kirsch and A. M. Brown, *J. Gen. Physiol.* **93**, 85 (1989).

7. B. E. Scanley, D. A. Hanck, T. Chay, and H. A. Fozzard, *J. Gen. Physiol.* **95**, 411 (1990).

8. D. DiFrancesco, A. Ferroni, S. Visentin, and A. Zaza, *Proc. R. Soc. Lond. Ser. B* **223**, 475 (1985).

9. M. F. Sheets and D. A. Hanck, *J. Physiol.* **454**, 299 (1992).

10. J. Satin, J. W. Kyle, M. Chen, P. Bell, L. L. Cribbs, H. A. Fozzard, and R. B. Rogart, *Science (Washington, D.C.)* **256**, 1202 (1992).

11. P. Backx, D. Yue, J. Lawrence, E. Marban, and G. Toaselli, *Science (Washington, D.C.)* **257**, 248 (1992).

12. C. J. Cohen, B. P. Bean, T. J. Colatsky, and R. W. Tsien, *J. Gen. Physiol.* **78**, 383 (1981).

13. D. A. Hanck, M. F. Sheets, R. Rogart, D. Doyle, S. Lustig, and H. Fozzard, *Biophys. J.* **53**, 534a (1988).

14. L. A. Alpert, H. A. Fozzard, D. A. Hanck, and J. C. Makielski, *Am. J. Physiol.* **257**, H79 (1989).

15. T. Powell, and V. W. Twist, *Biochem. Biophys. Res. Commun.* **72**, 327 (1976).

16. R. E. Ten Eick, *in* "Cardiac Electrophysiology: A Textbook" (M. R. Rosen, M. J. Janse, and A. L. Wit, eds.), pp. 3–27. Futura, Mount Kisco, 1990.

17. N. A. Burnashev, F. A. Edwards, and A. N. Verkhratsky, *Pfluegers Arch.* **417**, 123 (1990).

18. D. A. Hanck, T. Chay, and H. A. Fozzard, *J. Gen. Physiol.* **95**, 411 (1990).

19. L. H. Silver, E. L. Hemwall, T. H. Marino, and S. R. Houser, *Am. J. Physiol. (Heart Circ. Physiol).* **245**, H891 (1983).

20. M. L. Decker, M. Behnke-Barclay, M. G. Cook, J. J. LaPres, W. A. Clark, and R. S. Decker, *J. Mol. Cell. Cardiol.* **23**, 817 (1991).

21. W. M. Lue and P. A. Boyden, *Circulation* **85**, 1175 (1992).

22. Y. Sakakibara, J. A. Wasserstrom, T. Furukawa, H. Jia, C. E. Arentzen, R. Hartz, and D. H. Singer, *Circ. Res.* **71**, 535 (1992).

23. T. Mitsuiye and A. Noma, *Pfluegers Arch.* **410**, 7 (1987).

24. T. J. Colatsky, *J. Physiol.* **305**, 215 (1980).

25. R. Sato, A. Noma, Y. Kurachi, and H. Irisawa, *Circ. Res.* **57**, 553 (1985).

26. O. P. Hamill, A. Marty, E. Neher, B. Sakmann, and F. J. Sigworth, *Pfluegers Arch.* **391**, 85 (1981).

27. J. C. Makielski, M. F. Sheets, D. A. Hanck, C. T. January, and H. A. Fozzard, *Biophys. J.* **52**, 1 (1987).

28. M. F. Sheets, B. E. Scanley, D. A. Hanck, J. C. Makielski, and H. A. Fozzard, *Biophys. J.* **52**, 13 (1987).

29. B. Hille, Ed., "Ionic Channels of Excitable Membranes 2nd Ed., Vol. 3–4. Sinauer, Sunderland, MA, 1992.

30. J. Z. Yeh and G. S. Oxford, *J. Gen. Physiol.* **85,** 603 (1985).
31. E. E. Susanni, D. E. Vatner, and C. J. Homcy, *in* "The Heart and Cardiovascular System: Scientific Foundations" (H. A. Fozzard, R. B. Jennings, E. Haber, A. M. Katz, and H. E. Morgan, eds.), 2nd Ed., Vol. 2. Raven Press, New York, 1991.
32. A. E. Matell and R. M. Smith, *in* "Critical Stability Constants." Vol. 1. "Amino Acids." Plenum Press, New York, 1974.
33. B. Schubert, A. M. VanDongen, G. E. Kirsch, and A. M. Brown, *Science (Washington, D.C.)* **245,** 516 (1989).
34. G. S. Oxford, *J. Gen. Physiol.* **77,** 1 (1981).
35. E. C. Keung, *in* "Methods in Neurosciences," Vol. 4, pp. 30–44. Academic Press, San Diego, 1991.
36. D. A. Hanck and M. F. Sheets, *Am. J. Physiol.* **262,** H1197 (1992).
37. C. W. Clarkson, *Pfluegers Arch.* **417,** 48 (1990).
38. K. Benndorf, *Biomed. Biochim. Acta* **48,** 287 (1989).
39. T. Kimitsuki, T. Mitsuiye, and A. Noma, *Am. J. Physiol.* **258** (*Heart Circ. Physiol.* **27**), H247 (1990).
40. K. Ono, H. A. Fozzard, and D. A. Hanck, *Circ. Res.* **72,** 807 (1993).
41. D. J. Wendt, C. F. Starmer, and A. O. Grant, *Circulation* **86,** I (1992).
42. C. M. Armstrong and F. Bezanilla, *Nature (London)* **242,** 459 (1973).
43. C. M. Armstrong, *Physiol. Rev.* **61,** 644 (1981).
44. B. P. Bean and E. Rios, *J. Gen. Physiol.* **94,** 65 (1989).
45. R. W. Hadley and W. J. Lederer, *J. Physiol.* **415,** 601 (1989).
46. H. A. Fozzard and D. A. Hanck, *in* "The Heart and Cardiovascular System: Scientific Foundations" (H. A. Fozzard, ed.), 2nd Ed., Vol. 1, pp. 1091–1120. Raven Press, New York, 1991.

[10] Measurement of Dihydropyridine Modulation of Cardiac L-Type Calcium Channels: Molecular and Cellular Implications

Robert S. Kass and Ramesh Bangalore

Introduction

At least four calcium-channel subtypes (P, T, N, and L) have been detected based on their pharmacological and/or biophysical properties (1–3). Both T- and L-type channels have been studied in the heart (4), but the L-type channel is the target of the most extensively developed calcium-channel pharmacology. Drugs that regulate Ca^{2+} influx via voltage-gated Ca^{2+} channels have been called "calcium antagonists" or "calcium-channel blockers" but whether they block channels and/or compete for Ca^{2+} binding sites has not been uniquely determined. The drugs that have received the most attention belong to three distinct chemical classes (i) phenylalkylamines (verapamil, D-600); (ii) benzothiazepines [(+) *cis*-diltiazem]; and (iii) the 1–4 dihydropyridines (PN 200-110, nitrendipine, nifedipine, nisoldipine). These drugs bind to distinct, but allosterically coupled receptors, on the channel protein (5). Of all the chemical compounds that interact with calcium channels, the dihydropyridines (DHPs) have proved to be the most interesting, because these drugs bind to the channel with highest affinity and specificity (6).

Calcium-channel antagonists, in general, and the dihydropyridine derivatives, in particular, regulate calcium influx by modulating channel gating (7–10). Drug-induced gating changes resemble modes of gating that can occur under drug-free conditions (11, 12) supporting the view that molecular perturbations induced by these compounds promote indigenous conformational changes of the channel proteins. Since it has been shown that the α_1 subunit contains the specific binding sites for all three major classes of calcium-channel blockers described above (5, 13, 14), it is very likely that drug-induced conformational changes of the α_1 subunit underlie these gating changes.

Because voltage-dependent calcium channels contribute to the control of many aspects of cellular physiology including activation of contractile proteins and synaptic transmission, it is not surprising that drugs that modulate

Methods in Neurosciences, Volume 19

the entry of calcium, the channel blockers, have been found to be powerful therapeutic tools in the treatment of a variety of disorders including hypertension, angina, and some forms of cardiac arrhythmias. Therapeutic applications of these drugs have been expanded recently to the treatment of congestive heart failure, cardiomyopathy, atherosclerosis, and cerebral and peripheral vascular disorders (15). In addition to the major contributions calcium-channel blockers have made to the treatment and management of clinical disorders, these drugs have been equally important to the development of our understanding of the molecular properties of calcium channels. An understanding of the molecular mechanisms of action of this class of drugs will provide scientists with basic information that can be used in the synthesis of drugs designed to target specific tissue and pathophysiological conditions.

The aim of this chapter is to review the methodology needed to investigate the voltage-dependent modulation of L-type calcium-channel activity by dihydropyridine (DHP) derivatives. Experiments are summarized in which the voltage dependence of channel activity as well as the location of the binding site for dihydropyridines on the L-channel protein is probed. The experiments summarized were carried out on native channels in ventricular myocytes, but the experimental strategy can be applied to the investigation of DHP modulation of recombinant channel activity as discussed briefly in the last section of the chapter. Experimental strategy is described and results that point to an extracellularly located binding site for these important compounds are summarized.

Basic Methodology

Most of the experiments described as examples in this chapter are carried out in single cardiac cells isolated from either ventricle of adult guinea pigs using a method similar to that of Mitra and Morad (16) which has been previously described (17). Ionic currents are recorded using the patch-clamp procedure and recording methods are as described by Hamill *et al.* (18) for the whole-cell configuration. In some experiments, patch-clamp procedures are applied to Chinese Hamster Ovary cells that have been stably transfected with cDNA encoding the α_1 subunit of either the cardiac or the smooth muscle L-type channel.

Solutions for whole-cell calcium-channel experiments are chosen to minimize potassium-channel currents. As such, the standard pipette solution contains (in mM) 60 CsCl, 50 aspartic acid, 68 CsOH, 1 MgCl$_2$, 1 CaCl$_2$, 11 ethylene glycol bis(2-aminoethyl)ether N,N,N',N'-tetraacetic acid (EGTA), 5 K$_2$ATP, 10 HEPES (pH 7.4). The standard bath solution contains (in mM)

130 NaCl, 4.8 CsCl, 5 glucose, 5 HEPES (pH 7.4). In experiments designed to measure divalent ion currents, 2 mM MgCl$_2$ and divalent charge carriers at indicated concentrations are added to the basic solution. NaCl can be replaced by Tris–Cl or some other impermeant ion substitute such as N-methyl-D-glucamine in order to eliminate currents through Na channels. In experiments designed to measure monovalent currents through (L-type) calcium channels, the following are added to the basic extracellular solution: 2 mM EGTA plus 50 μM MgCl$_2$ (for the measurement of Na currents) or 2 mM ethylenediaminetetraacetic acid (EDTA) and 500 μM MgCl$_2$ (for the measurement of outward currents).

Drugs Used

Amlodipine and its quarternary derivative (QA) are dissolved in water as concentrated stock solutions. Amlodipine (and QA) concentrations were chosen as previously described (19–21). SDZ-207-180 was a gift of Sandoz, Ltd. (CH-4002 Basle, Switzerland); nisoldipine was a gift from Miles Laboratories (New Haven, CT); amlodipine and quarternary amlodipine were gifts from Pfizer Central Research (Sandwich, UK).

Experimental Approach

Voltage Protocols

The voltage protocols discussed in this chapter have been previously described in detail and illustrated in several papers beginning with Sanguinetti and Kass (8) through Kass and Arena (21). There are several procedures that are useful in assaying voltage-dependent drug activity, and each procedure is designed to provide information about access to and state dependence of interactions with the DHP receptor. Steady-state modulation of drug activity by membrane potential can be assayed by applying infrequent voltage pulses from different holding potentials (Fig. 1).

Drug onset can be measured conveniently by applying a depolarizing "train" protocol in which the holding potential is changed from a negative potential (−80 mV) to a moderately depolarized potential like −40 or −50 mV. Test voltage pulses can then be applied at regular intervals (once every 5 sec) to assay drug-induced changes in current amplitude. Control, drug-free, data must be obtained to ensure a minimum decrease in current amplitude due to slow inactivation. Application of this procedure in the presence of drug provides a means of evaluating the concentration dependence of the time

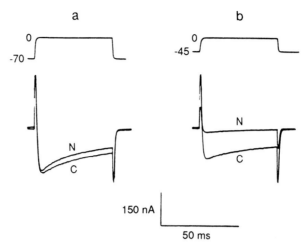

FIG. 1 Membrane potential modulates nisoldipine inhibition of calcium-channel currents in the Purkinje fiber. Currents measured in the absence (C) and presence (N) of 200 nM nisoldipine from −70 mV (a) and −40 mV (b) holding potentials. Nisoldipine completely blocks currents recorded from −40 mV but not from −70 mV. Reprinted with permission (8). Copyright © 1984, American Heart Association.

course of drug block as a function of conditioning potential amplitude. An example of this procedure used to measure the onset of a newly synthesized, neutral DHP compound is shown in Fig. 2. Sanguinetti and Kass (8) found that channel inhibition by the neutral DHPs nisoldipine and nitrendipine was markedly dependent on the magnitude of the conditioning potential, but not very sensitive to the test pulse amplitude or duration. This was taken as evidence that these neutral drugs did not require channel openings before blocking L-type channels. Similar experiments with the charged phenylalkalamine verapamil did show a strong dependence on pulse voltage and duration and suggested that charged calcium-channel blockers might require open channels in order to access the respective binding sites for each drug.

The contrast in actions of neutral and charged drugs was interesting, but conclusions about the access to and location of the DHP binding site were difficult to infer from them, because different classes of drugs were used and the number of charged DHPs available for investigation was limited. Experiments designed to probe the location of the DHP binding site thus had to await the synthesis of appropriate compounds.

Characteristic of the actions of previously investigated neutral DHP compounds is the rapid and relatively complete recovery from block at negative membrane potentials. Rapid recovery from channel inhibition at negative

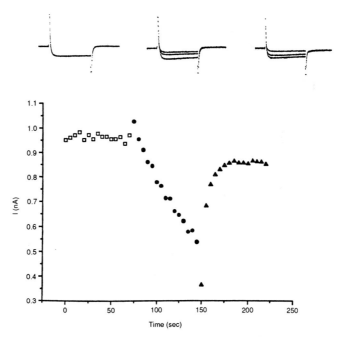

FIG. 2   Onset of and recovery from calcium-channel inhibition by a neutral dihydro-pyridine compound. Train protocols described in the Basic Methodology section were applied to promote block and recovery from block of calcium-channel activity in a single ventricular cell using whole-cell recording conditions. Barium (5 mM) currents were recorded in the absence (open symbols) and presence (filled symbols) of the compound DT2N (500 nM) (kindly synthesized by Dr. David Triggle). This compound had previously been found to bind with high affinity to DHP binding sites in rat heart cell membranes. In the absence of drug, there is no change in currrent amplitude with the depolarizing train protocol, but, in the presence of drug, current is inhibited in a time-dependent manner. Recovery from block is rapid after being returned to a negative (−80 mV) holding potential. Bangalore and Kass, unpublished data.

voltages explains the minimal use dependence caused by these drugs in cardiac preparations that have normally fully polarized diastolic potentials. Recovery from block can also be measured using train protocols, but, in this case, the holding potential is changed from a positive voltage (−40 mV) to a more negative potential (−80 mV). Pulses are applied to the same test potentials used to measure onset of drug block, but test pulse duration is brief (20 to 40 msec) to minimize pulse-induced inactivation. In experiments that require current measurements from potentials negative to −60 mV, prepulses (50 to 100 msec in duration) should be applied to -40 mV in order to inactivate T-type calcium-channel currents (22) if they are measurable in the preparation under investigation. Figures 3 and 4 illustrate the recovery of L-type channels in ventricular myocytes from block by two neutral DHP compounds.

Figure 3, which shows the time course of recovery from block by a neutral compound, illustrates another characteristic of the kinetics of unblock of calcium channels by this family of drugs. The experiment in this figure was carried out by measuring the recovery of channel activity after applying one long (30-sec) conditioning voltage step to 0 mV. In the absence of drug, this

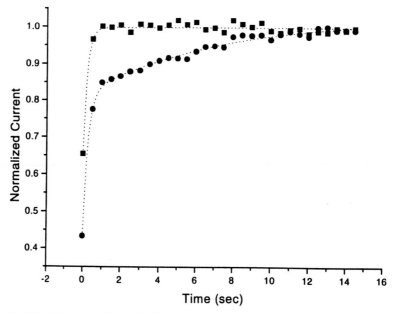

FIG. 3   Slow kinetics of drug "unblock" can identify the fraction of channels blocked. Calcium-channel currents measured from a -80 mV holding potential after 30 sec prepulses were applied to 0 mV in the absence (square) and presence (circles) of 500 nM DTN2. In the absence of drugs, channels recover from conditioning pulse-induced inactivation with a time constant on the order of 500 msec. In the presence of drug, the same protocol induces a slow component of channel recovery that is easily distinguished from the rapid drug-free channel kinetics. In this experiment, the slow time constant is on the order of 7 sec. The relative amplitude of the slow vs fast component of recovering current can be used to determine the fraction of channels inhibited as functions of drug concentration, conditioning pulse amplitude, and conditioning pulse duration. Bangalore and Kass, unpublished data.

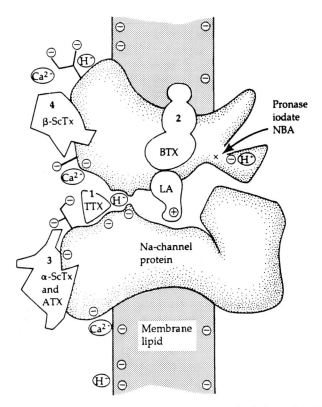

FIG. 4   Drug receptors on the sodium channel. Hypothetical view of the Na-channel macromolecule in the membrane. Receptors are labeled as follows: (1) tetrodotoxin (TTX), (2) batrachotoxin (BTX), (3) scorpion (α-ScTx) and anemone peptide toxins (ATX); and (4) β-scorpion toxin (β-ScTx). Reprinted with permission (23) by copyright permission of the Rockefeller University Press.

depolarization inactivates L-type channels even when Ba^{2+} is the charge carrier. However, under these conditions, the subsequent recovery from inactivation occurs within 1 sec at -80 mV. In contrast, in the presence of neutral drug, there is marked slow component of channel recovery that, presumably, reflects unbinding of drug molecules from the DHP receptor at this negative voltage. The relative fraction of current that recovers with this slow time course is a measure of the fraction of channels that had been blocked after 30 sec at 0 mV. Comparison of the effectiveness of structurally distinct DHP molecules at generating this slow component of recovery from block provides a systematic means of comparing relative potencies of the compounds. These procedures have provided strong evidence that DHP molecules interact with the DHP receptor in a strongly voltage-dependent manner and that channel openings are not required for access to the DHP binding site and, further, that once bound and blocked, L-type calcium channels recover readily from channel inhibition when membrane potential is returned to negative values.

These data demonstrate how neutral DHPs can regulate L-type calcium-channel activity, but they were not useful in providing a functional profile of the location of the site at which these molecules interact, because neutral drug molecules are not restricted in their access to the receptor binding site. In order to assay the binding site location, another approach has been necessary.

How to Test for Binding Site Location

The basic experimental strategy used to provide functional evidence for the location of the high-affinity binding site for dihydropyridines has relied on experiments previously carried out to probe local anesthetic binding site in sodium channels (Fig. 4). Based on ideas and experimental results described by Hille and others (23, 24), it was shown that the ionization state of a drug molecule could be a very useful tool in testing for access and binding to a channel-associated receptor. Because the channel is a protein inserted into a lipid bilayer membrane, ionized drug access to the receptor binding site is restricted to hydrophilic pathways. In contrast, as described above, neutral drugs are not so restricted and can follow either hydrophilic (channel pore) or hydrophobic (lipid membrane) pathways. Voltage protocols that limit or recruit the number of times channels open can discriminate between these possibilities. Furthermore, because ionized drugs cannot freely diffuse across the lipid bilayer, they are useful in determining the "sidedness" of access to the receptor binding site. Experiments using these approaches to probe the DHP binding site location in heart cells are summarized next.

Almost all previously investigated dihydropyridine derivatives are neutral drugs which are not restricted in their access to the DHP binding site (25) but, recently (21, 26–28), a limited number of charged DHPs have been made available for the investigation of ion channel properties, and these drugs have been used in studies designed to map the access to the DHP binding site via hydrophobic or hydrophilic pathways as the charge on the drug of interest was varied. This approach has provided functional evidence that the binding site for DHPs is physically distinct from the binding site for phenylalkylamines (and for benzothiazepine calcium-channel blockers). Thus, functional (and, indeed, molecular) data now clearly support the view of biochemical experiments that these drugs interact at unique receptor sites to regulate channel activity. The experiments using charged DHPs to map the DHP and phenylalkylamine binding sites are described next.

Internal Block of Calcium Channels by Phenylalkylamines

Charged calcium-channel blockers had previously been shown to inhibit channel activity via intracellular application (29, 30). In order to address the question of the binding site location, Heschler *et al.* (29) applied a quaternary phenylalkylamine to the intracellular surface of heart cells by dialysis with a patch-recording pipette. Their results indicated that the quaternary form of D600 (D890) blocked calcium channels from the intracellular membrane surface, consistent with drug access gained via open channels for extracellular application of tertiary compounds (such as D600). This previous work had two implications: first, it showed that intracellular dialysis with a molecule with a similar molecular weight was possible and, second, it provided evidence that the phenylalkylamine site was intracellularly accessible.

The location of the phenylalkylamine receptor site has also been investigated using a combination of photoaffinity labeling and immunoprecipitation with several sequence-specific antibodies. The results of this work restrict the site of photolabeling to the transmembrane helix S6 of domain IV and the beginning of the long intracellular C-terminal tail (31). Taken together with the functional studies of Heschler *et al.* (29) these data provide strong evidence for an intracellular location of the phenylalkylamine binding site.

Catterall and Striessnig (32) have proposed that the binding of phenylalkylamines to a receptor site formed in part by sequences near the intracellular end of the transmembrane segment IVS6 could inhibit channel activity by actually occluding the channel pore. Thus, in this view, the actions of these drugs result from an actual physical blockade of channel conductance pathways and not by regulation of indigenous channel gating.

Channel Block by Charged Dihydropyridines: Evidence for External Access

Before experiments could be carried out to determine the location of the DHP binding site relative to the membrane surface, it was first necessary to determine whether charged DHP compounds such as amlodipine, quaternary amlodipine, and SDZ-207-180 block channels in a characteristic voltage-dependent manner. This voltage dependence can therefore be taken as functional evidence for an interaction between the drug molecule and its receptor. The unique voltage dependence of inhibition (and recovery from inhibition at negative potentials) can therefore be used as a fingerprint by which DHP–channel interactions can be identified. As shown above, most neutral DHP compounds reversibly inhibit channel activity in this voltage-dependent manner. We found that this was not the case for some charged drugs.

One ionized drug that had unusually slow recovery kinetics was amlodipine, a tertiary compound with $pK_a = 8.6$. In contrast to nifedipine and nisoldipine, this compound is thus almost completely ionized at physiological pH (94% of amlodipine molecules are ionized at pH 7.4). In alkaline solutions a greater fraction of the drug molecules is neutral (at pH 10.0, 96% of amlodipine is neutral). Different forms of the drug can thus be selected by changing external pH (pH_o) and pH-induced changes in drug activity can be used to assay drug–channel interactions that may depend on the ionization state of the drug molecule.

Ionized amlodipine applied externally also blocked calcium-channel current (I_{Ca}) in a voltage-dependent manner that resembled block by neutral DHPs, but at a comparably slower rate. However, on repolarization, channels did not recover from block but appeared to be stabilized in a nonconducting state. Unblock was not promoted by further hyperpolarization nor was it enhanced by pulse application. Thus it is not likely that the drug remains trapped in the pore, but, instead, stabilizes a nonconducting state of the channel. Interestingly, this action of amlodipine was not seen when experiments were carried out with the neutral form of the drug molecule (pH 10.0). Neutral amlodipine (pH_o 10.0) blocks I_{Ca} in a voltage-dependent manner closely resembling other neutral DHP compounds. Block is promoted by depolarization and rapidly relieved when membrane potential is returned to negative values.

This striking difference in the recovery from block by neutral and charged forms of the drug was then used as an indirect assay for the location of the binding site for the drug molecule. Channels were first blocked by ionized drug (pH 7.4) and recovery was tested for but not observed, as expected (Fig. 5). The external solution was then rapidly changed to pH 10.0, and recovery protocols were repeated. Within the limitations of the kinetics of the change in external solution, it was found that recovery from block rapidly followed the change in external pH (Fig. 5). Because neutral, but not ionized, forms of the drug molecule reversibly block the channel, these data were consistent with the drug-bound receptor being accessible to rapid changes in external hydrogen ion concentrations. This suggested, but certainly did not prove, that the drug binding site must be near the external surface of the cell membrane.

In order to test more directly for the distinction between an extracellular and intracellular location of the DHP binding site, the sensitivity of calcium-channel current to internal and external drug application had to be measured. Internal application can be accomplished by including a high drug concentration in a patch pipette and waiting sufficiently long times for these molecules to diffuse out of the pipette and reach steady-state concentrations in the

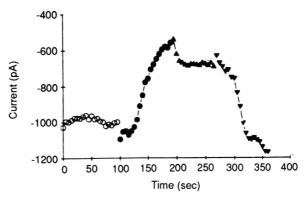

FIG. 5 Influence of external pH on recovery from amlodipine block. Current amplitude is plotted against time before (open symbols) and after application of 3 μM amlodipine (filled symbols). Onset of block, measured at pH 7.4, is shown as filled circles. Recovery in pH 7.4 is slow and incomplete (▲) but rapid and complete when external pH is changed to 10.0 (▼). Reprinted with permission from Kass and Arena, 1989.

cell under investigation. Drug–channel interactions can, in principle, be distinguished from channel rundown in the dialysis experiments by the voltage dependence of channel inhibition. However, this experimental approach must be used with great caution. First, great care must be taken to avoid leakage of drug from the patch pipette as the cell is approached. Particularly because the pipette drug concentrations are so high, a small leakage can result in large effects on the cells. Additionally, DHP compounds, including the ionized drugs, partition markedly into cell membranes and are difficult to wash out. Thus, leakage of drug from the approach pipette can cause channel inhibition via an externally applied pathway. Application of negative pressure while approaching the cell to be studied can minimize this problem. Second, sufficient times must be allowed for diffusional filling of intracellular compartments. Considering the cardiac ventricular myocyte as a cylinder, this diffusional delay can be substantial (up to 20 min), based on total cell capacitance (33). Tertiary amlodipine, although 94% ionized at pH 7.4, is not an optimum drug for these studies because the small fraction of neutral molecules can diffuse across the cell membrane and offset the dialysis of drug into the cell via the patch pipette. A permanently charged compound is better suited for these purposes and experiments have been carried out with quaternary amlodipine and SDZ-207-180(27). The ionization groups on the permanently charged molecules restrict them from diffusing out of the cell across the lipid membrane (23).

External Access to the DHP Binding Site: Evidence for Distinct Binding Domains

The ionized DHPs studied to date have been found to have little effect when applied internally to heart cell calcium channels. Figure 6 illustrates one of these experiments. It compares pulse-dependent block after 20 minutes of dialysis with 50 μM intracellular amlodipine to a 3-min exposure to 10 μM amlodipine applied externally to the same cell (28). Similar results were obtained for intracellularly and extracellularly applied SDZ-207-180 (27) and, more recently, quaternary amlodipine (Kass and Kwan, unpublished observations). In all cases intracellular application failed to inhibit Ca-channel

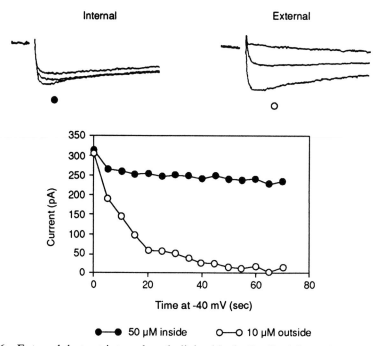

FIG. 6   External, but not internal, amlodipine blocks I_{Ca}. Peak inward currents were measured during depolarizing voltage pulse trains after 20 min dialysis with 50 μM amlodipine (filled symbols) and then after an additional 3 min exposure to 10 μM amlodipine applied externally (open symbols). Little indication of voltage-dependent block of I_{Ca} is seen for internally applied drug compared with externally applied drug despite the fivefold increase in internal drug concentration. Charge carrier is calcium. Reprinted with permission from Kass and Kwan, 1992.

activity when compared to control, drug-free dialysis. In order to control for the long diffusion times discussed above for intracellular loading of cells, experiments were also carried out in much smaller GH_4C_1 pituitary tumor cells for which the predicted diffusional time constant for cell dialysis is 40 sec (33), and similar results were obtained.

Experiments which compared intracellular and extracellular application of ionized amlodipine and SDZ-207-180 in heart and GH_4C_1 cells provided evidence that these two charged DHP compounds cannot reach the DHP receptor via an intracellular pathway. This suggests that the DHP binding site is closer to the extracellular rather than the intracellular face of the cell membrane. Because of the long hydrocarbon chain that separates the pyridine ring from the charge group on SDZ-207-180 (27), we can only estimate the location of the DHP binding site within the resolution of the length of this chain; however, our electrophysiological data are consistent with recent binding studies that provide structural evidence for opposite localization of the phenylalkylamine and DHP binding sites (31, 34, 35). Future experiments using different distances between pyridine rings and charge groups in other permanently charged DHP derivatives will improve the resolution of the location of this binding site.

Biochemical Experiments: Binding and Subunit-Specific Antibodies

The location of the binding domain for dihydropyridines has also been investigated by two groups using biochemical techniques on isolated and purified channel protein subunits. In the first published study the dihydropyridine binding site on the rabbit skeletal muscle α_1 subunit was investigated using tritiated azidopine and nitrendipine as ligands (36). Several peptides were found to be labeled using this approach, but the two major labeled peptides were shown to be located on the putative *cytosolic* domain of the calcium channel. Interestingly, this domain adjoins a possible calcium binding site (37) suggesting the possibility of conformation-dependent interactions between DHP and calcium binding.

In another approach, Catterall and colleagues (34, 35) used a combination of radioligand binding and peptide-specific antibodies raised against targeted fragments of the L-type channel α_1 subunit to also identify the DHP binding domain on the channel protein. This approach yielded two additional sites that appeared to be located on the extracellular domain of the channel. The electrophysiological data summarized above remain the principal means of distinguishing between these two possibilities, and additonal work is planned in the future to use probes specifically designed to improve the electrophysiological estimate of the binding site location.

In the view of Catterall and Streissnig (32), the action of dihydropyridines is to control channel gating by causing allosteric interactions among channel subunits and thus inhibit current by controlling the opening and closing of channels and not by occluding the channel pore.

Thus, although calcium binding sites are clearly important to channel function and modulation by DHPs and other drugs, it is clear that more work needs to be done in order to clarify the structural basis of the activity of these drugs and to determine the roles, if any, of channel subunits in causing the important regulatory actions of the calcium-channel antagonists.

Calcium Ions and Dihydropyridines

It has been known since the discovery of organic calcium-channel blockers by Fleckenstein (38) that there exists an interrelationship between the blocking activity of these drugs and divalent ion concentration. Divalent ions inhibit the binding of radiolabeled phenylalkylamines to membrane-bound L channels, but high-affinity dihydropyridine labeling of L-type channels in brain, cardiac, or smooth muscle membranes depends on the presence of divalent ions (5, 39) as does the binding of dihydropyridines to the purified DHP receptor (40). Binding of calcium also has important regulatory roles in the permeability and gating properties of drug-free native L-type channels. Calcium influx and the high calcium selectivity of skeletal muscle and heart L-type channels are best explained by a multiple ion (at least two)-binding model in which ion–ion repulsion is seen as the key to overcoming a strong intrapore cation binding (41). Divalent ion binding has been shown to induce protein conformational changes that, in turn, could affect ion permeation and channel gating (42, 43).

The structural predictions of a close physical relationship between the DHP binding site and a regulatory calcium binding site (36, 37) would suggest a molecular mechanism for this interrelationship between drug action and divalent ion binding, and this view has been strengthened by a report that positive allosteric regulators of DHP binding increase the calcium affinity of the L-type calcium channel and, simultaneously, alter DHP receptor kinetics (44). However, the location of this putative regulatory calcium binding site remains to be determined as conflicting evidence has been presented for extracellular (45) and cytoplasmic (36, 37) sites. Clearly knowledge of the drug- and calcium-binding domains on the α_1 subunit will contribute to our understanding of the molecular mechanisms of channel gating as well as the basis for allosteric drug interactions.

Summary

Calcium-channel antagonists are unique drugs in that they have gained prominence in their major contributions to clinical medicine and the basic science of calcium-channel proteins. Because of the multiple roles of calcium ions in cell function, communication, and biochemical regulation, the control of calcium entry by these drugs has given us therapeutic tools for a wide variety of clinical disorders. In addition, the high degree of selectivity and great sensitivity of these drugs for the L-type channel allow the cloning and molecular fingerprinting of the channel protein. Work is presently directed at combining the approaches of molecular biology and electrophysiology to provide further information about the molecular basis of tissue selectivity of these drugs and to allow the design of structurally related drug molecules that will be able to target and control specific regions of the channel protein.

References

1. B. P. Bean, *Ann. Rev. Physiol.* **51,** 367–384 (1989).
2. P. Hess, *Annu. Rev. Neurosci.* **13,** 337–356 (1990).
3. R. W. Tsien, P. T. Ellinor, and W. A. Horne, *TIPS* **12,** 349–354 (1991).
4. D. Pelzer, S. Pelzer, and T. F. McDonald, *Rev. Physiol. Biochem. Pharmacol.* **114,** 108–207 (1990).
5. H. Glossmann, and J. Striessnig, *Rev. Physiol. Biochem. Pharmacol.* **114,** 1–105 (1990).
6. R. A. Janis, and D. J. Triggle, "The Calcium Channel: Its Properties, Function, Regulation and Clinical Relevance." Telford Press, West Caldwell, NJ.
7. B. P. Bean, *Proc. Natl. Acad. Sci. U.S.A.* **81,** 6388–6392 (1984).
8. M. C. Sanguinetti, and R. S. Kass, *Circ. Res.* **55,** 336–348 (1984).
9. M. C. Sanguinetti, D. S. Krafte, and R. S. Kass, *J. Gen. Physiol.* **88,** 369–392 (1986).
10. P. Hess, J. B. Lansman, and R. W. Tsien, *Nature (London)* **311,** 538–544 (1984).
11. D. Pietrobon, and P. Hess, *Nature (London)* **346,** 651 (1990).
12. C. R. Artalejo, M. A. Ariano, R. L. Perlman, and A. P. Fox, *Nature (London)* **346,** 651 (1990).
12. C. R. Artalejo, M. A. Ariano, R. L. Perlman, and A. P. Fox, *Nature (London)* **348,** 239–242 (1990).
13. W. A. Catterall, M. J. Seagar, and M. Takahashi, *J. Biol. Chem.* **263,** 3535–3538 (1988).
14. M. M. Hosey, and M. Lazdunski, *J. Membr. Biol.* **104,** 81–105 (1988).
15. D. J. Triggle, *J. Cardiovasc. Pharm.* **18,** S1–S6 (1991).
16. R. Mitra, and M. Morad, *Am. J. Physiol.* **249,** H1056–H1060 (1985).
17. R. S. Kass, and J. P. Arena, *Abstr. Bayer Symp. Calcium Channel* (1988).

18. O. P. Hamill, A. Marty, E. Neher, B. Sakmann, and F. J. Sigworth, *Pfluegers Arch.* **391,** 85–100 (1981).

19. R. A. Burges, A. J. Carter, D. F. Gardiner, and A. J. Higgins, *Br. J. Pharmacol.* **85,** 281P (1985).

20. R. A. Burges, D. G. Gardiner, M. Gwilt, A. J. Higgins, K. J. Blackburn, S. F. Campbell, P. E. Cross, and J. K. Stubbs, *J. Cardiovasc. Pharmacol.* **9,** 110–119 (1987).

21. R. S. Kass and J. P. Arena, *J. Gen. Physiol.* **93,** 1109–1127 (1989).

22. B. P. Bean, *J. Gen. Physiol.* **86,** 1–30 (1985).

23. B. Hille, *J. Gen. Physiol.* **69,** 497–515 (1977).

24. L. M. Hondeghem and B. G. Katzung, *Biochim. Biophys. Acta* **472,** 373–398 (1977).

25. R. Rodenkirchen, R. Bayer, and R. Mannhold, *Prog. Pharmacol.* **5,** 9–23 (1982).

26. H. Valdivia, and R. Coronado, *J. Gen. Physiol.* **95,** 1–27 (1990).

27. R. S. Kass, J. P. Arena, and S. Chin, *J. Gen. Physiol.* **98,** 63–75 (1991).

28. R. S. Kass, and Y. W. Kwan, *J. Cardiovasc. Pharmacol.* **20**(Suppl. A), S6–S13 (1992).

29. J. Hescheler, D. Pelzer, G. Trube, and W. Trautwein, *Pfluegers Arch.* **393,** 287–291 (1982).

30. K. S. Lee, and R. W. Tsien, *J. Physiol.* **354,** 253–272 (1984).

31. J. Striessnig, H. Glossmann, and W. A. Catterall, *Proc. Natl. Acad. Sci. U.S.A.* **87,** 9108–9112 (1990).

32. W. A. Catterall, and J. Striessnig, *TIPS* **13,** 256–262 (1992).

33. M. Pusch, and E. Neher, *Pfluegers Arch.* **411,** 204–211 (1988).

34. H. Nakayama, M. Taki, J. Striessnig, H. Glossmann, W. A. Catterall, and Y. Kanaoka, *Proc. Natl. Acad. Sci. U.S.A.* **88,** 9203–9207 (1991).

35. J. Striessnig, B. J. Murphy, and W. A. Catterall, *Proc. Natl. Acad. Sci. U.S.A.* **88,** 10769—10773 (1991).

36. S. Regulla, T. Schneider, W. Nastainczyk, H. W. Meer, and F. Hofmann, *EMBO J.* **10,** 45–49 (1991).

37. J. Babitch, *Nature* (*London*) **346,** 321–322 (1990).

38. A. Fleckenstein, *Ann. N.Y. Acad. Sci.* 1–15 (1988).

39. H. Glossmann, and J. Striessnig, *Vitam. Horm.* **44,** 155–328 (1988).

40. V. Flockerzi, H. J. Oeken, F. Hofmann, D. Pelzer, A. Cavalie, and W. Trautwein, *Nature* **323,** 66–68 (1986).

41. R. W. Tsien, P. Hess, E. W. McCleskey, and R. L. Rosenberg, *Annu. Rev. Biophys. Biophys. Chem.* **16,** 265–290 (1987).

42. R. S. Kass, and M. C. Sanguinetti, *J. Gen. Physiol.* **84,** 705–726 (1984).

43. D. T. Yue, P. H. Backx, and J. P. Imredy, *Science* (Washington, D.C.) **21,** 1735–1738 (1990).

44. R. Staudinger, H. G. Knaus, and H. Glossmann, *J. Biol. Chem.* (1991).

45. H. Ebata, J. S. Mills, K. Nemcek, and J. D. Johnson, *J. Biol. Chem.* **265**(1), 177–182 (1990).

[11] Potassium Channels of Cardiac Myocytes

Michael C. Sanguinetti

Introduction

The study of ionic currents in cardiac tissue lagged behind that of squid giant axons and cultured neuronal cells for many years due to the lack of suitable methods to isolate viable, Ca^{2+}-tolerant cardiac myocytes from intact tissue. Until about 1980 our understanding of the ionic basis of cardiac action potentials was based largely on studies of isolated cardiac Purkinje fibers using the two-microelectrode voltage-clamp technique or of trabeculae and papillary muscles using the sucrose-gap-clamp technique. While these early studies formed the experimental foundations of cardiac electrophysiology, the techniques used were technically difficult and fraught with artifacts which made neurophysiologists shudder at the inadequacies of the experimental data. Luckily for those interested in continuing the exploration of cardiac currents, advances in cell isolation and voltage-clamp techniques have eliminated most of these inadequacies. The development of suitable cell isolation techniques (1–5) and the concurrent introduction of patch-clamp methodology (6, 7) have led to a blossoming of studies directed at unraveling the complexities of cardiac electrophysiology.

Arguably the most interesting finding in the field of cardiac electrophysiology within the last 5–10 years, made possible by the development of these new methodologies, has been the discovery of the plethora of K^+-channel types present in cardiac myocytes. Years of continued study will be required before we understand the exact role these various K^+ channels play in shaping the configuration of action potentials from different species and regions of the heart. Further studies are also needed to define how differential K^+-channel expression can account for developmental changes in action potential repolarization.

While the diversity of cardiac K^+ channels represents a treasure chest for future study, it also complicates the efforts to characterize any one specific current in isolation from other K^+ currents that overlap it in time and/or voltage dependence of activation. In recognition of this problem, this chapter emphasizes cell isolation techniques and the use of pulse protocols and pharmacological agents to separate multiple K^+ currents from one another. The reader is referred to the many excellent papers and reviews covering the details of single-cell voltage-clamp techniques, which for the most part are common to all isolated cell types (6–9).

Methods in Neurosciences, Volume 19

TABLE I   Cardiac Potassium Currents Characterized by Electrophysiological Techniques

Potassium current	Abbrev.	Distinctive properties	Single-channel conductance $(\gamma)^a \cdot$ pS	Refs
Inward rectifier	I_{K1}	Triple-barrel structure	32 $\gamma \propto ([K^+]_o)^{0.22}$	10, 11
Delayed rectifier	I_{Kr}	Moderately rapid activation; very rapid inactivation	10–11	12–14
Delayed rectifier	I_{Ks}	Very slow activation; no inactivation; cloned from rat, mouse, and human heart.	<1–3	15–17
Delayed rectifier	I_{RAK}	Fast activation; no inactivation; cloned from rat atria	?	18, 19
	(HK2)	Cloned from human heart	?	20, 21
Plateau K current	I_{Kp}	Rapid activation; no inactivation	14	22
Transient outward (voltage activated)	I_{to1} (I_{eo})	Rapid activation and rapid inactivation	14–17 (5.4 K) 12 and 27 (5.4 K, mice)	23–25
([Ca^{2+}]$_i$ activated)	I_{to2}	Activation dependent on ↑ in [Ca^{2+}]$_i$	120 (10.8 K)	26–28
Acetylcholine activated	I_{KACh}	Similar to I_{K1}, but only found in atria and Purkinje fibers	$\gamma = 13.3 \, ([K^+]_o)^{0.23}$	29–31
[ATP]$_i$ inhibited	I_{KATP}	Closed under "normal" conditions	$\gamma = 23.6 \, ([K^+]_o)^{0.24}$ 35 (5.4 K)	32, 33
Arachidonic acid activated	I_{KAA}	Activated by AA	124	34
Phosphatidylcholine activated	I_{KPC}	Activated by PC	60	34
[Na]$_i$ activated	I_{KNa}	Activated by [Na$^+$]$_i$ > 20 mM; K_D for Na = 66 mM	210	35, 36
Sarcoplasmic reticulum	SR$_K$	Blocked by Ca^{2+}	154	37, 38

^{a} Measured in presence of 150 mM [K$^+$]$_o$.

Diversity of Cardiac Potassium Channels

Nomenclature Based on Electrophysiology

At least 13 distinct types of mammalian cardiac K$^+$ currents have been described using standard whole-cell voltage-clamp techniques. Table I summarizes the distinctive properties and major physiological roles of each channel type. Also listed is the single-channel conductance of the channels.

Undoubtedly, many other cardiac K^+ channels will be characterized once experimental methods exist to differentiate new channels from those already described.

Cloned Cardiac Potassium Channels

The diversity of K^+ channels is also demonstrated by the increasing number of channels (seven at the time of this writing) that have been cloned from mammalian heart. The cDNA encoding several types of I_{to} and I_K channels has been cloned and sequenced from rat and human cDNA libraries. Most of the cloned cardiac K^+ channels activate relatively rapidly and inactivate only very slowly, requiring several seconds for significant, but incomplete inactivation. These include the Shaker homologs, Kv1.1 (= RK1) (see Refs. 39, 40), Kv1.2 (= RAK, RK2) (see Refs. 19, 40), Kv1.5 (= RK4, HK2, Kv1) (see Refs. 20, 40), and the Shab homolog Kv2.1 (= drK1) (see Ref. 39). When these channels are expressed in *Xenopus* oocytes the currents most resemble I_{RAK}, the delayed rectifier K^+ current recorded from rat atrial myocytes (18). The channel ("I_{sK}") that underlies the very slowly activating delayed rectifier K^+ current (I_{Ks}) was originally expression cloned from rat kidney and was later cloned from rat (17) and mouse (41) myocardium. The Shaker analog Kv1.4 (= RHK1, RCK4, RK3, HK1) (see refs. 20, 42) and the Shal homolog Kv4.2 (= RK5) (see Ref. 43) are rapidly (a few milliseconds) activating and inactivating I_{to} channels.

Unfortunately, expression cloning of additional cardiac K^+ channels has been hindered by the very poor expression of channels in *Xenopus* oocytes injected with polyadenylated mRNA isolated from heart tissue. The seven cardiac K^+ channels listed above were all isolated by screening cDNA libraries.

Cell Isolation Techniques

Several methods for isolation of Ca^{2+}-tolerant myocytes from adult mammalian heart have been described in recent years. Isolation of these cells requires a variable period of exposure to nominally Ca^{2+}-free salt solutions. Lowering of extracellular Ca^{2+} is necessary for proper dissociation of the cells, causing disruption of desmosomes and permitting optimal activity of collagenase. Prior to about 1976, isolation of cells from cardiac tissue was relatively simple, but the cells would hypercontract on reexposure to normal levels of extracellular Ca^{2+}. This Ca^{2+} intolerance prevented the use of isolated adult mammalian myocytes in any kind of experiment that called for bathing the cells in normal Ca^{2+}. The probable cause of Ca^{2+}-overload caused by reexposure of myocytes to normal Ca^{2+} levels is discussed in a review of cell isolation methods prior to 1983 (44).

Langendorff Perfusion of Small Mammalian Hearts

The first successful and relatively simple technique for isolating viable cells was introduced in 1976 by Powell and Twist (1). A breakthrough in isolation techniques was the introduction of procedures that only exposed the tissue to low $[Ca^{2+}]_o$ solutions for relatively short periods (<20 min) or allowed for recovery of the cells in "relaxing" solutions that contain high levels of K^+ and low concentrations of $CaCl_2$ (3). Many investigators today preincubate isolated cells in a recovery medium first described by Isenberg and Klockner (3), a solution they called "KB" medium (*KraftBruhe* = "power soup"). Usually cells are allowed to recover in KB medium for at least 1 hr before being reexposed to solutions containing normal $[Ca^{2+}]$. The composition of KB medium is as follows (mM): KCl, 85; K_2HPO_4, 30; $MgSO_4$, 5; Na_2ATP, 5; pyruvate, 5; succinate, 5; β-hydroxybutyrate, 5; creatine, 5; pH is adjusted to 7.2 with KOH and pCa is adjusted to 7.5 with EGTA.

An even simpler cell isolation method was introduced by Mitra and Morad (5) which yielded Ca^{2+}-tolerant cells from several small vertebrate species. This technique with slight modifications has been used for many years in our laboratory to consistently yield guinea pig ventricular myocytes suitable for voltage-clamp experiments. An outline of the technique is presented below:

1. Anesthetize animal, remove heart, and place in warmed HEPES-buffered saline (solution A, Table II) containing 1.8 mM $CaCl_2$ and continuously gassed with 100% O_2.

2. Cannulate aorta to perfusion apparatus. Perfuse heart with prewarmed (37°C) nominally Ca^{2+}-free (no added $CaCl_2$) HEPES-buffered saline (solution B) for 5 min at a rate of 10 ml/min.

3. Perfuse with Ca^{2+}-free HEPES-buffered saline + enzymes (solution C) for 8 min at 10 ml/min (without recirculation).

4. Perfuse with HEPES-buffered saline containing 0.2 mM $CaCl_2$ (solution D) for 5 min at 10 ml/min.

5. Remove heart from perfusion cannula. The heart should still be intact at this time, but should be quite flaccid. Cut ventricle and atria separately into small (5 × 5 mm) pieces and place into beakers containing 15 ml of solution D. Shake in water bath (37°C) for 2–3 min until tissue just begins to disintegrate. This procedure allows the more superficial cells of the tissue

TABLE II   Solutions Used in the Isolation of Myocytes from the Hearts of Small Mammals

Solution[a]	NaCl	KCl	CaCl$_2$	MgCl$_2$	HEPES	Glucose	Enzymes
A	132	4	1.8	1.2	10	10	—
B	132	4	—	1.2	10	10	—
C	132	4	—	1.2	10	10	Worthington Type 2 collagenase (150 units/ml)
							Sigma Type 14 protease (0.35 units/ml)
D	132	4	0.2	1.2	10	10	—

[a] All solutions are adjusted to pH 7.2 with NaOH.

pieces to dissociate from the less-damaged cells of the interior of the tissue mass.

6. Remove pieces of tissue from beakers and place in separate beaker containing 15 ml of fresh solution D. Continue shaking for 5 min to release cells from disintegrating tissue.

7. Remove undigested pieces of tissue from beaker, and then filter cell suspension through a piece of woven polypropylene mesh (210 μm opening, Spectra/Mesh) into a 50-ml plastic centrifuge tube to separate single cells from incompletely dissociated clumps of cells. Allow cells to settle to the bottom of the centrifuge tube for 20 min. Aspirate off solution to remove suspended cells (most dead), and then resuspend cells pellet with 15 ml of solution A.

Cells kept at room temperature in this solution are suitable for electrophysiological recordings for up to about 10 hr. When cells are kept in this solution for periods >8–10 hr, we have noted a distinct decline in the magnitude of delayed rectifier K$^+$ current (I_{K_s}).

A similar, but improved version (80–90% Ca^{2+}-tolerant rod-shaped cells) of this myocyte isolation method was reported by Yazawa et al. (45). The improvements were (i) cannulation of the aorta in situ while the guinea pig is artificially respired; (ii) perfusion of the isolated heart with 60–100 units/ml collagenase (Yakult); (iii) subsequent treatment of isolated cells with protease (0.05–0.2 mg/ml) and deoxyribonuclease (0.02 mg/ml) for 10–15 min to improve gigaohm seal formation.

These techniques are useful for dissociating cells from small hearts, but are impractical for larger hearts. To isolate cells from mammalian species such as dogs, cows, and sheep, most investigators have enzymatically digested minced tissue (3). An alternative approach is to perfuse a wedge of ventricular free wall via a coronary artery. A relatively high yield of Ca^{2+}-tolerant cells from canine left ventricle is obtained by perfusion of a wedge of left ventricular free wall through a branch of the left circumflex artery (46). After perfusion of the tissue wedge with a solution containing collagenase, the partially digested tissue is minced and further digested by continuous trituration in a fresh enzyme solution.

Methods for long-term maintenance of rat (47) and rabbit (48) myocytes in primary cell culture have been described. Enzymatically isolated cells are cultured in a media supplemented with amino acids, 5% fetal bovine serum, antibiotics, and cytosine arabinoside (10 μM). Cytosine arabinoside is an antimetabolite that suppresses proliferation of interstitial cells that would otherwise overwhelm the nondividing myocytes (48).

Nonperfusion Method

The chunk method of myocyte isolation has been used to obtain viable cells from frog hearts by several investigators (49–51). The technique has also been tried with less success using chunks of mammalian heart (3). The basic procedure developed for bullfrog (*Rana catesbiana*) myocyte isolation (50) is outlined below:

1. Remove heart and place in normal Ringer's saline (160 mM NaCl, 5 mM KCl, 1 mM CaCl$_2$, 1 mM MgCl$_2$, 10 mM glucose, 5 mM HEPES; adjust pH to 7.6 with NaOH). Dissect atrium away from ventricle; place each in nominally Ca^{2+}-free Ringer's saline.

2. Mince tissues into small (approx. 2 mm^2) pieces. Transfer pieces to 5 ml (atria) or 10 ml (ventricle) of nominally Ca^{2+}-free Ringer's saline containing 0.15% Sigma type I collagenase (200 units/mg) and 0.1% Sigma type III trypsin (10,000 units/mg).

3. Stir solution slowly (approx. 60 rpm) for 45 min using a small stir bar.

4. Pipette off enzyme solution; resuspend tissue in 5 ml (atria) or 10 ml (ventricle) of nominally Ca^{2+}-free Ringer's plus 0.1% bovine serum albumin (BSA). Continue stirring for 5 min.

5. Pipette off BSA solution, and resuspend tissue in 5 ml (atria) or 10 ml (ventricle) of nominally Ca^{2+}-free Ringer's saline plus 0.05% Sigma type I collagenase. Stir atrial tissue for at least another 30 min. Stir ventricular tissue for 2 hr; triturate with Pasteur pipette and then continue digestion for another 2–3 hr.

6. When tissue is fully digested, dilute saline 1 : 4 with fresh normal Ringer's saline.

Muscle Slices

Treatment of cells with proteolytic enzymes may alter some properties of sarcolemmal channels. Using a technique similar to that first developed for brain tissue, Burnashev *et al.* (52) have successfully recorded single-channel currents from thin slices of newborn rat hearts. Ventricles were immobilized, sliced with a vibrating microslicer parallel to the long axis of the tissue, and then allowed to incubate in normal physiological saline at room temperature before electrophysiological recording. This technique may prove useful in future studies of the effects of enzymatic treatment on single-channel properties.

Voltage-Clamp Techniques

Electrode Glass

Many types of glass have been used to record single-channel and whole-cell currents from isolated cardiac myocytes. Most investigators have chosen to use borosilicate glass capillaries (e.g., Kimax-51) for standard single-channel or whole-cell recording due to the relative ease of seal information.

The type of glass used to make recording pipettes is important not only with respect to ease of gigaseal formation, but also because some types of glass have been reported to release divalent cations that can block ionic currents (53, 54). In particular, the lead potash glass (8161) was found to induce slow inactivation of K^+ currents in pituitary cells (53), and several other types of glass were reported to release cations that blocked cGMP-activated channels from rod photoreceptors (54). There have been no reports of possible effects of patch pipette glass composition on kinetics or magnitude of cardiac currents; however, it is likely that similar problems have affected the recording of these currents as well.

Pipette Solutions

Electrodes used for standard whole-cell voltage-clamp experiments are filled with solutions that vary according to the type of currents that are to be recorded. To record cardiac K^+ currents, a typical intracellular solution would contain (in m*M*) potassium aspartate or potassium gluconate, 110; KCl, 20; $MgCl_2$, 0–2; EGTA or BAPTA, 1–10; $CaCl_2$, varies depending on [EGTA] or [BAPTA]; HEPES, 5–10; glucose, 0–10; K_2ATP, 2–5; adjust pH to 7.2 with KOH. To record single K^+ channels in a cell-attached or inside-out configuration, pipettes are typically filled with either a standard extracellular solution (e.g., solution A of Table II) or a high $[K^+]$ solution, typically composed of (in mM) KCl, 150; $CaCl_2$, 1.8; HEPES, 5; adjust pH to 7.4 with KOH.

One of the problems associated with studying whole-cell currents using perfusion pipettes or channels in excised patches is the dilution or loss of endogenous modulators that may dramatically alter current magnitude and kinetics (whole-cell) or single-channel behavior. Methods have recently been developed to circumvent some of these potential problems. Whole-cell currents can be recorded using pipettes filled with a standard intracellular solution plus nystatin. Nystatin is an ionophore that permeabilizes the patch of membrane under the pipette, allowing the contents of the pipette to mix freely with the cytoplasm, but without allowing large macromolecules to pass from the cell interior into the pipette (55). Thus, reasonable control of intracellular ion composition can be achieved without dilution of larger endogenous modulators. The other major advantage afforded by the use of nystatin-containing pipettes is the prevention of current "rundown" that commonly occurs during perfusion of a cell with normal whole-cell pipette solutions. A modification of the standard cell-attached patch configuration, the "open-cell-attached" (OCA) mode, has been used to study I_{KATP} channels under conditions in which intracellular ion concentrations can be controlled without loss of large cytoplasmic molecules (56, 57). The OCA patch configuration is formed by an on-cell patch, and then the entire cell is permeabilized with a brief exposure to 1% saponin. This allows the composition and concentration of internal ions to be manipulated by simply bathing the cell in the desired intracellular solution, with the distinct advantage of not "washing out" vital intracellular molecules such as soluble kinases.

A hybrid technique of the suction pipette method that minimizes internal dialysis of voltage-clamped cells was introduced by Hume and Giles (50) and is described in more detail in a methods paper (58). To minimize perfusion of cells with the pipette solution, negative pressure (applied via an air-tight syringe) is attached to a side port of the microelectrode holder and very small-tipped electrodes (pulled from 1-mm o.d. square-bore glass, Glass Co. of America, Bargaintown, NJ) are filled with 0.5–2 M potassium gluconate (58). Electrodes are pulled in two stages similar to standard patch pipettes, but do not require fire polishing to achieve consistently tight seals. We have used this technique routinely to measure delayed rectifier K^+ currents in guinea pig myocytes (12). The great advantages of the method is twofold. First, it permits the recording of currents under more physiological conditions than possible if the cell contents are exchanged for a simple pipette solution.

Second, it prevents the normal rundown of I_K that is encountered with the use of standard patch pipettes (58). The disadvantages of the method are also twofold. The investigator has no control of the cytoplasmic concentration of ions or regulatory factors (e.g., ATP, GTP, pH), and only relatively small currents can be recorded due to the large series resistance (R_s) associated with use of high-resistance electrodes. We have found that the ideal electrode resistance is between 3 and 6 MΩ for 1-mm-o.d. square-bore pipettes filled with 0.5 M potassium gluconate, 25 mM KCl, and 5 mM K$_2$ATP. This leads to an uncompensated R_s of 9–18 MΩ. Even with 90% R_s compensation, this leaves an uncompensated R_s of 0.9–1.8 MΩ.

Cell Chambers

Most investigators fabricate their own single-cell recording chambers based on their particular needs. The main problems facing the cellular cardiac electrophysiologist in this regard have been design of recording chambers compatible with inverted microscopes that allow for rapid changes in flow and stable temperature control of perfusion solutions. Relatively simple designs have been described by Datyner *et al.* (59) and improved on by Cannell and Lederer (60). Even faster methods of solution changes ("concentration jump") have since been developed. A single cell can be immersed in a microdrop (<0.1 μl), isolated from a larger volume of solution contained in an outer bath by a ring of Sylgard (61). This arrangement is reported to allow changes in external bathing media in less than 10 msec. A different method, called the oil-gate concentration method has been used to rapidly (half-time of 6.5 msec) switch the solution bathing inside-out patches of cardiac cells (62). Detailed descriptions of these chambers can be found in the cited references.

Separation of Currents in Whole-Cell Voltage-Clamp Experiments

Obviously, the ideal way to study a single K$^+$-channel type is with a single-channel recording technique (6, 7). This technique is the only reasonable way to investigate many aspects of intracellular regulation of channel function. However, channels do not always behave normally in excised membrane patches, and the effects of endogenous modulators and drugs may be altered (57). For many experimenters, whole-cell currents are the only practical means to study the physiological or pharmacological regulation of a particular K$^+$ current. In this case, it is extremely important to eliminate or at least minimize the contribution of other overlapping currents.

TABLE III   Activators and Blockers of Cardiac Potassium Channels[a]

Channel type	Activators	Blockers	Refs
I_{K1}	Voltage; high $[K^+]_e$; platelet-activating factor (10–100 pM)	Ba^{2+} (1–10 μM), Cs^+, Mg^{2+}, Ca^{2+}, low $[K^+]_e$	11, 63
I_{Kr}	Voltage	E-4031 (300 nM), dofetilide (32 nM), d-sotalol (10 μM)], La^{3+} (10 μM), low $[K^+]_e$	12, 14 63a 64, 66
I_{Ks}	Voltage; catecholamines; Stimulation of protein kinase A or C; endothelin (3 nM)	Quinidine (10 μM), tedisamil (2.5 μM), Co^{2+} (1 mM),	67–72
I_{RAK}	Voltage	4AP[b] (1 mM)	18–21
I_{Kp}	Voltage	Ba^{2+} (<0.5 mM)	22
I_{to1}	Voltage	α_1 Adrenoceptor agonists (phenylephrine, 30 μM) (methoxamine, 0.23 mM), tedisamil (6 μM), tedisamil (6 μM), 4AP (1–3 mM), Ba^{2+}	23, 69 72–74
I_{to2}	Voltage, Ca^{2+}	Ca^{2+} channel blockers, e.g., Co^{2+}; ryanodine (3 μM), caffeine (10 mM) (via inhibition of Ca^{2+} transients)	28, 75
I_{KATP}	Cromakalim (3 μM), nicorandil (74 μM), RP 49356 (3–300 μM), SR 44866 (2–10 μM), metabolic inhibitors (cyanide, 2,4-DNP)	$[ATP]_i$ (122 μM), glyburide (5–20 μM), tolbutamide (400 μM), 5-hydroxydecanoate (0.16 μM)	33, 57 75a–80
I_{KACh}	Voltage; ACh (0.1 μM), CGRP (0.3 μM), somatostatin (1 μM), adenosine (0.8 μM), endothelin (1 nM), α_K subunit of G_k	Ba^{2+} (3 μM), Cs^+, Mg^{2+}, quinidine (10 μM)	29–31 81–85
I_{KAA}	Arachidonic acid (10 μM) 6-keto PGF$_1$, 12-HHT, low pH$_i$		34
I_{KPC}	Phosphatidylcholine (10 μM), 12-HHT, PGF$_2$		34
I_{KNa}	$[Na^+]_i$ > 10 mM, ouabain, low $[K^+]_e$ (via inhibition of Na/K-ATPase)	R56865 (0.1–1 μM)	35, 86
SR$_K$ channel		Ca^{2+} (15–21 mM), neomycin B (0.2 mM)	87

[a] Approximate EC$_{50}$ (activators) or IC$_{50}$ (blockers) values are given in parentheses.
[b] 4-Aminopyridine

Most cardiac K^+ currents can be adequately separated from "contaminating" currents by careful selection of appropriate ionic conditions, voltage protocols, and specific activators or blockers. A list of cations and drugs that activate or block cardiac K^+ channels is given in Table III. Unfortunately, very few specific blockers of cardiac K^+ currents are available. Note that several drugs and cations are relatively nonspecific blockers of K^+ currents (e.g., quinidine, 4-aminopyridine, tetraethylammonium, Ba^{2+}, Cs^+). Specific blockers have been described for I_{Kr} (methanesulfonanilides) and I_{KATP}) (sulfonylureas, 5-hydroxydecanoate), but not for the other 11 K^+ currents characterized by electrophysiological techniques. Likewise, there are very few cardiac K^+-channel activators. The first seven channel types listed in Table III are activated by changes in membrane potential, whereas the other channels are activated by intracellular metabolites (I_{KAA}, I_{KPC}), intracellular Na^+ (I_{KNa}), acetylcholine (I_{KACh}), or a decrease in $[ATP]_i$ (I_{KATP}). I_{KATP} channels are specifically activated by a structurally diverse group of agents collectively referred to as K^+-channel openers.

It is oftentimes possible to isolate a particular cardiac K^+ current from other currents by taking advantage of specific blockers, different rates of activation and deactivation, variable threshold potentials for voltage-dependent activation, or variable degrees of voltage-dependent rectification. As an example, these principles can be used to isolate I_{Ks} from other whole-cell currents in isolated guinea pig ventricular myocytes. Under standard conditions (normal $[Na^+]_i$, $[ATP]_i$), depolarization of a guinea pig myocyte from -80 mV to voltages >-20 mV activates inward Ca^{2+} current (T-and L-types), Na^+ current, and multiple outward K^+ currents: I_{K1}, I_{Kp}, I_{Kr}, and I_{Ks}. To study a single current (e.g., I_{Kr}) in isolation from other currents, specific blockers and voltage protocols must be utilized. L-type I_{Ca} can be blocked with dihydropyridines (e.g., nisoldipine). T-type I_{Ca} and I_{Na} can be voltage inactivated by holding potentials positive to -40 mV. Each of the remaining K^+ currents has a characteristic threshold and rate of activation. I_{K1} and I_{Kp} activate fully within a few milliseconds, whereas I_{Kr} activates over 10's to 100's of milliseconds, and I_{Ks} requires many seconds to reach steady-state activation. Outward I_{K1} is activated almost instantly at potentials ranging from its reversal potential ($= E_K$, equilibrium potential for K^+) to about -20 mV (the threshold of activation). At potentials >-20 mV, I_{K1} is not activated. I_{K1} requires the presence of external K^+ for activation and, therefore, can be eliminated by bathing cells in a K^+-free saline. I_{Kr} is specifically inhibited by certain methanesulfonanilides, such as dofetilide (Table III). Therefore, bathing a cell in a K^+-free saline containing dofetilide (1 μM) and nisoldipine (0.4 μM) and applying test pulses from a holding potential of -40 mV result in specific activation of only I_{Kp} and I_{Ks}. Since I_{Kp} is essentially time independent, the only *time-dependent* current elicited during long test pulses to depolarized potentials would be I_{Ks}. The magnitude of I_{Ks} is extremely temperature sensitive. This current is much larger and activates much faster at elevated temperatures. For this reason we usually study I_{Ks} at 35°C.

Inorganic blockers of I_{Ca} can also be used when studying I_K, but it must be appreciated that these cations have effects in addition to blocking I_{Ca}. For example, Cd^{2+} has been shown to increase the magnitude and shift the voltage dependence of I_K activation to more positive potentials in cat ventricular cells (88). I_K of cat cells most resembles I_{Kr} with respect to its marked inward rectification and sensitivity to block by methanesulfonanilides. L-type I_{Ca} is also routinely blocked with inorganic cations such as La^{3+} or Co^{2+}; however, both of these cations are potent blockers of I_{Kr} and thus should not be used when studying this current. Co^{2+} at 3 mM (89) and La^{3+} at >10 μM (65) block I_{Kr} in guinea pig ventricular myocytes. In addition, Co^{2+} blocks I_{Ks} in these cells with an IC_{50} of 1 mM (90). On the other hand, La^{3+} is useful to block both I_{Ca} and I_{Kr} when one wishes to study I_{Ks} in isolation from these other currents.

I_{Kr} can be measured in the presence of I_{Ks} either as a methanesulfonanilide-sensitive current (12, 65, 66, 89) or as time-dependent outward current within a narrow window of potentials negative to that required for I_{Ks} activation. This range of potentials can be widened by bathing cells in a Ca^{2+}-free saline. In the absence of extracellular Ca^{2+}, the half-point for the voltage dependence of I_{Kr} activation is shifted by about -15 mV (to -40 mV), whereas I_{Ks} activation is shifted by $+9$ mV (to $+22$ mV) (see Ref. 65). For example, in Ca^{2+}-free saline, only I_{Kr} is activated during 550-msec pulses to -10 mV from a holding potential of -50 mV (63a).

Considerable differences exist between species with regard to the type and magnitude of K^+ currents expressed within a given cell type. These inherent differences should be exploited when one wishes to study a particular K^+ current. For example, adult rat ventricular cells have a very large I_{to} and I_{RAK}, but little or no I_{Kr} or I_{Ks} (91). Rat atrial cells have a large I_{RAK}, but only a small I_{to} and no I_{Ks} (18). Rabbit and cat cardiac myocytes have a large I_{to} and I_{Kr}, but no measurable I_{Ks} (13, 24, 88, 92). Frog myocytes have a very large I_{Ks} (15), but no I_{Kr} or I_{to}. Guinea pig myocytes have a large I_{Ks} and I_{Kr}, but no measurable I_{to} (12).

References

1. T. Powell and V. W. Twist, *Biochim. Biophys. Res. Commun.* **72,** 327 (1976).
2. T. Powell, *J. Mol. Cell Cardiol.* **11,** 511 (1979).
3. G. Isenberg and U. Klockner, *Pfluegers Arch.* **395,** 6 (1982).
4. M. F. Sheets, C. T. January, and H. A. Fozzard, *Circ. Res.* **53,** 544 (1983).
5. R. Mitra and M. Morad, *Am. J. Physiol.* **249,** H1056 (1985).
6. O. P. Hamill, A. Marty, E. Neher, B. Sakmann, and F. J. Sigworth, *Pfluegers Arch.* **391,** 85 (1981).

7. B. Sakmann and E. Neher, eds., "Single Channel Recording." Plenum Press, New York, 1983.
8. D. Noble and T. Powell, eds., "Electrophysiology of Single Cardiac Cells." Academic Press, London, 1987.
9. T. G. Smith, H. Lecar, S. J. Redman, and P. W. Gage, eds., "Voltage and Patch Clamping with Microelectrodes." American Physiological Society, Bethesda, 1985.
10. Y. Kurachi, *J. Physiol. (London)* **366,** 365 (1985).
11. H. Matsuda, *J. Physiol. (London)* **435,** 83 (1991).
12. M. C. Sanguinetti and N. K. Jurkiewicz, *J. Gen. Physiol.* **96,** 195 (1990).
13. T. Shibasaki, *J. Physiol. (London)* **387,** 227 (1987).
14. M. Horie, S. Hayashi, and C. Kawai, *Jpn. J. Physiol.* **40,** 479 (1990).
15. J. R Hume, W. Giles, K. Robinson, E. F. Shibata, R. D. Nathan, K. Kanai, and R. Rasmusson, *J. Gen. Physiol.* **88,** 777 (1986).
16. J. R. Balser, P. B. Bennett, and D. M. Roden, *J. Gen. Physiol.* **96,** 835 (1990).
17. K. Folander, J. S. Smith, J. Antanavage, C. Bennett, R. B. Stein, and R. Swanson, *Proc. Natl. Acad. Sci. U.S.A.* **87,** 2975 (1990).
18. W. A. Boyle and J. M. Nerbonne, *Am. J. Physiol.* **260,** H1236 (1991).
19. M. Paulmichl, P. Nasmith, R. Hellmiss, K. Reed, W. A. Boyle, J. M. Nerbonne, E. G. Peralta and D. E. Clapham, *Proc. Natl. Acad. Sci. U.S.A.* **88,** 7892 (1991).
20. M. M. Tamkun, K. M. Knoth, J. A. Walbridge, H. Kroemer, D. M. Roden, and D. H. Glover, *FASEB J.* **5,** 331 (1991).
21. D. J. Snyders, K. M. Knoth, S. L. Roberds, and M. M. Tankun, *Mol. Pharmacol.* **41,** 322 (1992).
22. D. T. Yue and E. Marban, *Pfluegers Arch.* **413,** 127 (1988).
23. I. R. Josephson, J. Sanchez-Chapula, and A. M. Brown, *Circ. Res.* **54,** 157 (1984).
24. R. B. Clark, W. R. Giles, and Y. Imaizuni, *J. Physiol. (London)* **405,** 147 (1988).
25. K. Benndorf, F. Markwardt, and B. Nilius, *Pfluegers Arch.* **409,** 641 (1987).
26. S. A. Siegelbaum and R. W. Tsien, *J. Physiol. (London)* **299,** 485 (1980).
27. G. Callewaert, J. Vereecke, and E. Carmeliet, *Pfluegers Arch.* **406,** 424 (1986).
28. G.-N. Tseng and B. F. Hoffman, *Circ. Res.* **64,** 633 (1989).
29. B. Sakmann, A. Noma, and W. Trautwein, *Nature* **303,** 250 (1983).
30. E. Carmeliet and K. Mubagwa, *J. Physiol. (London)* **371,** 219 (1986).
31. J. Codina, A. Yatani, D. Grenet, A. M. Brown, and L. Birnbaumer, *Science (Washington, D.C.)* **236,** 442 (1987).
32. A. Noma, *Science (Washington, D.C.)* **305,** 147 (1983).
33. C. G. Nichols and W. J. Lederer, *Am. J. Physiol.* **261,** H1675 (1991).
34. M. A. Wallert, M. J. Ackerman, D. Kim, and D. E. Clapham, *J. Gen. Physiol.* **98,** 921 (1991).
35. M. Kameyama, M. Kakei, R. Sato, T. Shibasaki, H. Matsuda, and H. Irisawa, *Nature* **309,** 354 (1984).
36. Z. Wang, T. Kimitsuki, and A. Noma, *J. Physiol. (London)* **433,** 241 (1991).
37. B. Tomlins, A. J. Williams, and R. A. P. Montgomery, *J. Membr. Biol.* **80,** 191 (1984).

38. J. A. Hill, R. Coronado, and H. C. Strauss, *Biophys. J.* **55,** 35 (1989).
39. S. L. Roberds and M. M. Tamkun, *Proc. Natl. Acad. Sci. USA.* **88,** 1798 (1991).
40. S. L. Roberds and M. M. Tamkun, *FEBS Lett.* **284,** 152 (1991).
41. E. Honore, B. Attali, G. Romey, C. Heurteaux, P. Ricard, F. Lesage, M. Lazdunski, and J. Barhanin, *EMBO J.* **10,** 2805 (1991).
42. J. C. L. Tseng-Crank, G.-N. Tseng, A. Schwartz, and M. A. Tanouye, *FEBS Lett.* **268,** 63 (1990).
43. T. A. Blair, S. L. Roberds, M. M. Tamkun, and R. P. Hartshorne, *FEBS Lett.* **295,** 211 (1991).
44. B. B. Farmer, M. Mancina, E. S. Williams, and A. M. Watanabe, *Life Sci.* **33,** 1 (1983).
45. K. Yazawa, M. Kaibara, M. Ohara, and M. Kameyama, *Jpn. J. Physiol.* **40,** 157 (1990).
46. K. Hewett, M. J. Legato, P. Danilo, and R. B. Robinson, *Am. J. Physiol.* **245,** H830 (1983).
47. R. E. Weishaar and R. U. Simpson, *Cell. Biol. Intl. Rep.* **10,** 745 (1986).
48. J. Haddad, M. L. Decker, L.-C. Hsieh, M. Lesch, A. M. Samarel, and R. S. Decker, *Am. J. Physiol.* **255,** C19 (1988).
49. M. Tarr and J. W. Trank, *Experientia* **32,** 338 (1976).
50. J. R. Hume and W. Giles, *J. Gen. Physiol.* **78,** 19 (1981).
51. R. Fischmeister and H. C. Hartzell, *J. Physiol.* (*London*) **376,** 183 (1986).
52. N. A. Burnashev, F. A. Edwards, and A. N. Verkhratsky, *Pfluegers Arch.* **417,** 123 (1990).
53. G. Cota and C. Armstrong, *Biophys. J.* **53,** 107 (1988).
54. R. E. Furman and J. C. Tanaka, *Biophys. J.* **53,** 107 (1988).
55. S. J. Korn, A. Marty, J. A. Cannor, and R. Horn, this series, volume 4, [11].
56. M. Kakei, A. Noma, and T. Shibasaki, *J. Physiol.* (*London*) **363,** 441 (1985).
57. C. G. Nichols and W. J. Lederer, *J. Physiol.* (*London*), **423,** 91 (1990).
58. J. R. Hume and R. N. Leblanc, *Mol. Cell. Biochem.* **80,** 49 (1988).
59. N. B. Datyner, G. A. Gintant, and I. S. Cohen, *Pfluegers Arch.* **403,** 318 (1985).
60. M. B. Cannell and W. J. Lederer, *Pfluegers Arch.* **406,** 536 (1986).
61. S. Hering, D. J. Beech, and T. B. Bolton, *Pfluegers Arch.* **410,** 335 (1987).
62. D. Qin, M. Takano, and A. Noma, *Am. J. Physiol.* **257,** H1624 (1989).
63. Y. Imoto, T. Ehara, and H. Matsuura, *Am. J. Physiol.* **252,** H325 (1987).
63a. N. K. Jurkiewicz and M. C. Sanguinetti, *Circ. Res.* **72,** 75 (1993).
64. E. Carmeliet, *J. Pharmacol. Exp. Ther.* **232,** 817 (1985).
65. M. C. Sanguinetti and N. K. Jurkiewicz, *Am. J. Physiol.* **259,** H1881 (1990).
66. M. C. Sanguinetti and N. K. Jurkiewicz, *Pfluegers Arch.* **420,** 180 (1992).
67. K. Yazawa and M. Kameyama, *J. Physiol.* (*London*) **421,** 135 (1990).
68. K. B. Walsh and R. S. Kass, *Science* (*Washington, D.C.*) **242,** 67 (1988).
69. I. A. Dukes, L. Cleemann, and M. Morad, *J. Pharmacol. Exp. Ther.* **254,** 560 (1990).
70. J. R. Balser, P. B. Bennett, L. M. Hondeghem, and D. M. Roden, *Circ. Res.* **69,** 51971. (1991).
71. Y. Habuchi, H. Tanaka, T. Furukawa, Y. Tsujimura, H. Takahashi, and M. Yoshimura, *Am. J. Physiol.* **262,** H345 (1992).

72. Z. Fan and M. Hiraoka, *Am. J. Physiol.* **261,** C23 (1991).

73. D. Fedida, Y. Shimoni, and W. R. Giles, *J. Physiol. (London)* **423,** 257 (1990).

74. X.-L. Wang, E. Wettwer, G. Gross, and U. Ravens, *J. Pharmacol. Exp. Ther.* **259,** 783 (1991).

75. D. Escande, A. Coulombe, J.-F. Faivre, E. Deroubaix, and E. Coraboeuf, *Am. J. Physiol.* **252,** H142 (1987).

75a. D. Escande, D. Thuringer, S. Le Guern, J. Courteix, M. Laville, and I. Cavero, *Pfluegers Arch.* **414,** 669 (1989).

76. K. Nakayama, Z. Fan, F. Marumo, and M. Hiraoka, *Circ. Res.* **67,** 1124 (1990).

77. M. Takano and A. Noma, *Naunyn-Schmiedeberg's Arch. Pharmacol.* **342,** 592 (1990).

78. D. Thuringer and D. Escande, *Mol. Pharmacol.* **36,** 897 (1989).

79. J.-F. Faivre and I. Findlay, *Biochim. Biophys. Acta* **1029,** 167 (1990).

80. T. Notsu, I. Tanaka, M. Takano, and A. Noma, *J. Pharmacol. Exp. Ther.* **260,** 702 (1992).

81. D. Kim, *Pfluegers Arch.* **418,** 338 (1991).

82. D. L. Lewis and D. E. Clapham, *Pfluegers Arch.* **414,** 492 (1989).

83. Y. Kurachi, T. Nakajima, and T. Sugimoto, *Pfluegers Arch.* **407,** 264 (1986).

84. D. Kim, *Circ. Res.* **69,** 250 (1991).

85. Y. Kurachi, T. Nakajima, and T. Sugimoto, *Nauyn-Schmiedeberg's Arch. Pharmacol.* **335,** 216 (1987).

86. H.-N. Luk and E. Carmeliet, *Pfluegers Arch.* **416,** 766 (1990).

87. Q.-Y. Liu and H. C. Strauss, *Biophys. J.* **60,** 198 (1991).

88. C. H. Follmer, N. J. Lodge, C. A. Cullinan, and T. J. Colatsky, *Am. J. Physiol.* **262,** C75 (1992).

89. M. C. Sanguinetti and N. K. Jurkiewicz, *Am. J. Physiol.* **260,** H393 (1991).

90. Z. Fan and M. Hiraoka, *Am. J. Physiol.* **261,** C23 (1991).

91. M. Apkon and J. M. Nerbonne, *J. Gen. Physiol.* **97,** 973 (1991).

92. E. Carmeliet, *J. Pharmacol. Exp. Ther.* **262,** 809 (1992).

[12]   Airway Smooth Muscle Ion Channels

Jerry M. Farley

Introduction

Airway smooth muscle plays an important role in the control of airway diameter, particularly in disease states such as asthma. The importance of the muscle in asthma has meant that considerable effort has been expended in development of drugs that relax the muscle and in understanding the transduction processes that mediate stimulus-induced muscle contraction and relaxation. One aspect of the transduction process, the membrane flux of ions through ion channels, has been studied for many years. Until recently, the primary electrophysiological methods for the study of ion channel function in smooth muscle were indirect, for example, through the measurement of contraction or membrane potential utilizing microelectrodes or sucrose gap. Radioisotope flux experiments permitted a more direct measurement ion flow through the membrane, but were cumbersome to perform and only able to measure slow events (typically seconds). The recent developments of the patch-clamp technique and fluorescent ion-sensing dyes have permitted examination of ion channels and changes in intracellular ion concentration in airway and other smooth muscles with much better time resolution (milliseconds). A short review below describes several ion channels that have been found in airway smooth muscle, the mechanisms by which their activity is controlled, and their importance in the function of the muscle. A discussion of methods used to measure membrane properties or ion content follows.

General Properties

Airway smooth muscle is electrically quiescent and action potentials are not initiated after depolarization of the tissue (1–7). Slow waves are observed in some airway smooth muscle most notably from guinea pig (8, 9). The voltage–current relationship obtained using either microelectrodes or sucrose gap shows that the membrane acts like a rectifier (1, 4, 5). Depolarization of the tissue by a constant current pulse results in smaller changes in membrane voltage compared with hyperpolarizing currents than would be expected for an ohmic resistor.

Methods in Neurosciences, Volume 19

The application of contractile agonists causes the muscle to depolarize in a concentration-dependent manner. For example, both acetylcholine (2, 7) and histamine (3, 6) depolarize airway smooth muscle. The depolarization is not accompanied by action potentials, although slow wave activity can be induced, particularly after histamine (6). The cell is depolarized, but the depolarization is small, 20 to 25 mV (3, 7, 10). The agonist-induced depolarization has been suggested to arise from activation of nonspecific cation channels (11, 12), the inhibition of potassium channels (13–15), the activation of a chloride channel (12), and the activation of voltage-gated calcium channels (2, 15–18).

Potassium Channels

A significant role for potassium channels in the normal function of airway smooth muscle has been proposed (19). Blockade of potassium channels with tetraethylammonium (TEA) causes the tissue to contract rhythmically (20–23). Regenerative action potentials also occur in the presence of TEA and the rectification of the voltage–current relationship is reduced (4, 6, 21). The spontaneous contractions/action potentials that occur in the presence of TEA are blocked by calcium-channel antagonists (e.g., verapamil) or removal of extracellular calcium (6, 20, 21), suggesting that the spontaneous action potentials and contractions result from the opening of voltage-gated calcium channels. Thus, it is clear that airway smooth muscle cells have a substantial permeability to potassium ions as well as having voltage-gated calcium channels. The large potassium conductance acts as a current shunt such that the activation of voltage-gated calcium channels does not lead to a regenerative action potential.

The large-conductance calcium-activated potassium channel (K_{Ca}) has a conductance in symmetrical intracellular and extracellular potassium concentrations of ~220–260 pS (13, 24, 25). The density of K_{Ca} is high in each cell making it difficult to study single-channel properties (13, 24). Recently airway smooth muscle was used as a source for isolation and purification of K_{Ca} channels with their subsequent introduction into lipid bilayers (26). The probability of channel opening is increased by elevation of intracellular calcium ion concentration and depolarization (25). The channel is sensitive to inhibition by TEA (13, 24, 27), but is insensitive to inhibition by 4-aminopyridine, apamin, and glybenclamide (14, 27). K_{Ca} are transiently activated by the spontaneous release of calcium from the SR. The currents in the whole cell that are generated by this activation have been termed spontaneous transient outward currents, STOC (13, 25) [or I_{oo} (28)]. Acetylcholine acting at muscarinic receptors can both increase the activity of K_{Ca} and inhibit the activity of the channels. The former occurs through the release of calcium from the SR (25, 27) and the latter through a G protein-mediated process at the level of the channel (29). These channels are also activated by β adrenoceptor activation through phosphorylation by a cAMP-dependent protein kinase (30). Therefore these channels are probably important in relaxation of the muscle during β adrenoceptor activation and are also involved normally in moderating the activity of the muscle.

A voltage-dependent potassium permeability, the delayed rectifier (27, 31), is also present in airway smooth muscle. It is sensitive to inhibition by 4-aminopyridine, is much less sensitive to inhibition by TEA, and is not blocked by glybenclamide. The relative insensitivity of the delayed rectifier channel to TEA, when compared with that of K_{Ca}, suggests that the rectification of smooth muscle membrane may arise from the presence of a delayed rectifier and is not due to K_{Ca} since high concentrations of TEA are required to cause spontaneous activity in the muscle (27, 31, 32).

Calcium Channels

Calcium influx through the cell membrane occurs through two types of channels: voltage gated and receptor operated. For a general discussion of these channels see reviews by Kitamura *et al.* (33), Coburn and Baron (10), and Bolton (34). Voltage-gated calcium channels are activated by depolarization and are of two types, L-type and T-type. L-type calcium channels are inhibited by dihydropyridine and other calcium-channel blockers and do not quickly inactivate. T-type calcium channels are insensitive to organic calcium-channel antagonists and inactivate rapidly during a depolarizing voltage step. Both have been found in airway smooth muscle (17, 35). Receptor-operated channels (ROC) on the other hand are not considered to be voltage gated, although they may be voltage sensitive, but are activated by agonists acting at receptors (11). They are insensitive to blockade by organic calcium-channel antagonists (11, 16). Receptor-operated channels are not generally considered to be selective for calcium and are usually viewed as nonselective cation channels (12) although they may not be permeant to barium (11). Receptor-operated channels are important particularly in the sustained phase of agonist-induced contractions (16).

Conclusion

Most of the information we have concerning the activity of ion channels has come through organ bath measurements of tension, electrophysiological techniques (microelectrode, patch clamp, sucrose gap), radioisotope flux measurements, and the use of ion-sensitive fluorescent dyes. The methods associated with microelectrode, patch-clamp, and fluorescence measurements as they relate to tracheal smooth muscle are now discussed.

Methods

Recent developments of ion-sensitive fluorescent dyes and the patch-clamp technique have greatly expanded our knowledge of how airway and other smooth muscles work. We first discuss the use of the fluorescent calcium dye Fura-2 {1-[2-(5-carboxyoxazol-2-yl)-6-aminobenzofuran-5-oxy]-2-(2'-amino-5'-methylphenoxy)ethane-N,N,N',N'-tetraacetic acid} to measure calcium in airway smooth muscle (36, 37) and then methods associated with the study of the electrophysiological properties of the muscle (2, 13, 25, 38).

General Dissection

The methods are described for use of porcine tracheal airway smooth muscle from young, 6- to 8-week-old animals. The trachea is quickly removed and transferred into normal Ringer solution (pH 7.4) containing (mM) NaCl, 113; KCl, 4.8; KH_2PO_4, 1.2; $MgSO_4$, 1.2; $CaCl_2$, 2.5; HEPES, 10; and glucose, 5.5. Antibiotics are included in the transport and storage solution to sterilize the tissue: penicillin, 100 units/ml, and streptomycin, 100 μg/ml. For all of the procedures described, the tracheal smooth muscle is cleaned of all extraneous tissue. The dissection is performed by cutting through the anterior aspect of the trachea, i.e., through the cartilage rings. The trachea is then opened and the epithelial tissue is removed by pulling it off starting at the left proximal end. Care should be taken when pulling the muscle from over the muscle. Unlike canine trachea where the muscle lies under the cartilage rings, in pig the muscle lies just under the epithelial layer luminal to the cartilage rings. The trachea can then be pinned out in a sterile dissection dish (sterilized with 70% ethanol or by autoclaving) with the luminal side down. The bottom of the dissection dish is coated with Sylgard (Dow 182 silicone) to permit the tissue to be pinned out. The connective tissue from the exterior surface of the posterior of the trachea is removed using a pair of iridectomy scissors to cut away connective tissue. The cartilage of the rings is cut as close to the insertion of the muscle as possible. Once cleaned of all connective tissue, overlying cartilage, and any gland cell acini, the trachea is turned over lumen-side up and pinned out. The surface of the tracheal muscle is then cleaned of all mucus glands. The illumination used should be low-angle or dark-field illumination since it is easier to see the glands. The glands are removed by picking each off with a pair of forceps such as Dumont Medical (5/45 or curved end) fine forceps. Once cleaned the muscle from these young animals should be translucent. The muscle is less than 200-μm thick in young animals (36). For fluorescence and microelectrode studies, rings are cut from the tissue with a small (~1 mm length) piece of cartilage left attached at each insertion. A silk suture (8 O) can be tied to the cartilage at one or both ends. For dissociation the muscle sheet is cut free of the cartilage.

Fluorescence Methodology

Tsien *et al.* have developed a series of calcium-sensitive fluorescent dyes which have been employed to measure intracellular calcium concentration [see (39)]. Fura-2 has a high fluorescence yield and is sensitive to calcium but has a relatively low affinity for calcium. Thus Fura-2 senses changes in calcium with little buffering effect on calcium. When Fura-2 binds with calcium, the fluorescence of the excitation spectrum at 340 nm will increase and that at 380 nm will decrease. The wavelength shifts permit calcium concentrations to be calculated from the ratio of amplitudes at a pair of excitation wavelengths and in principle cancel out variations in dye loading (39).

Fura-2 Loading

In order to load Fura-2 into muscle strips, the strips are exposed to Fura-2-acetoxymethyl ester (Fura-2/AM), a relatively apolar molecule that is permeable to the plasma membrane. It is trapped inside the cells after deesterification by cytosolic esterases to Fura-2.

A concentrated stock solution (2 mM) of Fura-2/AM in dimethyl sulfoxide (DMSO, Sigma, in sealed vials) is used to make the solutions for loading. The stock solution is stored at -20 to $-80°C$ in single-experiment aliquots. The muscle is loaded with Fura-2/AM by incubation of each strip in 2 ml normal Ringer solution containing 2–8 μM Fura-2/AM for 2–3 hr at 37°C in the dark and room air. The low concentrations of DMSO ($<1\%$) used during the loading procedure have no effect on the muscle. We have found that stretching the muscle slightly by pinning it to a block of plastic facilitates loading. The reason for this is not known. Muscles are loaded in cuvettes to minimize the volume of loading solution. After loading, the smooth muscle strip is mounted vertically on a mounting block (described below and shown in Fig. 1) and placed in a cuvette. The cuvette is continuously perfused with Krebs bicarbonate solution which is maintained at 35–36°C and bubbled with

95% O_2 and 5% CO_2. The solution consists of (mM) NaCl, 113; KCl, 4.8; KH_2PO_4, 1.2; $MgSO_4$, 1.2; $CaCl_2$, 2.5; $NAHCO_3$, 18; and glucose, 5.5. The tissue is washed continuously for 30–50 min prior to the experiment to wash out Fura-2/AM and DMSO and to permit complete deesterification of intracellular Fura-2/AM. Loading of the muscles is staggered in time such that when one experiment is finished the next muscle is coming out of the loading procedure and is ready to be mounted.

One end of the muscle must be fixed within the cuvette to measure tension. To do this a special mounting block was constructed in our machine shop and is shown in Fig. 1. The base, shaped like a square cap, fis tightly over the cuvette holder in the spectrofluorometer. The cuvette can be placed into the holder through the center hole and protrudes slightly above the base. The base forms a platform to which two inflow and one suction lines are attached as well as the muscle holder. The muscle holder is mounted on a diagonal to the cuvette at one corner to permit positioning of the muscle within the cuvette to obtain maximal fluorescence. The vertical portion of the muscle holder inserts ino the cuvette to near its bottom. The muscle is pinned to a small piece of rubber glued to the back at the bottom of the muscle holder. A silk suture, previously tied to the cartilage at the free end of the muscle, is threaded through a small hole (0.060 in.) drilled in the cuvette chamber door directly above the center of the cuvette. A small wire with a loop on one end is used to pull the suture through the hole in the door as it is being closed. The muscle can then be attached to the transducer by the silk suture. The isometric force transducer (Harvard Bioscience although others could be used) is attached to a rack and pinion gear mounted on the door such that the transducer lever is centered over the hole. The tension on the muscle can be set during the 30-min wash to 3 g (38) using the rack and pinion gear. Initially the muscle is stretched to a tension of 6–8 g to stretch the elastic elements in the muscle. Usually the muscle will relax to the baseline tension of 3 g during the equilibration period with little extra adjustment needed.

The cuvette is perfused continuously throughout the experiment with a Krebs bicarbonate solution to which drugs and other compounds are added. If it is important to rapidly exchange the solution, the cuvette is emptied by suction through one inflow line and then rapidly refilled by injection with a syringe through this same line. Rapid exchange is important to capture the fast changes that occur when agonists are applied to the muscle. Draining the cuvette for a short time does not alter the fluorescence signal to any extent. Small movements of the muscle do not alter the fluorescence signal either (36). If the rate of drug addition is not important then inflow solution need only be switched to a reservoir containing the drug.

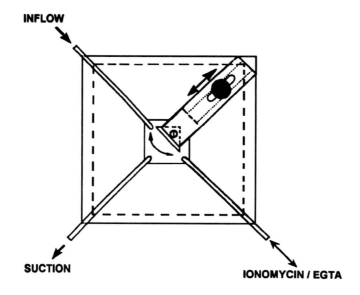

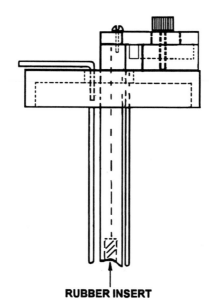

FIG. 1   This is a drawing of the muscle mounting block used to hold a tracheal muscle strip vertically in a cuvette in an SLM SPF 500C spectrofluorometer. Top and side views are shown. The block is described in the text. The block we have used is made out of aluminum that is painted with flat black enamel. On the bottom of the long piece which extends below the square base, an arrow indicates the location

As a practical matter we have found that exposure to an agonist such as acetylcholine can bring about alterations in the response of the muscle to subsequent exposure to acetylcholine for long periods of time, e.g., 30 min (36). Therefore, in many of our experiments the muscle is exposed to only one concentration of acetylcholine, unless of course the experimental design was to use or study the changes in muscle function caused by multiple application of agonist.

Measurements of Fluorescence and Isometric Tension

The fluorescence is monitored in a computer-controlled Aminco SPF-500C spectrofluorometer. The functions of the spectrofluorometer are controlled by a computer via an RS232 interface. The software to do this was kindly supplied by Dr. Marcy Petrini (Department of Medicine, UMC). The program takes emission data (510 nm) at any two specified excitation wavelengths (340 and 380 nm) by alternately moving the excitation monochromator to each wavelength and averaging the fluorescence emission intensity (five samples) at each wavelength for approximately 1 sec. The slew rate for this monochromator is approximately 50 nm/sec, so the minimum sample interval is a little less than 3 sec. The final sample interval is about 4 sec.

Calculations

Tension is measured using an isometric force transducer interfaced to the computer via a Metrabyte Dash-16 A/D converter and also displayed on a chart recorder. Tension is sampled immediately following the acquisition of the fluorescence at both excitation wavelengths. Tension is measured to a precision of approximately 1 mg. Tension should be normalized to the cross-sectional area of the muscle. This can be estimated according to the method of Murphy (40) and reported as Newtons/cm^2. The length of the muscle is measured while it is in the cuvette under basal tension. A standardized procedure should be developed to measure the weight of the muscle. In the case of airway smooth muscle we blot the muscle on a laboratory tissue (Kimwipe), three times on each side, to remove extracellular solution clinging to the tissue. The cartilage attached to each end of the muscle is carefully removed before weighing. The average cross-sectional area is then computed as muscle weight/(length $\times$ density). A density of 1.060 g/cm^3 is assumed for tracheal smooth muscle (40).

---

of a piece of rubber to which the muscle can be pinned. The arrows on the top view indicate the adjustable portions of the holder. The inflow and outflow tubing is made of 304 stainless.

The ratio of the fluorescence intensity (R) emitted at 510 nm obtained by alternately exciting at 340 and 380 nm at each time interval is used to estimate changes in intracellular free Ca^{2+} concentration. The autofluorescence of the tissue is subtracted from the fluorescence data. Cartilage fluoresces more than the signal from the muscle, thus it is critical to assure that the cartilage is not in the light path of the fluorometer. The fluorescence and absorption spectrum of all compounds to which the muscle is to be exposed should also be determined. Fluorescence of added compounds will alter the estimated ratios and cause errors in the calculated calcium concentration. A simple example is phenol red, a pH-sensitive dye included in most culture media. If the compound you add has a significant absorption at 340 or 380, the efficiency of excitation of Fura-2 will be altered and cause errors in the calculated calcium concentration.

At the end of each experiment the maximum ratio (R_{MAX}) and minimum ratio (R_{MIN}) for Fura-2 are determined by first permeabilizing the membrane using ionomycin (20 μM) in Krebs bicarbonate solution containing 2.5 mM Ca^{2+}, to saturate the dye, followed by EGTA solution (125 mM), to chelate all of the calcium in the muscle. Typically 5 min in each is sufficient to obtain a stable ratio. The intracellular calcium concentration for each time point is then calculated using the following formula: $[Ca^{2+}]_i = (R - R_{MIN})/(R_{MAX} - R) \times S_{f2}/S_{b2} \times K_d$. S_{f2} and S_{b2} are the intensities of fluorescence emission (510 nM) of Fura-2 obtained at an excitation wavelength of 380 nm of the minimum and maximum calcium concentrations, respectively. The K_d for Fura-2 is assumed to be 224 nM (39).

Advantages and Disadvantages

The advantages of this methodology are that you can estimate the intracellular calcium concentration at the same time you are measuring tension permitting correlation of these processes. Information can also be gained about transduction processes that are involved in the rise in intracellular calcium, their time course, and their relative importance. However, it should be remembered that these are tissue-level calcium measurements and will only approximate the actual cytosolic calcium concentrations. Non-smooth muscle cells will contribute to the signal as will cell organelles that may contain Fura-2 (e.g., the nucleus). Another practical disadvantage with these measurements is that they require a lot of time to make. The dissection and loading of the muscle take up to 6 hr, followed by an experimental protocol that may last 2–3 hr. Even with careful planning, complete experiments (five to six muscle strips) can take 12 hr. Finally, the use of this type of fluorometer is cumbersome. The position of the transducer outside the cuvette chamber makes positioning of the muscle in the cuvette difficult. However, the information that can be obtained warrants overcoming these difficulties.

Electrophysiological Measurements

Microelectrode Measurements

Measurements of membrane potential are important in the study of airway smooth muscle since tissue level measurements are made. There are several important requirements for successful microelectrode measurement of membrane potential in airway smooth muscle. The first requirement is patience. Tracheal smooth muscle cells are small (5–10 μm in diameter) and are tightly wrapped in multiple layers of connective tissue. Many attempts may be required to obtain stable recordings. Also needed are (i) a good micromanipulator, (ii) an electrode puller, (iii) a microelectrode amplifier, (iv) a muscle chamber, (v) a vibration-free surface, and (vi) data recording capability.

A stable micromanipulator is required. Kopf, Haer, or other hydraulic motorized micromanipulators are excellent since they allow precise rapid movements of the microelectrode. Manual hydraulic or mechanical manipulators (e.g., Huxley type) can also be used. The characteristics of all useful micromanipulators are that they do not drift and are capable of a very fine movement (less than 2 μm) and rapid acceleration of the pipette without backlash. We have used all of the types of micromanipulators listed above successfully (2, 37, 38).

A microelectrode puller capable of pulling high-resistance microelectrodes (70–100 MΩ when filled with 3 M KCl or citrate) is also needed. We have used an Industrial Science Associates Model M1 horizontal puller. However, numerous pullers are available which can pull these type of pipettes. The pipettes are held in a pipette holder. These are also commercially available (WPI New Haven, CT; Warner Instruments Hamden, CT; etc.) or they can be made out of a piece of acrylic plastic tube, one end of which is covered with a small sheet of rubber (e.g., like that found in capped injection vials glued with instant glue) through which a small 0.8-mm hole has been punched. A male gold pin that will fit in the headstage of a high-impedance amplifier (e.g., WPI Model M-707) to which a silver wire or sintered silver pellet has been soldered is glued into the other end (5-minute Epoxy works well). The solder joint must be completely insulated with the glue and should never come in contact with the solution filling the holder.

The muscle chamber can be very simple: a slot in a piece of acrylic, ~1.5" × 0.5" × 0.5" deep. Inflow and outflow lines, one at each end of the slot, can be placed in the chamber from the top or through the bottom. The slot is filled 0.25 deep with Sylgard (Dow 182), to which the muscle can be pinned. Since the muscle should be maintained at 37°C it is necessary to heat the chamber and the incoming solutions. We do this by mounting the chamber as the lid of a plastic box into which heated water is pumped from a circulating water bath.The inflow solutions can be gravity fed into the chamber and are preheated within the plastic box by circulation through coiled polyethylene tubing. The solutions should be gassed with either 100% oxygen or an oxygen/carbon dioxide mixture if a bicarbonate buffer is utilized. In order to hold the chamber so that it does not move, we strap the chamber onto a lead brick. We illuminate the chamber with a fiber optic wand positioned to provide low-angle or right-angle illumination of the muscle surface. This permits the tip of the electrode to be seen in the solution and the surface of the muscle to be visualized.

The muscle chamber and lead brick are placed on a metal surface that does not vibrate. A metal plate (1″- steel 30″ × 48″ is a good size) laid on a partially inflated 26″ bicycle inner tube can be used or if one is available a vibration isolation table (e.g., we use a Technical Manufacturing Corp., Micro-g) is best. We have used both successfully (2, 37, 38). The microscope, a simple stereomicroscope will do, and the micromanipulator are also mounted firmly on the metal plate. The plate should be grounded.

It is convenient to have a means of recording membrane potential in some permanent form. A computer or a simple strip chart recorded can be used. We have used a strip chart recorder, Gould Model 2200, to observe changes in potential as penetration of the muscle tissue occurs. An oscilloscope and/or a digital voltmeter are also useful to display the membrane potential.

The characteristics of the pipettes are important for successful impalement of airway smooth muscle cells. The microelectrodes should have resistances between 70 and 100 MΩ when filled with 3 M KCl or citrate. The tips should be approximately 1 cm long when pulled from 0.8- to 1-mm glass. The tip of the microelectrodes should be very sharp and the shank should have a shallow taper near the tip. We have used several methods to fill the pipettes. Initially (2) we used a method developed by Tasaki et al. (41). Kimax-51 melting point capillaries were pulled in batches of 30–50 and were mounted on glass microscope slides with rubber bands. The settings to pull electrodes should previously be determined before large numbers are pulled. The microelectrodes are then completely filled with filtered methanol (0.2 μm) under vacuum. Following this they are placed in filtered distilled water for approximately 4 hr to replace the methanol. The microelectrodes can also be stored in this solution. Approximately 4 hr before the microelectrodes are to be used they are put into filtered 3 M KCl. These electrodes have very good mechanical and electrical characteristics. However, the ease of using fiber-filled capillaries has largely supplanted these more cumbersome techniques

(37, 38). Glass fiber-filled capillaries (e.g., FMG 15, Dagan, and others) can be pulled and then filled directly with KCl either by backfilling through a 32-gauge needle inserted down near the tip or according to the manufacturer's suggestions.

To impale tracheal smooth muscle cells these steps should be followed.

1. The cleaned muscle is pinned lumen-side up in the chamber. Two short insect pins are inserted through the cartilage at each end of the smooth muscle strip. A small amount of tension, just enough to stretch the muscle slightly, is placed on the strip. This is important. Too much or too little tension will decrease the possibility of impaling a cell.

2. The microelectrode is lowered into the solution. The tip potential is balanced to zero and the resistance is checked. The tip is then lowered to just above the muscle surface. This is performed under $20\times$ to $40\times$ magnification.

3. The microelecrode is then moved down with fine adjustments until the tip potential deflects to positive voltages (typically 1–3 mV). At this point the tip is resting against the muscle surface.

4. The tip is then backed up 10–30 μm from the surface and rapidly accelerated into the muscle tissue, approximately 50 μm, while the recording of the potential is watched. Typically the following occurs. There is a positive deflection of a few millivolts followed by a rapid transient negative deflection of 50–60 mV with a return of the tip potential to near 0 mV. The transient negative deflection occurs because a cell is penetrated, but the electrode tip goes completely through the cell. The rapid acceleration can be obtained by using the highest steps/second setting on a motorized micromanipulator. If a mechanical manipulator is used, rapidly rotating the fine adjustment knob with a "flick of your wrist" will work.

5. The microelectrode is very slowly withdrawn from the tissue using the fine movement control. If the impalement is successful another sharp negative deflection will occur that will reach to a potential equal to or greater than the transient observed. The membrane potential measured should remain stable for 30 to 60 sec before it is accepted.

Stable recordings that last for many minutes can be obtained using this method. The microelectrodes must have very sharp tips with nearly parallel sides near the tip, since this minimizes the damage of completely penetrating the cell and permits the membrane to rapidly reseal as the microelectrode tip is pulled back into the cell. Using this method, measurement of membrane potential in quiescent muscle is reasonably easy. Application of chemicals which cause movement of the muscle generally dislodges the electrode tip from the cell. The techniques given here can be used to measure membrane potential after the muscle has reached a steady-state tension such as that induced by acetylcholine. Recording stable potentials from cells during contraction is more difficult, however.

Advantages and Disadvantages

The microelectrode technique permits measurement of a very important cell parameter, that is, membrane potential. This is the variable that controls the sensitivity of many conductances (e.g., voltage-gated calcium channels) to activation. However, the limitation of the method is that the membrane potential can be altered by many factors. If the conductance of the cell increases or decreases membrane potential may change; however, if a nonstoichiometric pump changes its activity membrane potential may also change. Therefore, it is important to not know only whether the conductance of the membrane changes. This can be done by coupling the microelectrode technique with a technique that permits the entire tissue to be hyperpolarized or depolarized. These techniques are described by (42). Another technique that may be used is the sucrose-gap technique (43).

Patch-Clamp Methodology

Cell Isolation

The use of the patch-clamp technique requires that the membrane of cells from which recordings are to be made be free of all connective tissue. In general this means that cells must be isolated from the tissue. It is important that the isolation procedures minimize damage to the cells. Isolated tracheal smooth muscle cells should be in an elongate, relaxed state and contract in response to stimuli. We basically try to remove everything that is not part of the cell so that the cells will just fall out of the connective tissue matrix. The muscle is first cleaned of epithelia, gland cells, adherent connective tissue, and cartilage as noted above. The following methods are adapted from other sources (25, 38, 44). To dissociate the cells the following protocol is performed.

1. The sheet of muscle is placed in 5 ml of dispersing solution in a 60-mm plastic petri dish. This solution is composed of 0.4–0.8 units/ml protease type XIV (Sigma), dissolved in a HEPES buffered solution (HBS) containing (mM); NaCl, 140; KCl, 5.5; CaCl$_2$, 0.9; glucose, 5.5; and Hepes, 10, pH = 7.4. This solution is sterilized by filtration through a 0.2 μm sterile syringe filter.

2. The petri dish is then placed in an incubator at 37°C on a rotary shaker and rotated at a slow speed (~10 rev/min) for 0.5 hr, until the tissue begins to loosen. The tissue is then removed with forceps and rinsed in HBS.

3. The tissue is pulled into strips with forceps and placed in a second enzyme solution containing: 1 mg/ml (~300 units/ml) Type I collagenase (Sigma), 0.2 mg/ml type IV elastase (Sigma); 0.1 mg/ml type I deoxyribonuclease (Sigma), 1500 units/ml hyalouronidase (type 5, Sigma); and 1 mg/ml bovine serum albumin in HBS. This is placed on the rotary shaker for 30–35 min at 37°C.

4. Remaining small tissue pieces and free cells of the dissociation solution are gently pipetted or poured into a sterile 50 ml tube through sterile nylon mesh (Falcon 2350, Cell Strainer). The cell suspension is diluted to 50 ml with HBS by pouring or pipetting the solution through the mesh.

5. The cells are removed from the dissociation medium by low speed centrifugation, 60 g for 15 min at 10°C. The remaining enzyme solution is carefully decanted since the cells form a very loose pellet at the bottom of the tube.

6. The cell pellet is resuspended in HBS (6–10 ml) containing 5 mM glucose without serum and plated onto acid cleaned sterile glass coverslips (25 mm^2) in 35-mm plastic dishes. The coverslips are cleaned previously by soaking in 1 N HCl overnight , rinsed with distilled water, allowed to dry, placed in petri dishes, and sterilized by uv irradiation within 12 hr of use.

We allow the plates to sit at room temperature in the laminar flow hood for ~0.5 hr. This permits the cells to attach to the coverslip before the dishes are flooded. The dishes are then flooded with Medium 199 containing 10% fetal calf serum (Hyclone), 18 mM bicarbonate, and penicillin/streptomycin. After being plated, the cells are placed in an incubator and maintained at a temperature of 37°C and in an atmosphere containing 5% CO_2. They can be used immediately and are always used within 48 hr of plating. After 48 hr the cells flatten and begin to divide. Other investigators do not put the cells in culture, but rather leave them in a physiological saline solution and store the cells at ~4–5°C until needed (17). The cells in this case are pipetted into a glass-bottomed recording chamber and allowed to lightly attach before use.

Patch Clamp

The patch clamp originally developed by Neher and Sakmann (45, 46) permits the current flow through single ion channels to be measured. The method electrically isolates a small patch of cell membrane from the cell and bath by pressing a small-tipped (approx. 1 μm), fire-polished, glass pipette against the membrane and then applying gentle suction to form a seal with very high electrical resistance. The glass–membrane seal effectively isolates the cell membrane patch from the rest of the bath so that the currents flowing through this patch can be accurately measured. Because the noise levels are exceptionally low, it is possible to measure currents of less than 1 pA (10^{-12} amperes) in size. This resolution is sufficient to measure single-channel open and closed times of the channel as well as the conductance of the channel. A variation of this technique called whole-cell recording permits the control of both the intra- and extracellular environment and the recording of currents from single small cells. A book (47) and several reviews (48–53) of this topic are available. This review considers only those methods we have used for airway smooth muscle.

To form a patch with tracheal smooth muscle we use pipettes that are short tapered, <6 mm total, with a bullet-shaped tip. The tip opening appears to be ~1 μm. The pipettes are fabricated on a vertical puller (Kopf) with a standard tungsten coil. We use melting point capillaries made of Kimax-51 glass, 0.8–1 mm external diameter. This is a borosilicate glass. The tubing is inexpensive and has been found to have adequate electrical properties. Since it is a borosilicate glass it does not leach aluminum (as do the alumina silicates) or lead as some of the other glasses do. A two-stage pull is used to form the tip. The solenoid setting is zero. The first pull is 5–6 mm at a current of ~20 A. The glass is recentered and pulled again at a heat setting of ~14.5 A. Small changes in the second setting are made to form the exact type of tip desired. The pipettes can be used immediately without fire polishing if desired. We make 20–30 pipettes at once, however, and store them in standard pipette storage jars. The stored pipettes work well but must be fire polished before use. Also, we coat the tips with Q-dope (GC Electronics), a polystyrene plastic, although we have used clear nail polish as well. The tips are coated by pushing the tip into the side of a small brush loaded with Q-dope with only the very tip of the pipette sticking out below the brush. The pipette is then rotated. The pipettes are placed tip down in a pipette storage jar and and the coating is allowed to dry for several hours. If the orifice of the pipette accidentally becomes coated, the pipette can still be used since the coating will be vaporized during fire polishing. The coating increases the dielectric of the pipette tip and makes the tip hydrophobic. The latter characteristic keeps solution from wetting the outside of the pipette; both reduce the capacitance of the pipette and subsequent noise of the recording. Fire polishing is performed under a microscope (a simple Bausch and Lomb student model) at 900×. The tip is brought near (three to five tip diameters) a small (100 μm) glass-coated platinum wire heated electrically. The tip is observed as the wire is heated (dull red heat) and is withdrawn as soon as a small change in the refraction of light occurs at the tip (1–3 sec). The tip of the pipette is filled by applying suction to the back end of the pipette with the tip inserted into the solution. The solution should be filtered with a 0.2-μm filter (e.g., syringe filters are good for this). Extracellular solution is used for attached patch or inside-out recording and intracellular solution for whole-cell or outside-out recording. We fill the tip quickly into the shank with solution then backfill with filtered solution 1–2 mm above the tip. We have made a needle for backfilling out of fused silica tubing (used for capillary electrophoresis) glued with 5-min Epoxy into a luer fitting taken from a disposable needle. The fused silica is durable and small, and will not interact with compounds used in intracellular solutions (particularly EGTA) as does stainless steel. Minimize getting solution on the interior of the pipette above the tip since this will increase the capacitance of the pipette. The pipette is then cut to length with a diamond scribe and inserted into the pipette holder. Borosilicate and harder glasses can be broken in this way without too much difficulty. However, when using softer glasses, Corning 8161 for example, you should be as gentle as possible in cutting and breaking the glass. The tips shatter easily. After connecting the pipette holder to the patch-clamp headstage the pipette can be lowered into the chamber. The chamber solution should be kept as low as possible, 0.1–0.3 mm above the cells again to reduce capacitance and as stable as possible to eliminate changes in tip capacitance during the experiment.

Whole-Cell Recording

The following method is used to record from a whole cell.

1. As the tip is lowered through the air–water interface, a positive pressure should be applied to the pipette to keep debris on the solution surface from fouling the tip. Once the tip is through the interface the positive pressure may be released. The resistance of the tip should be measured and the capacitance of the pipette balanced while a small 0.1-mV pulse is applied through the pipette. The resistances of the tips we use range from 3 to 5 MΩ when filled with normal ionic strength solutions.

2. Initially an attached patch is formed by pressing the pipette against the cell. The resistance should increase 2–4$\times$. Gentle suction (by mouth) is then applied to the pipette. The suction is slowly increased until the resistance begins to increase (10–20$\times$).

3. The suction should be released since often a seal will form. The resistance may also drop and the suction will have to be reapplied and often increased. Typically a "gigaseal" will form rapidly once the resistance has increased by 20–30$\times$. Final seal resistances range from 10 to 30 GΩ for tracheal smooth muscle.

4. After the seal is formed, the patch of membrane inside the pipette is ruptured by applying slightly greater suction than needed to form a seal and

simultaneously applying 1 to 3 short (1 msec) but large voltage pulses (1 V) to the membrane. This usually results in the rupturing of the membrane within the pipette lumen, a decrease in resistance, and a prolongation and increase in size of the capacity transients.

5. The access and cell resistance should be checked periodically since the membrane can reseal over the tip during the experiment. The resting potential of tracheal smooth muscle cells measured in the current clamp mode when the pipette contains normal internal K^+ ($\sim$140 mM) is around -50 to -60 mV (25), a value which compares favorably with the -60 mV value obtained using the microelectrode technique (2).

The intracellular medium rapidly exchanges with the pipette solution such that the internal ionic composition is easily controlled and drugs, second messengers, etc. can be applied intracellularly. However, soluble metabolic components that may be important for the activity of the channels to be studied can be lost or diluted. One method which has been used to minimize the loss of intracellular components is the perforated patch technique (54–56). In this method nystatin, a pore-forming antibiotic, is introduced into the membrane patch within the lumen of the pipette. The formation of nonselective ionic pores permits a low electrical resistance pathway to the cell interior to form, but larger soluble intracellular components cannot leave the cell. This technique has recently been used to record from tracheal smooth muscle cells (12).

The choice of intracellular and extracellular solutions will be dictated by the experimental design. However, the solutions should be osmotically balanced, or the intracellular solution should be slightly hypotonic to the extracellular solution. This will keep the cell from swelling after rupture of the membrane. A calcium buffer should be used in the intracellular medium to reduce the calcium concentration. However, the choice of calcium-buffering capacity should be carefully considered in the design of solutions. For example, in our experiments with minimal calcium buffering (0.1 to 1 mM EGTA) in the intracellular solution, STOC, due to the activation of many calcium-activated potassium channels, are observed that can be several hundred picoamperes in amplitude at 0 mV. If the EGTA concentration is increased to 4–10 mM these currents are reduced in size and can be abolished (25). The inhibition occurs presumably due to the rapid chelation of calcium released from the sarcoplasmic reticulum. Obviously, if you are interested in studying the STOC or another current dependent on increases in intracellular calcium the choice of an appropriate EGTA concentration is important. The addition of metabolically active substances may also be important. For example, ATP may alter the activity of channels in airway (57), thus its inclusion or deletion will affect the types of channels that are observed. We usually use a relatively normal ionic medium that is buffered with HEPES for most of our experiments (13, 25). This solution is simple and has the following composition (mM): 1–20 NaCl, 140 KCl, 10 HEPES, 0.2–10 EGTA, 4 MgCl$_2$, and 3 ATP. The variable NaCl concentrations offset changes in osmolality caused by changes in EGTA concentration. The pH of the solution is 7.1.

Advantages and Disadvantages

The great advantage of the patch-clamp technique is that it permits measurements of conductances in smooth muscle cells. This has permitted numerous conductances to be identified and their properties as well as the mechanisms by which they are controlled to be determined. This information has tremendously advanced our understanding of how smooth muscle works. Several things should be considered in the interpretation of patch-clamp data. The data obtained from patch clamp are of necessity from cells isolated from their normal cell-to-cell connections. This will alter their function to some degree. Enzymatic digestion of the tissue may also alter the properties of the cells. Recording from the whole cell will change the intracellular environment and lead to changes in cell physiology. In addition, even though a conductance exists in the membrane of airway smooth muscle cells it is important to show that it is physiologically relevant in processes (e.g., contraction) that normally occur *in situ*. Even considering all of the possible problems that are caused by the use of the patch-clamp technique, the benefits far outweigh the difficulties that may occur.

Acknowledgments

This work was supported in part by NIDA 05094 and the Mississippi Lung Association. Martha Wood provided expert technical assistance. Joe Ed Smith provided the machining skills and built the muscle and pipette holders described.

References

1. C. T. Kirkpatrick, *J. Physiol.* (*London*) **244**, 263 (1975).
2. J. M. Farley and P. R. Miles, *J. Pharmacol. Exp. Ther.* **201**, 199 (1977).
3. R. F. Coburn and T. Yamaguchi, *J. Pharmacol. Exp. Ther.* **201**, 276 (1977).
4. H. Suzuki, K. Morita, and H. Kuriyama, *Jpn. J. Physiol* **26**, 303 (1976).
5. A. R. Cameron and C. T. Kirkpatrick, *J. Physiol.* (*London*) **270**, 733 (1977).
6. C. T. Kirkpatrick, *in* "Smooth Muscle: An Assessment of Current Knowledge"

(E. Bulbring, A. F. Brading, A. W. Jones, and T. Tomita, Eds.), p. 385. University of Texas Press, Austin, TX, 1981.

7. R. F. Coburn, *Am. J. Physiol.* **236,** C177 (1979).
8. R. C. Small, *Br. J. Pharmacol.* **77,** 45 (1982).
9. F. Ahmed, R. W. Foster, R. C. Small, and A. H. Weston, *Br. J. Pharmacol.* **83,** 227 (1984).
10. R. F. Coburn and C. B. Baron, *Am. J. Physiol. Lung Cell. Mol. Physiol.* **258,** L119 (1990).
11. R. K. Murray and M. I. Kotlikoff, *J. Physiol. (London)* **435,** 123 (1991).
12. L. J. Janssen and S. M. Sims. *J. Physiol. (London)* **453,** 197 (1992).
13. H.-M. H. Saunders and J. M. Farley, *J. Pharmacol. Exp. Ther.* **260,** 1038 (1992).
14. H. Kume and M. I. Kotlikoff, *Am. J. Physiol. Cell Physiol.* **261,** C1204 (1991).
15. M. I. Kotlikoff, H. Kume, and M. Tomasic, *Biochem. Pharmacol.* **43,** 5 (1992).
16. J. M. Farley and P. R. Miles, *J. Pharmacol. Exp. Ther.* **207,** 340 (1978).
17. M. Kotlikoff, *Am. J. Physiol.* **254,** C793 (1988).
18. M. Tomasic, J. P. Boyle, J. F. Worley III, and M. I. Kotlikoff, *Am. J. Physiol. Cell. Physiol.* **263,** C106 (1992).
19. J. Black and P. J. Barnes, *Thorax* **45,** 213 (1990).
20. E. A. Kroeger and N. L. Stephens, *Am. J. Physiol.* **228,** 633 (1975).
21. R. W. Foster, R. C. Small, and A. H. Weston, *Br. J. Pharmacol.* **79,** 255 (1983).
22. Y. Ito and T. Itoh, *Br. J. Pharmacol.* **83,** 667 (1984).
23. S. L. Allen, J. Cortijo, R. W. Foster, G. P. Morgan, R. C. Small, *et al., Br. J. Pharmacol.* **88,** 473 (1986).
24. J. D. McCann and M. J. Welsh, *J. Physiol. (London)* **372,** 113 (1986).
25. H.-M. H. Saunders and J. M. Farley, *J. Pharmacol. Exp. Ther.* **257,** 1114 (1991).
26. D. Savaria, C. Lanoue, A. Cadieux, and E. Rousseau, *Am. J. Physiol. Lung Cell. Mol. Physiol.* **262,** L327 (1992).
27. M. I. Kotlikoff, *Am. J. Physiol. Lung Cell. Mol. Physiol.* **259,** L384 (1990).
28. T. Hisada, Y. Kurachi, and Sugimoto, *Pfluegers Arch.* **416,** 151 (1990).
29. M. Yamaya, W. E. Finkbeiner and J. H. Widdicobme, *Am. J. Physiol. Lung Cell. Mol Physiol* **261,** L491 (1991).
30. H. Kume, A. Takai, H. Tokuno, and T. Tomita, *Nature (London)* **341,** 152 (1989).
31. J. P. Boyle, M. Tomasic, and M. I. Kotlikoff, *J. Physiol. (London)* **447,** *329* (1992).
32. K. Okabe, K. Kitamura, and H. Kuriyama, *Pfluegers Arch.* **409,** 561 (1987).
33. K. Kitamura, N. Teramoto, M. Oike, Z. Xiong, S. Kajioka, *et al., Adv. Exp. Med. Biol.* **304,** 209 (1991).
34. T. B. Bolton, *Physiol. Rev.* **59,** 606 (1979).
35. J. F. Worley, III, and M. I. Kotlikoff, *Am. J. Physiol. Lung Cell. Mol. Physiol.* **259,** L468 (1990).
36. C. C. Shieh, M. F. Petrini, T. M. Dwyer, and J. M. Farley, *J. Pharmacol. Exp. Ther. Exp.Ther.* **256,** 1 (1991).
37. C. C. Shieh, M. F. Petrini, T. M. Dwyer, and J. M. Farley, *J. Pharmacol. Exp. Ther.* **260,** 261 (1992).
38. P. M. Mohan, C. M. Yang, H. M. Saunders, T. M. Dwyer, and J. M. Farley, *J. Auton. Pharmacol* **8,** 93 (1988).

39. G. Grynkiewicz, M. Poenie, and R. Y. Tsien, *J. Biol. Chem.* **260,** 3440 (1985).
40. R. A. Murphy, *in* "Handbook of Physiology: The Cardiovascular System." Vol. 2. D. F. "Vascular Smooth Muscle" Bohr, A. P. Somylo, and H. V. Jr. Sparks, eds.), p. 325. *Am. Physiol. Soc., Bethesda, Maryland, 1980.*
41. I. Tasaki,E. H. Polley, and F. Orrego, *J. Neurophysiol.* **17,** 454 (1954).
42. Y. Abe and T. Totila, *J. Physiol (London)* **196,** 87 (1968).
43. T. B. Bolton, *J. Physiol. (London)* **250,** 175 (1975).
44. R. Mitra and M. Morad, *Am. J. Physiol.* **249,** H1056 (1985).
45. E. Neher and B. Sakmann, *Nature (London)* **260,** 799 (1976).
46. Deleted in press.
47. "Single-Channel Recording." Plenum Press, New York 1983.
48. E. Neher and B. Sakmann, *Sci. Am.* **266,** 44 (1992).
49. J. M. Farley, *in* "Neuromethods." Vol. 9. "The Neuronal Environment" (A. A. Boulton, G. B. Baker, and W. Walz, eds.), p. 363. Humana Press, Clifton, NJ, 1988.
50. T. M. Dwyer and J. M. Farley, *in* "Physical Methods in the Study of Cellular Biophysics" (M. Dino, ed.), p. 233. A. R. Liss, New York, 1983.
51. O. P. Hamill, A. Marty, E. Neher, B. Sakmann, and F. J. Sigworth, *Pfluegers. Arch.* **391,** 85 (1981).
52. M. Cahalan and E. Neher, *Methods Enzymol.* **207,** 3 (1992).
53. E. Neher, *Neuroscience* **26,** 727 (1988).
54. R. Horn and A. Marty, *J. Gen. Physiol.* **92,** 145 (1988).
55. R. Horn and S. J. Korn, Methods Enzymol. **207,** 149 (1992).
56. S. Sala, R. V. Parsey, A. S. Cohen and D. R. Matteson, *J. Membr. Biol.* **122,** 177 (1991).
57. K. Groschner, S. D. Silberberg, C. H. Gelband, and Van Breemen, *Pfluegers Arch.* **417,**517 (1991).

Section IV

Molecular Biology and Electrophysiology

[13] Potassium Channels in Mammalian Brain: A Molecular Approach

John A. Drewe, Hali A. Hartmann, and Glenn E. Kirsch

Introduction

Within the past 5 years more than 20 cDNA gene products that encode different voltage-gated K^+ channels have been cloned from brain cDNA libraries. The electrophysiological characteristics of the cloned channels have been determined by *in vitro* expression in *Xenopus* oocytes. The channels share delayed rectifier features including sigmoidal onset of activation on depolarization, steep voltage dependence of activation gating, and strong K^+ selectivity over Na^+, but show considerable variability in gating kinetics and sensitivity to blockers. Much progress has been made in identifying structural domains that specify these functional characteristics. This review describes some of the molecular techniques that we have used in the structure–function analysis of rat brain K^+ channels (1–5).

Deletion Mutations

As shown in Fig. 1 the coding region of Kv2.1, a typical representative of rat brain K^+ channels, contains six transmembrane domains (S1–S6) flanked by long C- and N-termini. We have used two techniques to make deletions in the amino and carboxy ends that help define the minimal core protein necessary for a functional channel. The first method uses restriction endonuclease sites, and the second technique uses exonucleases to achieve nested deletions.

Amino Terminus Deletions Using Restriction Enzyme Sites

Kv2.1 contains in-frame methionines at positions 17, 102, and 140 in the primary sequence. In accordance with the "scanning model" of eukaryotic translation initiation, we assume that protein synthesis initiates at the most upstream AUG. Hence, deletions can be made by removing upstream AUG to shift the initiation point. As shown in Fig. 1, upstream segments of Kv2.1 are removed by cuts in the 5′ polylinker of pBluescript and at a restriction

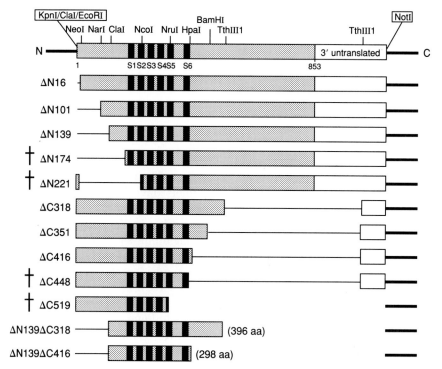

FIG. 1 Deletion mutations of Kv2.1 at the amino and carboxy terminus. The delayed rectifier K^+-channel Kv2.1 is shown as an insert in the phagemid vector pBluescript SK (−). Relative positions of restriction enzyme sites in the coding region (shaded) are indicated; restriction sites in the polylinker of pBluescript are boxed. Six putative transmembrane α-helical segments (S1–S6) in the central core region are shown as vertical black bars. Deletions are indicated by thin horizontal lines. The Δ notation on the left-hand side indicates the number of amino acid residues deleted from the N- and C-termini. The crosses indicate deletions that did not express K^+ channels in oocytes.

site within the coding region of Kv2.1. In the coding region *NcoI*, *NarI*, *ClaI*, and the second *NcoI* sites are used to create ΔN16, ΔN101, ΔN139, and ΔN221 deletions, respectively (the Δ notation indicates the number of residues deleted from either the N- or C-termini). The cDNA is then blunt ended and religated.

Clones lacking multiple methionines in the amino terminus can still be cut in the 5′ polylinker of the plasmid and at available restriction enzyme sites within the coding region of the cDNA. Synthetic linker pairs of oligonucleotides containing an in-frame AUG can then be ligated upstream of the point of restriction cut.

Carboxy Terminus Deletions Using Restriction Enzyme Sites

The C-terminus of the cDNA region to be transcribed requires both a stop codon at the end of the coding region and, to increase stability of the message, a poly(A) tail. Kv2.1 contains a TthIII1 restriction endonuclease site in the 3′ untranslated region of the cDNA, which is followed within eight amino acids by termination codons in each of the three reading frames. A stretch of DNA with stop codons in all three reading frames will be rare in most cDNA molecules since there are only three stop codons of 64 possible combinations of three nucleotides. In most cases, therefore, a synthetic oligonucleotide linker containing stop codons in all three reading frames can be ligated between the restriction cut in the 3′ untranslated region and the cut in the carboxy coding region of the cDNA.

Kv2.1 contains only two naturally occurring restriction sites in the carboxy end of the coding region, TthIII1 and a second BamHI, providing deletions ΔC318 and ΔC351, respectively (Fig. 1). For ΔC318, Kv2.1 is cut with TthIII1, blunt ended, and religated. For ΔC351 Kv2.1 is cut with BamH1 and TthIII1, blunt ended, and religated. Also, a silent restriction enzyme site for HpaI is introduced at the 3′ end of the S6 transmembrane region for later use in chimera construction. For ΔC448, Kv2.1 is cut with HpaI and TthIII1, blunt ended, and religated.

Construction of Deletions without Restriction Enzyme Sites

Deletions in both the amino and carboxy termini, beginning just upstream of the S1 and just downstream of the S6 transmembrane regions, are made using exonuclease III followed by treatment with S1 nuclease and blunt ending with T4 DNA polymerase. The rate of exonuclease III action is slowed by the addition of 150 mM NaCl such that two new deletions are obtained, ΔN174 and ΔC416, in the amino and carboxy termini, respectively.

A preferred alternative method would be to use polymerase chain reaction (PCR) techniques. Synthetic oligonucleotide PCR primers for the amino terminus could be designed to incorporate one of the restriction sites present in the 5′ polylinker of the plasmid followed by an initiating methionine plus the sequence of the cDNA where the deletion is to begin. For the carboxy terminus antisense primers would be designed to incorporate a restriction site from the 3′ untranslated region of the cDNA followed by a stop codon plus the sequence of the carboxy terminus where the deletion is desired. Due to the possibility of spontaneous mutations during the PCR reaction, the region spanning the PCR reaction product would have to be completely sequenced in the final construct.

Combining Amino and Carboxyl Termini Deletions

Two double-mutant constructs, ΔN139-ΔC318 and ΔN139-ΔC416, are constructed by combining the longest deletion mutations at the amino and carboxy terminus that give functional channels (Fig. 1). The procedure makes a ΔN139 mutation with a *Cla*I digest in the N-terminus of the ΔC318 and ΔC416 deletion constructs. The smallest channel construct that gives a measurable K$^+$ current in oocytes was found to be the ΔN139–ΔC416 construct (1, 6). Thus a core region spanning 33 amino acids upstream of the S1 transmembrane region and 30 amino acids downstream of the S6 is essential for function. The following sections describe procedures used in analysis of the core region of the protein.

Construction of a Chimeric Potassium Channel

Structure-function analysis based on point mutations has provided useful information. An alternative approach, however, is to make large-scale cassette mutations that swap functional domains between different channels. This method has the advantage of simplifying later mutagenesis in the region of interest. We have used this approach to identify a region (P-region) which forms part of the lining of the aqueous pore in K$^+$ channels (3). The P-region is in the link between transmembrane segments S5 and S6. Figure 2 shows the alignments of the amino acids of Kv2.1 (host) and Kv3.1 (donor) in the P-region. The segment marked by a cross-hatched bar was transplanted from Kv3.1 to Kv2.1. The following section describes construction of the chimeric cDNA.

Identification and Creation of Recognition Sequences

The first consideration in chimera construction is the location of restriction sites in the cDNA flanking the region to be cut out. In the absence of naturally occurring unique restriction sites, "silent" sites are engineered based on degeneracy of the amino acid code. The nucleotide sequence of the channel is compared with the nucleotide sequences recognized by various restriction enzymes to determine which base changes would allow recognition by restriction enzymes without altering the encoded amino acid sequence (i.e., silent sites). The required base changes are then made by site-directed mutagenesis.

Silent *Bsp*mI and *Stu*I sites are engineered into Kv2.1 as indicated in Fig. 3. The *Bsp*mI recognition sequence, ACCTGC, requires two base changes,

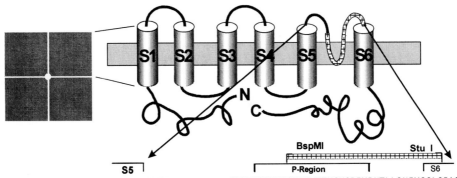

```
                                                              BspMI                    Stu I
                        S5                         P-Region                    S6
Kv2.1:  FFAEKDEDDTK.........FKSIPASFWWATITMTTVGYGDIYPKTLLGKIVGGLCIAG
Kv3.1:  YY--RIGAQPNDPSASEHTH--N--IG----VV----L----M--Q-WS-ML--A--ALT-
```

FIG. 2 Chimeric transplant of the P-region. A popular model of the folding pattern of transmembrane segments S1–S6 of a single K^+-channel subunit embedded in a membrane is shown at the top of the figure. A top view of the aggregation of four subunits around a central pore to form a complete channel is shown on the left. The P-region lies between S5 and S6. A sequence alignment of the S5–S6 linkers in Kv2.1 and Kv3.1 is shown in the lower part of the figure. The dashes represent residues identical to Kv2.1. The dots represent interruptions that maintain the alignment with Kv3.1. The cross-hatched bar shows the fragment transplanted from Kv3.1 to Kv2.1. Silent *Bsp*MI and *Stu*I sites were engineered into Kv2.1 for the construction of the chimera as described in the text.

(C → A and C → T) without changing the coded amino acids isoleucine and proline. Changes of G → A and G → C are necessary for creating the *Stu*I recognition sequence (AGGCCT) while maintaining the code for two glycines. These mutations are made using an oligo-directed mutagenesis system (7). A fragment containing the bases to be changed is subcloned into M13 phage for mutagenesis and a single-strand template DNA is synthesized by host TG1 cells. Complementary oligonucleotides of approximately 25 bases with the appropriate base changes for the mutations are synthesized such that the mismatched bases reside near the center of the strand. When mixed with template DNA the bases at the ends of the oligo pair up exactly with the template to allow low-stringency annealing of the oligonucleotide to the template. After isolation of the mutated clones and sequencing to verify that the correct bases have been mutated, the smallest segment containing the mutated area is selected for religation into the parent cDNA. Although the recovery of small fragments of DNA (<600 bp) is lower than the recovery of larger fragments, less of the smaller-sized fragments is required for ligation back to the parent vector. Also, smaller inserts reduce the amount of double-

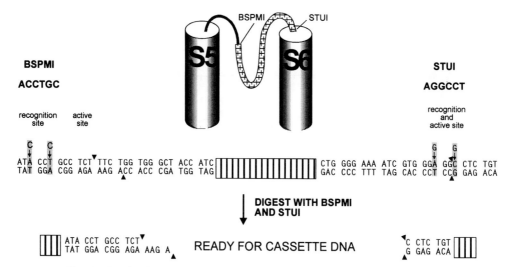

FIG. 3   Design for cassette DNA swapping in Kv2.1. The double-stranded nucleotide sequence of Kv2.1 after mutagenesis of the four bases (shaded boxes) needed for the *Bsp*mI and *Stu*I silent restriction sites is shown. Inverted triangles (▼) show enzyme cleavage points. The Kv2.1 nucleotide sequence between the *Bsp*mI to *Stu*I sites is represented by a striped box. The nucleotide sequence at the bottom of the figure shows the cassette insertion region of Kv2.1 after restriction enzyme digestion. The rest of Kv2.1 is represented by striped boxes.

stranded DNA sequencing necessary to confirm that the correct insert has been ligated into the parent cDNA. Thus, in Kv2.1 the *Bsp*mI and *Stu*I sites are between two *Bam*H1 sites and the *Bam*H1 fragment is 650 base pairs. For a 2 : 1 molar ratio of insert to vector DNA, 20 ng of a 650-bp fragment is needed for ligation into 100 ng of a 6.1-kb parent vector. This requires 8 μg of the mutated insert DNA to be digested by the appropriate restriction enzymes in a 300-μl volume reaction. The DNA is precipitated with 0.1 volume of 3 M sodium acetate and 2 volumes of 100% ethanol at $-20°C$. The fragment is gel-purified in a high-quality preparative agarose gel, cut out, and electroluted from the agarose by dialysis with 10 mM Tris–EDTA. A total of 10 μg of the Kv2.1 parent cDNA also is cut with *Bam*H1. After precipitation of the linear 6.1-kb fragment, the ends of the fragment are dephosphorylated to prevent recircularizing in the presence of T4 DNA ligase and thereby increase the amount of linear parent fragment (or vector) available for ligation with the insert. The ligated product is amplified in an appropriate host cell by standard techniques (8). In addition to sequencing the entire ligated fragment, it is digested with the flanking restriction enzymes to confirm that the ligation restored the correct bases.

Design of Donor DNA Cassette

The donor DNA cassette is prepared using PCR methods (9). PCR primers are designed such that the final cassette product contains *Bsp*mI and *Stu*I sites which, after enzyme digestion, provide compatible ends for ligation into the host Kv2.1 DNA. Figure 4 shows a schematic diagram of the forward and reverse PCR primers used to prime Kv3.1 DNA. Both *Bsp*mI and *Stu*I recognition sites are included in the forward and reverse primer, respectively. *Bsp*mI recognizes the sequence ACCTGCC; it cuts the DNA four bases down on the sense strand and eight bases down on the antisense strand to create a 3' recessive end (only the sense strand is shown in Fig. 4). This type of restriction enzyme provides great flexibility. The bases of the recognition sequence will be cleaved from the final cassette product thereby providing a site in the primer for incorporation of bases mismatched with the template. At the cleavage site four and eight bases downstream, however, the donor DNA can be primed from the template DNA without any other base changes, unless the point of ligation in the final cassette DNA would produce a change in the amino acid or a frameshift.

Unlike the recognition sequence for *Bsp*mI, the *Stu*I recognition and active site sequence are the same and provide a blunt end to the cassette DNA. Five to six additional bases of Kv3.1 DNA are then included in the primer directly preceding the restriction enzyme recognition sites. In Fig. 4, the bases are ATAT (preceding the *Bsp*mI recognition sequence) and GGAC (preceeding the *Stu*I recognition site). These are included in the primer so that the PCR product will contain extra bases before the recognition sequence to stabilize the restriction enzyme as it digests the PCR product to produce the cassette DNA. This combination of endonuclease sites, one sticky ended (or 3' recessive) and the other blunt ended, provides directional cloning of the cassette DNA (Fig. 4) back into the parent cDNA (Fig. 3), as the fragment can only insert itself into the parent DNA in one direction.

The total length of the primers are approximately 30 to 40 bases. Obviously, too many mismatched bases between the primers and the template DNA may prevent annealing in the PCR reaction and contribute to false priming (primers annealing to an incorrect segment of the template DNA). To facilitate the most specific annealing conditions and decrease the likelihood of false priming, the template DNA should be the smallest fragment possible which includes the desired template sequence. For example, the full-length cDNA of Kv3.1 is 4.3 kb; but, when used for template in the PCR, it is first

FLOW DIAGRAM FOR PREPARATION OF CASSETTE DNA

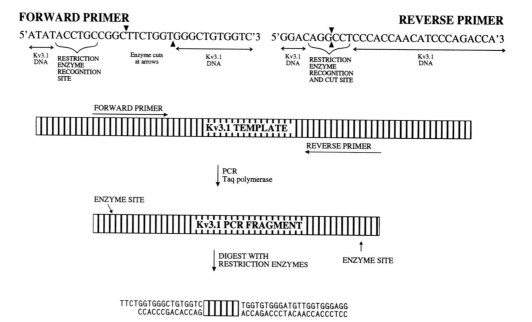

FORWARD PRIMER   **REVERSE PRIMER**

5'ATATACCTGCCGGCTTCTGGTGGGCTGTGGTC'3   5'GGACAGGCCTCCCACCAACATCCCAGACCA'3

TTCTGGTGGGCTGTGGTC   TGGTGTGGGATGTTGGTGGGAGG
CCACCCGACACCAG   ACCAGACCCTACAACCACCCTCC

"Kv3.1 CASSETTE DNA"

FIG. 4   Flow diagram for PCR amplification of Kv3.1 cassette. DNA primers were designed with the recognition sequences of *Bsp*mI and *Stu*I. The forward primer consists of codons for the sense strand of Kv3.1 DNA that will anneal to the antisense strand of the template DNA and, after the addition of Taq polymerase, will amplify the sense strand. The base changes (mismatches) from Kv3.1 DNA needed to create *Bsp*mI's recognition sequence are included in the recognition sequence part of the forward primer and will be amplified in the PCR fragment. The reverse primer consists of codons in the 5' → 3' direction of the antisense strand that will anneal to Kv3.1 sense strand and amplify the antisense strand of Kv3.1 DNA. The sequence of the primer includes the base changes for the *Stu*I recognition sequence. Note that the PCR fragment is only the length of the distance between the two primers. After digestion with *Bsp*mI and *Stu*I, the cassette DNA has compatible ends for ligation into the *Bsp*mI–*Stu*I cut Kv2.1.

digested with the enzyme *Nci*I to produce a 1.8-kb fragment which includes the segment to be amplified by PCR.

One characteristic of Taq polymerase is its persistent ability to end PCR products with an adenine base, even though the PCR products are extended.

We have found that our PCR products can be digested well with the respective enzymes if the PCR fragments are subjected to T4 DNA polymerase prior to enzyme digestion to ensure blunt ends.

The PCR products are prepared for ligation as described above for small DNA inserts. The ligated segment is sequenced and a restriction digest made. Confirmation of the ligated product by sequencing is essential to avoid the propagation of spontaneous, unintended mutations in constructs that serve as the base for later work.

In Vitro Transcription of RNA

The K^+-channel cDNA clones are propagated in either pSP72 (Promega) or pBluescript SK(−) (Strategene) plasmid vectors. These clones have different sequences corresponding to the three common phage RNA polymerase recognition sites, T7, T3, and SP6. In vitro translation is optimized for a high-quality yield of cRNA using either T7 or T3 RNA polymerase. The following steps are performed:

1. Linearization of the DNA template at the 3′ end of K^+-channel cDNA.
2. Removal of RNase contamination from the DNA template.
3. RNA transcription.
4. Quantification of RNA yield by radioactive tracer incorporation.
5. Qualification by denaturing agarose gel.

Linearization of cDNA Template

Plasmid DNA used as a template for RNA transcription is isolated by both a standard alkaline lysis and polyethylene glycol (PEG) purification, or more recently using the Maxiscript system (Promega). The orientation of the clone with respect to the RNA polymerase sites is determined by restriction analysis and sequencing. The RNA polymerase species corresponding to the region 5′ to the initiating methionine of the K^+ channel cDNA is used for translation. RNA polymerase, after attaching to the promoter site, can transcribe RNA in the 5′ to 3′ direction until it falls off either due to a termination sequence or, as in our case, a physical cut in the DNA template at a unique restriction site on the 3′ end of the K^+-channel clone. The RNA polymerase on reaching the cut will fall off and, if the transcription has gone to completion, all of the RNA molecules will be of the same length. The restriction enzyme used for linearization must leave either a 5′ overhang or blunt end because T3 and T7 RNA polymerases also will bind to 3' overhangs and begin synthesis of the antisense cRNA.

Removal of RNase Contamination from DNA Template

RNase A is used for isolating plasmid DNA because, unlike other types of RNase, it can be completely inhibited to avoid RNA degradation during transcription. Since RNase A also is present in human skin precautions against contamination must be observed in all steps after removal of the RNase by proteolysis. RNase is very stable and can only be inactivated by chemical reagents or by autoclaving. All solutions used in the handling of RNA are autoclaved and all glassware is baked. The following procedures are used to remove RNase from the plasmid DNA.

1. All restriction endonuclease digests used in the linearization of the cDNA template are made in H buffer (100 mM NaCl, 50 mM Tris, pH 7.5, 10 mM MgCl$_2$). If another buffer is used for restriction digest then the ionic concentrations should be modified to that of H buffer before continuing.

2. The DNA is treated with proteinase K under partially denaturing conditions: the restriction digest is incubated 1–3 hr at 37°C with 0.4% SDS plus proteinase K at 100 μg/ml to partially denature the proteins and inactivate contaminating RNase. From this point RNase-free precautions must be maintained (including use of latex gloves, separate containers from chemicals used in DNA work, and autoclaved solutions).

3. The proteins are removed by a sequence of phenol (preequilibrated to saturation with Tris, U.S. Biochemicals) extractions: (a) 1 volume phenol, (b) 1 volume phenol : chloroform : isoamyl alcohol (25 : 24 : 1 ratio), and (c) 1 volume chloroform : isoamyl alcohol (24 : 1 ratio)

4. DNA is precipitated with 0.1 volumes 3 M sodium acetate and 2.5 volumes of ethanol.

RNA Transcription

T7 and T3 RNA polymerase initiate RNA transcription at specific promoters and incorporate free ribonucleotides on to the RNA strand until it falls off of the 3' end of the K$^+$ channel where the cDNA had been previously linearized.

RNA transcribed *in vivo* is modified by the addition of a poly(A) tail and methylation of the 5' guanine (capping). Although a poly(A) tail is not essential in the *Xenopus* oocyte it may enhance RNA stability. Thus, for *in vitro* RNA transcription the poly(A) tail should be included in the cDNA at the

3′ end of the K^+-channel clone. To cap the 5′ end of the RNA, a methylated guanine derivative (m7 G(5′)ppp(5′)G) (Pharmacia or GIBCO BRL) is added in a 4 : 1 excess over GTP during the transcription. Capping is essential for high levels of channel expression in the oocyte. The use of the nonmethylated form of the cap analog (Pharmacia) leads to functional RNA molecules but the reliability of expression with the methylated form is higher. RNAsin (Promega) is included in the incubation mixture to specifically inactivate residual RNase A.

1. To a 25–150 μl reaction volume is added cDNA template, 0.2 μg/μl; ATP, 400 μM; CTP, 400 μM; UTP, 400 μM with 0.002–0.008 μCi per μl [^{32}P]UTP (Amersham); GTP, 240 μM; methylated GTP (cap analog), 1000 μM, RNAsin, 2 units/μl; either T7 or T3 RNA polymerase (Stratagene), 2 units/μl in 1x Stratagene buffer (40 mM Tris–Cl, pH 8.0, 8 mM MgCl$_2$, 50 mM NaCl, 2 mM Spermidine, and 30 mM dithiothreitol). The reaction must be mixed at room temperature, not at 4°C, or the spermidine will precipitate.

2. Incubation is for 40 min at 37°C. Longer incubation times do not result in a greater yield of RNA.

3. Two microliters are then removed for quantification of the RNA (see below).

4. Distilled water is added to the remaining 23 to 148 μl of reaction mixture to reach a final volume of 200 μl.

5. The sample is extracted twice with 200 μl of Tris-equilibrated phenol :- chloroform : isoamyl alcohol (25 : 24 : 1 ratio) and then once with chloroform : isoamyl alcohol (24 : 1 ratio)

6. The RNA is precipitated twice in 200 μl of ammonium acetate and 800 μl ethanol. (Free nucleotides do not precipitate with ammonium salts and 90% are removed with the first precipitation and 99% with two precipitations.)

Quantification of RNA

To quantify the RNA, the 2-μl sample is precipitated with 6% TCA and filtered. The RNA concentration can be calculated knowing the original specific activity of the UTP. The usual yield is 40–120 μg/ml. The sample is diluted to a stock solution of 250 μM with 100 mM KCl and aliquoted for injection.

Qualification by Denaturing Agarose Gel Electrophoresis

Since full-length cRNA is essential for channel expression, the following procedures are used to determine the quality of the transcription product. Denaturing agarose electrophoresis is used to evaluate the length of the transcribed RNA. A total of 125 ng of RNA is treated with 1 M glyoxal for 1 hr at 45°C and electrophoresed in a phosphate-buffered agarose gel. The RNA in the agarose gel is fixed with 7% TCA and the gel can be dried with a home blow dryer. The radioactive RNA is visualized by overnight exposure to X-ray film. A single sharp band corresponding to the full length of the clone should be observed. Longer products result from transcription of uncut template, but their presence will not affect the translation efficiency of the RNA. However, shorter products may not give functional channels. Two likely causes of shortened product are contamination by RNase or premature termination due to either secondary structure of the DNA or the presence of a cryptic termination site. Smearing of the RNA band indicates RNase degradation and a new DNA template should be made. A sharp band of less than full length indicates premature termination. The amount of functional RNA in the sample can be estimated by the relative intensity of the full-length band.

Oocyte Expression and Electrophysiology

Ionic currents through voltage-gated K^+ channels are usually recorded under voltage-clamp conditions where the membrane potential and external ionic concentration are controlled. Step changes in membrane potential activate the voltage-sensitive channels and the electrochemical driving force controls the net flow of current through open channels. The conductance of the voltage-clamped membrane varies with both time and voltage. We have used the voltage-clamped *Xenopus* oocyte to determine the biophysical properties of expressed K^+ channels. The oocyte is normally a nonexcitable cell but during development unfertilized oocytes readily synthesize protein from foreign mRNA introduced into the cytoplasm. In our work rat brain K^+-channel expression was readily detectable within 24 hr after microinjection of cRNA that encodes the K^+-channel protein. Functional K^+ channels in the oocyte plasmalemma were detected by an intracellular microelectrode voltage-clamp recording method. Compared with the level of expression of brain K^+ currents, endogenous channels present in the oocytes contribute negligible current, hence this system is advantageous for studying a homogenous population of K^+ channels.

Oocyte Isolation

In our earlier work oocyte injection (1–6) was performed by manual dissection of oocytes from the ovarian lobes and microinjection through the intact follicular cell layer using relatively small-caliber (10 μm opening) beveled tip micropipettes. We have changed this procedure after testing expression in two groups of oocytes injected with the same amount of Kv2.1 cRNA using our original procedure and an enzyme treatment described below, in which large-caliber (18 μm) unbeveled micropipettes were used to microinject defolliculated oocytes. The enzyme procedure gave three-to ninefold improvement in expression. A possible explanation for the difference is that the rough edges of the beveled micropipette may have damaged the cRNA during injection. We have standardized the following procedures.

Female *Xenopus laevis* (Nasco) are kept in tanks of circulating spring water at room temperature (20–23°C) and fed weekly with chopped beef liver. Stage V and VI oocytes are obtained 1 day prior to injection by the following procedures. Under general anesthesia (MS-222, 0.075%) several ovarian lobes are removed from the frog and placed in a filtered (0.2 μm) sterile Ca^{2+}-free solution (OR-2 (in mM) NaCl, 82.5; KCl, 2.5; $MgCl_2$, 1; HEPES 5; pH 7.4) at room temperature. The lobe is cut into pieces consisting of 20–30 oocytes/piece. At this point the oocytes are present in tightly bound clusters that must be treated enzymatically to digest the connective tissue and remove the follicular cell layer. A 1-ml aliquot of the oocyte cluster suspension is placed in a sterile 15-ml tube containing a solution of 24 mg of collagenase (GIBCO) in 12 ml OR-2. The tube is placed on a low-speed shaker (Vortex Genie 2 at 200 rpm) for 40 min. At the end of this period dissociation of oocytes from the cluster can be observed by gently shaking the tube. The oocytes are allowed to settle to the bottom of the tube and the supernatant is drawn off and replaced gradually with Ca^{2+}-containing solution (SOS, in mM, NaCl, 100; KCl, 2; $CaCl_2$, 1.8; $MgCl_2$, 1; HEPES, 5; sodium pyruvate, 2.5; and gentamycin, 50 mg/liter; pH 7.4). Defolliculated, healthy oocytes are selected under stereomicroscopic inspection, placed in fresh SOS solution, and again examined after 1 hr. Undamaged oocytes are then transferred to an 18°C incubator until injection.

Oocyte RNA Injection

Injection pipettes are prepared from capillary glass (Drummond 3-00-203-GT/X) on a micropipette puller (Sutter Instruments, P-87). The tips are broken under microscopic inspection to a tip opening diameter of 18 μm. Prior to injection frozen aliquots of cRNA are thawed and diluted to the desired concentration (generally 1–200 pg/nl) in sterile 100 mM KCl solution. The oocytes are injected with 46 nl of cRNA solution using an automatic injector system (Drummond, Nanoject) mounted on a micromanipulator (Marzhauser, MM-3) under stereomicroscopic observation.

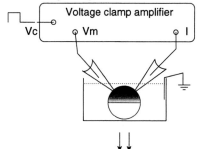

Oocyte voltage clamp

A. Recording configuration

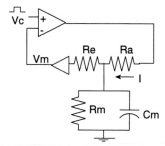

B. Electrical equivalent circuit

FIG. 5   Two-microelectrode oocyte voltage clamp. (A) Recording configuration. An oocyte impaled by two micropipettes. V_c, V_m, and I are, respectively, command potential, membrane potential, and current. (B) A simplified electrical equivalent circuit in which the voltage-clamp circuit is represented by two operational amplifiers; one compares V_c and V_m in a feedback configuration and the other measures V_m in a voltage-follower configuration. The voltage-measuring and current-injecting pipettes are represented by resistors R_e and R_a, respectively. The oocyte membrane is represented by the parallel resistive and capacitive elements R_m and C_m, respectively.

Whole-Cell Voltage Clamp

The principle of the whole-cell voltage clamp is shown in Fig. 5. An oocyte immersed in bathing solution [typically Na-Mes (in mM) NaOH, 100; methanesulfonic acid, 100; CaCl$_2$, 2; HEPES, 10; pH 7.3] is impaled with two micropipettes. One pipette was connected to a high input impedance voltage-measuring amplifier and the other pipette is connected to the output of the voltage-clamp feedback amplifier to inject current into the cell. Measured membrane potential and a command potential, usually in the form of a step voltage pulse, are compared at the differential input of the feedback amplifier

Two microelectrode clamp response time

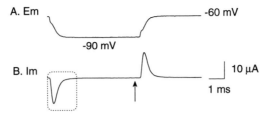

FIG. 6 Response time of two-microelectrode oocyte voltage clamp. An uninjected oocyte bathed in Na-Mes, standard solution (see text), is voltage clamped using low-resistance micropipettes (0.3 MΩ, 3 M KCl filled). The membrane potential (A) and current (B) responses to a 5.5 msec step pulse (not shown) to -90 from holding potential -60 mV are shown. The boxed region of the current trace was integrated to determine capacitive charge movement (Fig. 7). Steady-state ionic current was measured at the point indicated by the arrow.

and the difference is amplified. The output voltage of this amplifier drives sufficient current through the current injection pipette to nullify the difference between command signal and the membrane potential. The fidelity with which the membrane potential follows the command potential is limited by the gain of the feedback amplifier and its maximum output current. If the gain is too low the error will be significant. Too high a gain drives the feedback amplifier into oscillation because of phase differences between the frequency response of the voltage measuring system and the feedback amplifier. The tendency to oscillation can be overcome in several ways. First, phase lag in the voltage-measuring system can be minimized by reducing the product of the electrode resistance times stray capacitance. A stray capacitance of about 1 pF times an electrode resistance of 10 MΩ gives a time constant of 10 μs. The usual solution in microelectrode work is to compensate the stray capacitance by boosting the high-frequency components of the signal. This is probably unnecessary in the oocyte system since large microelectrode tips with resistance <1 MΩ are readily tolerated and should therefore be used to reduce the time constant of the recording system. Second, the capacitive coupling between voltage and current electrodes should be reduced through shielding. In oocyte recording we find that an insulated, grounded metal shield placed just above the bath surface between the two micropipettes provides sufficient decoupling to significantly reduce oscillation. Last, the bandwidth and phase characteristics of the feedback amplifier should be adjustable. Figure 6 shows the typical response time in an uninjected voltage-clamped oocyte. The rise time of the membrane potential in response to a

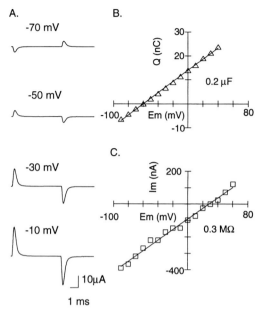

Passive electrical properties

FIG. 7   Passive electrical properties of uninjected oocyte. (A) Membrane current traces to the indicated test potentials from a holding potential of −60 mV. (B) The linear relationship between capacitative charge movement, obtained by integration of the area of the ON transient of the current trace (see Fig. 6), and test potential. The x-intercept at −60 mV was the holding potential. From the slope we obtain a value of 0.2 μF for the cell membrane capacitance. The time constant of the decay phase of the capacitative transient current divided by the membrane capacitance yields a maximum series resistance of 1.2 kΩ. (C) The steady-state relationship between ionic current and test potential. The x-intercept is the reversal potential for the leakage current that flows mainly through endogenous Cl$^-$ channels and reverses near the Nernst potential for Cl$^-$ under our standard test conditions (low chloride solution, methanesulfonate substitution). The slope gives 0.3 MΩ cell membrane resistance.

30-mV hyperpolarizing step is about 0.5 msec and the capacitative transient subsides in about 1 msec. This response time sets the limit of resolution of fast-gating transitions, especially the deactivation of tail currents at strongly negative test potentials and activation rise time at strongly positive test potentials.

In addition to feedback gain, the fidelity of the oocyte clamp is limited by the maximum output of the feedback amplifier. This is an especially critical problem in oocytes (average diameter, 1.2 mm) where the combined load of a large membrane capacitance and low input resistance can easily exceed the ± 10 V compliance of ordinary amplifiers. At high-output (± 100 V) amplifier together with low-resistance (<0.5 MΩ tip resistance) current injection pipette are required. Figure 7 illustrates the typical passive properties of an uninjected oocyte. A series of test pulses from -90 to $+60$ mV in 10-mV increments from a holding potential of -60 mV evoked passive electrical responses from the membrane of an uninjected oocyte. Membrane currents are shown in Fig. 7A. Note that the capacitative spikes can easily exceed 20 μA and would therefore require at least a 20-V output through and access resistance of 1 MΩ. Several commercial voltage-clamp amplifiers with 100 V output are now available.

In practice, during whole-cell recording oocytes are first impaled with two microelectrodes (filled with 3M KCl, 10 mM HEPES, pH 7.4). The pipette resistance is reduced by breaking back the tip to an opening diameter of about 12 μm. This is done in the recording chamber by gently rubbing the tip against a clean surface such as the tip of fine dissecting forceps. Impalement proceeds by first positioning the micropipette at the surface of the oocyte under stereomicroscopic observation and advancing it axially with the micromanipulator until visible dimpling of the surface appears. A gentle tap on the micromanipulator with a pencil causes penetration that can be detected visually by a slight rebounding of the oocyte surface at the impalement site. Successful impalement also can be determined electrically when the potential jumps from zero (bath reference ground) to the membrane resting potential (-20 to -70 mV). At the end of the experiment the micropipettes are withdrawn and can often be reused. Reuse is facilitated by attaching micrometer syringes through a length of plastic tubing to the side port of the micropipette holder. Pressure applied through the syringe clears cytoplasm that often plugs the micropipette tip. Alternatively, the tip can be filled with 2% agar in 3 M KCl and the pipette backfilled with 3 M KCl solution.

Acknowledgments

We acknowledge the support of Drs. A. M. Brown and R. H. Joho, in whose laboratories these methods were developed. This work was supported by National Institutes of Health Grants NS29473 (G. E. K.) and NS08895 (J. A. D.); Grants NS23877 and HI42267 to A. M. Brown and NS28407 to R. H. Joho; and the American Heart Association (Texas Affiliate) (H. A. H.).

References

1. A. M. J. VanDongen, G. C. Frech, J. A. Drewe, A. M. Brown, and R. H. Joho, *Neuron* **5,** 433 (1990).
2. M. Taglialatela, A. M. J. Vandongen, J. A. Drewe, R. H. Joho, A. M. Brown, and G. E. Kirsch, *Mol. Pharmacol.* **40,** 299 (1991).
3. H. A. Hartmann, G. E. Kirsch, J. A. Drewe, M. Taglialatela, R. H. Joho, and A. M. Brown, *Science* **251,** 942 (1991).
4. G. E. Kirsch, J. A. Drewe, H. A. Hartmann, M. Taglialatela, M. DeBiasi, A. M. Brown, and R. H. Joho, *Neuron* **8,** 499 (1992).
5. G. E. Kirsch, J. A. Drewe, M. Taglialatela, R. H. Joho, M. DeBiasi, H. A. Hartmann, and A. M. Brown, *Biophys. J.* **62,** 136 (1992).
6. J. A. Drewe, M. Taglialatela, G. E. Kirsch, H. A. Hartmann, A. M. Brown, and R. H. Joho, *Biophys. J.* **59,** 198a (1991).
7. J. W. Taylor, J. Ott, and F. Eckstein, *Nucleic Acids Res.* **13,** 8764 (1985).
8. J. Sambrook, E. F. Fritsch, and T. Maniatis, "Molecular Cloning: A Laboratory Manual" 2nd Ed. Cold Spring Harbor Laboratory, Cold Spring Harbor, NY, 1989.
9. H. A. Erlich, "PCR Technology: Principles and Applications for DNA Amplification." Stockton Press, New York, 1989.

[14] Methods for Expression of Excitability Proteins in *Xenopus* Oocytes

Michael W. Quick and Henry A. Lester

Introduction

The study of membrane proteins of excitable cells has been significantly enhanced by heterologous expression experiments (29). One of the most popular expression systems for these studies is the oocyte from the African frog *Xenopus laevis*. The oocyte has been widely used because (i) it is easily manipulated, (ii) it is capable of successfully translating exogenous RNA and DNA messages into membrane proteins, and (iii) it is amenable to a number of electrophysiological and biochemical techniques. This was first shown by Gurdon and colleagues 20 years ago for globin RNA (20) and since then has been important in the study of many components of excitable cells including voltage-gated ion channels, ligand-gated ion channels, and membrane transporters. To cite only two recent examples of its usefulness, the oocyte has been used extensively in the functional expression of a large number of cDNA clones and in elucidating structure–function relationships through assays of chimeric and mutant proteins. This chapter presents some of our laboratory's methods for studying these ion transport components and processes in oocytes. An important drawback of the oocyte system is the amount of variability in the quality of the oocytes. Therefore, we first detail our current procedures for minimizing variability by optimizing the maintenance of frogs and oocytes. Next, we discuss several different methods used in the lab for delivery of exogenous messages into oocytes. Finally, we present some of the methods we use to examine expression of excitability proteins in oocytes. For related techniques, we suggest a number of excellent papers (2, 6, 8, 26, 27, 39, 40).

Care and Maintenance of Frogs

We purchase sexually mature female, oocyte-positive *X. laevis* from Xenopus One (Ann Arbor, MI, Cat No. 4216), although they are available from a number of other sources. Frogs may be ordered that have been injected with human chorionic gonadotropin (HCG) to ensure that the ovaries are functioning properly and to induce development of new oocytes. However,

Methods in Neurosciences, Volume 19

we currently use frogs that are not HCG injected. Groups of frogs are housed in large fiberglass tanks (3 × 1 m) that have been filled to a height of approximately 20 cm with tap water. The tap water is filtered ("Super-changer," Tranter, Inc., Wichita Falls, TX) and the temperature is maintained between 18 and 20°C. The room light/dark cycle is currently 11/13 hr, although we have also used 12/12 hr and 13/11 hr. The tanks are cleaned and the water is completely changed three times per week. The frogs are fed once per week on a diet of either chopped beef heart or liver (approximately 10 g/frog/week). Commercially available frog food pellets are commonly used as well, although we have had better oocyte success using fresh beef.

Surgical Procedures

The frog is placed in a plastic container and anesthetized in 1 liter of 0.2% MS-222 (tricaine; 3-aminobenzoic acid ethyl ester, methanesulfonate salt; Sigma, St. Louis, MO, Cat No A5040) in deionized water. This procedure takes approximately 30 min and can be accelerated by (i) beginning with cold anesthetic and (ii) surrounding the container with ice. Successful anesthesia is indicated when the frog fails to move when touched on the legs or abdomen.

During this 30-min period, the surgical instruments can be prepared. These instruments include one pair of sharp scissors, one pair of blunt end forceps, two No. 4 forceps, one 18-gauge 1.5″ syringe needle, one curved suture needle with 6-0 silk suture (Ethicon, Somerville, NJ), and the "operating table." All metal tools, which have been previously cleaned and autoclaved, are immersed in 95% ethanol and dried immediately prior to use. The operating table is simply a 30 × 19 × 5-cm Pyrex dish, filled with ice, and covered with one layer of aluminum foil and one layer of paper towels. The frog is removed from the container and placed on its back on the paper towels. Some protocols suggest using completely sterile conditions (e.g., a hood) for the surgeries (39); however, we have not found this to be necessary.

The frog is rinsed with deionized water and the abdominal region is swabbed with ethanol. *Xenopus* are naturally marked with a stich-like pattern that runs down their sides and meets at the midline of the lower abdomen. The incision is typically located 1–2 cm rostral to these markings and 0.5–1 cm to the left or right of the midline. The incision is created by inserting the syringe needle through the skin, moving the needle tip just under the skin, and creating a second hole by pushing outward through the skin approximately 1 cm away from the initial insertion. The skin is cut between these two points with the scissors, using the needle as a guide. This procedure is repeated for the abdominal wall, which is located immediately below the skin. The ovarian lobes are gently pulled from the abdomen using the No.

TABLE I   Composition of Solutions Used for Oocyte Preparation, Injection, and Incubation[a]

OR-Mg		ND96	
Compound	Concentration (mM)	Compound	Concentration (mM)
NaCl	82	NaCl	96
MgCl$_2$	20	HEPES	5
HEPES	5	KCl	2
KCl	2	CaCl$_2$	1.8
		MgCl$_2$	1

ND96-PG		ND96-HS	
ND96		ND96-PG	
Sodium-pyruvate	2.5 mM	5% Horse serum	
Gentamycin	50 μg/ml		

[a] OR-Mg and ND96 are prepared as 10× stock solutions. All solutions are sterile filtered using 0.22-μm cellulose acetate filters. The pH is adjusted with 1 M NaOH to 7.4–7.5; osmolality is between 200 and 240 mOsmol/kg. Sterile aliquots of horse serum (EIA negative, Irvine Scientific, Santa Ana, CA, Cat. No. 4023) are kept at -20°C, heat inactivated at 56°C for 30 min, and added after the saline has been filtered.

4 forceps. Each lobe typically contains hundreds of oocytes. The amount needed is removed by cutting that portion of the lobe away from the remaining ovary (6). The cut ovary is placed into a petri dish with OR-Mg saline (see Table I), and the remaining, attached ovarian lobe can be gently shaken back into the frog. The incision is then sutured (usually with two stitches). We suture through both the skin and the abdominal wall with the same stitch. We often mark the frog at this time for future reference by discoloring its skin in unique patterns using a swab dipped in HCl. The frog is allowed to recover for several hours in a small container of its home tank water (making sure that its head can easily protrude above the water level) before it is returned to the group tank. We typically reuse our frogs every 2–4 months; many frogs are reused six to eight times.

Oocyte Defolliculation

The extracted ovarian lobes are rinsed several times with OR-Mg. Using No. 4 forceps, the lobe is gently torn into small clumps of approximately 1 cm in length (25–30 oocytes per clump). Each clump is transferred into a new petri dish containing OR-Mg. The clumps are then placed into collagenase (Type A, Boehringer-Mannheim, Indianapolis IN, Cat No. 1088 785; Type 1A, Sigma, Cat No. C9891). This enzymatic treatment (i) strips individual oocytes from the ovary and (ii) removes the follicle cell layer that surrounds each oocyte (15). We currently use a collagenase concentration of 2 mg/ml in OR-Mg. However, it has been our experience that different lots of collagenase vary considerably in their enzymatic properties; we recommend testing each new lot on a single oocyte batch to determine the appropriate concentration. Equal volumes of oocytes are placed into 15-ml plastic test tubes containing 5 ml collagenase solution. These tubes are placed at room temperature on a test tube rocker ("Speci-Mix," Thermolyne, Dubuque, IA) at a 45° angle from vertical. Typical defolliculation lasts 1.5 to 3 hr, with fresh collagenase replacement every hour. Starting at 1.5 hr, oocytes should be checked, as described below, every 15 min by removing several with a precut and flamed Pasteur pipette into a dish containing OR-Mg.

Oocyte Isolation

A crucial step in oocyte preparation is determining when to terminate enzyme treatment. If the treatment is too short, the oocyte remains covered by follicle cells which can render injection more tedious and, more importantly, can confound experimental results due to receptors and transporters endogenous to the follicle cells. If the treatment is too long, the enzyme can damage the integrity of the oocyte membrane. The presence of the follicle is indicated by small blood vessels surrounding the oocyte and a sheen across its surface. We typically stop when 50–75% of the oocytes are still surrounded by follicle. (After 24 hr of incubation, 90% of the follicle-laden oocytes will have sloughed off this layer.) The oocytes are rinsed quickly with OR-Mg three or four times to remove the collagenase and then transferred to a petri dish containing OR-Mg for selection.

Rapid selection of healthy oocytes is also crucial. Our selection criteria typically include (i) a uniformly dark animal hemisphere that is free from discolorations, (ii) a uniformly light vegetal hemisphere, and (iii) a well-defined color separation of these two hemispheres along the equator. Stage V and VI oocytes are usually selected (14), although for some applications stage II or III oocytes can be used. The advantages of these earlier stage oocytes are that their smaller size (i) results in lower membrane capacitances which can be critical for rapid voltage-clamp electrophysiology and (ii) requires less RNA per injection (25). We have not noted any differences in the translation ability of these earlier stage oocytes compared to that of later stages.

Selection of the desired oocytes is performed manually using a cut and fire-polished Pasteur pipette. The acceptable oocytes are transferred to a petri dish containing fresh OR-Mg. When selection is completed, the OR-Mg is removed and the oocytes are rinsed three times in ND96 (see Table I). A second selection is then performed to obtain a batch of oocytes more uniform in size and color. These oocytes are plated into non-tissue culture-treated petri dishes (Lux 60 × 15 mm, Nunc, Inc., Naperville IL, Cat No. 5220) in ND96-HS (see Table I). These types of dishes are preferred because they prevent the oocytes from becoming damaged by sticking to the bottom. The dishes are then placed in an 18°C incubator and are continuously and gently rotated (setting 2 or 3) on a tabletop shaker ("Orbital Shaker," Bellco Biotechnology, Vineland, NJ; "Red Rotor," Hoefer Scientific Instruments, San Francisco, CA).

Maintaining Oocytes

The oocytes are checked twice daily. The healthy oocytes are placed into new dishes with fresh ND96-HS while the unhealthy oocytes are discarded. Transferring the good oocytes is recommended because the dying oocytes secrete proteases that may act to accelerate the degradation of healthy oocytes (44). Dying oocytes can be distinguished by a significant change in their appearance from that described above. The most common changes are (i) a large white spot on the animal pole indicative of oocyte maturation, (ii) mottling of the animal hemisphere, or (iii) a white or gray coloration of the entire oocyte surface. A white scar on the oocyte often occurs at the injection site; these oocytes should not be considered unhealthy. We typically wait 24 hr after isolation to inject the oocytes. Injections can be performed on the day of isolation; however, there is usually a 5 to 10% attrition rate for oocytes during this initial 24-hr period. By waiting, one avoids injecting those oocytes that would otherwise die anyway.

One of the major problems in using oocytes for heterologous expression is the variability in the quality that can often occur across batches. This variability is often enhanced during certain times of the year, most noticeably during the summer months, and can greatly reduce the number of successful experiments. We have recently found that by incubating the oocytes with saline containing 5% horse serum, we can (i) increase the viability (time to death) of our oocyte batches, (ii) decrease the amount of variability both within and between oocyte batches, and (iii) increase our percentage of successful experiments (33). The effect of horse serum on the resting potential and leak currents (assessed using the two-electrode voltage clamp) and on viability is shown in Fig. 1 for water-injected oocytes. The data represent measurements taken on a total of 40 oocytes, 20 from each of two different frogs (circle and square symbols), and 10 oocytes from each frog under each

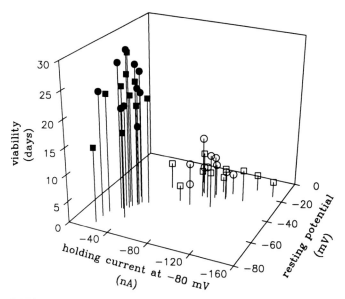

FIG. 1   Viability, resting potential, and leak current of water-injected oocytes incubated with (closed symbols) or without (open symbols) horse serum. The data represent 20 oocytes from each of two frogs on successive weeks (circle and square symbols, respectively). Non-serum-treated oocytes were incubated in ND96-PG. All oocytes were incubated in individual wells of a multiwell polystyrene plate for identification purposes. At 3 days postinjection, the resting potential and background current (holding potential −80 mV) were measured using a two-electrode voltage clamp. Measurements were made in ND96; serum-incubated oocytes were placed in ND96-PG 6 hr prior to measurement. Viability was determined by oocyte appearance.

condition. The mean resting potential, holding current at −80 mV, and viability for the serum-treated oocytes (open symbols) were −31 mV, −87 nA, and 6.4 days, respectively. For the serum-treated oocytes (closed symbols), the mean values were −58 mV, −39 nA, and 23 days. Because of these beneficial effects of horse serum, we routinely include it in our incubation medium. However, we place the oocytes in ND96-PG approximately 4–6 hr prior to measurement to ensure that the serum does not have a distorting effect on the oocyte responses (31).

Injection of Oocytes

The initial step in oocyte injection is to prepare injection needles. The glass (8″ Microdispenser Bores, Drummond Scientific, Broomall, PA, Cat No. 3-00-216G8) should be incubated at 200°C for 24 hr prior to use, in order to destroy any RNases that may be present (for this same reason, gloves should be worn throughout injection). Needles are pulled on a vertical pipette puller (David Kopf Instruments, Tujunga, CA, Model 720). The goal is to pull a tip that has a long, tapering shaft. The needle is then placed under a dissecting microscope and held in place by modeling clay. The tip of the needle should be broken to a diameter of approximately 20 μm. This can be accomplished by using fine forceps, a glass slide, or a grooved metal spatula. As has been previously noted (39), there is a trade-off in obtaining the appropriately sized needle. Needles with larger diameters can injure the oocyte, while smaller diameter needles can be easily clogged by oocyte cytoplasm, dirt, and even viscous injection solution. It is preferable to break the needle tip such that the edge is beveled. This provides a sharp point with which to penetrate the oocyte. Needle tips with a flat surface often greatly deform the oocyte surface before the plasma membrane is penetrated, and this can lead to excessive damage. Be sure to have several extra needles prepared in case the first ones become clogged or broken. We store our needles on clay in 150-mm covered petri dishes.

A syringe containing paraffin oil is inserted (using a 21-gauge needle) into the base of an injection needle and approximately 0.5–1 cm of oil is dispensed. The oil serves as a clean seal and interface between the injection solution and injector. The next step is to secure the glass injection needle to the injector (10 μl Digital Microdispenser, Drummond Scientific, Model 510X). This a positive displacement device in which a steel shaft drives the injection solution down the needle. The injector fitting is loosened, the steel injector shaft is slid into the base of the needle, and the fitting is tightened. The injector dispenser is advanced until all of the air is out of the needle and oil droplets emerge from the needle tip. The injector is then secured to a three-axis micromanipulator, which allows for precise control over injector position.

The third step is to backfill the needle with the desired injection solution (RNA, DNA, etc.). All solutions should be centrifuged prior to use in order to remove dirt and debris that could clog the needle. The amount of solution needed is calculated (volume injected × number of oocytes). An additional 0.5–1 μl should be added to this amount because not all the solution can be drawn into the needle. The solution is pipetted to form a drop on the surface of a sterile petri dish. Each small division represents 10 nl. Make sure that there is sufficient needle volume available to drawn in the desired amount of solution. By adjusting the micromanipulator, the needle tip can be positioned just above the drop. Be sure to eliminate all air bubbles from the needle before submerging the tip into the solution. Once the needle tip is in the solution, the injector dispenser should be slowly rotated to backfill the needle. It is important to avoid drawing air bubbles into the needle, for the bubbles will result in oocytes injected with air rather than solution. The air bubbles can be seen at the steel injector shaft/paraffin oil interface or at the paraffin oil/injection solution interface. If air bubbles are present, the dispenser is advanced to expel the air. One technique that we find useful is repeatedly drawing solution and expelling a small fraction of it. This minimizes clogs by constantly forcing debris away from the needle tip. The last resort is to change needles.

The fourth step is to prepare the oocytes for injection. If the oocytes have been incubated with ND96-HS, they should be removed 10–15 min prior to injection and placed in ND96-PG. We have found that the serum tends to make the oocytes slippery and more difficult to impale. Our injection dish is the underside of a 60 × 15-mm petri dish lid on which rows have been scratched using a hypodermic needle. The lid is filled with ND96-PG and the desired number of oocytes are placed in orderly rows between the lines. The lines (i) provide a barrier so that the oocytes do not roll during injection and (ii) keep the oocytes ordered such that injected oocytes can be distinguished from noninjected oocytes. In stage V and VI oocytes, the nucleus is within the animal hemisphere (14), and the oocyte typically orients on the dish with that hemisphere up. Although most authors suggest that the oocytes be injected on their vegetal hemisphere to avoid damaging the nucleus and killing the cell, we have found that there is less leakage when injections are performed on the animal hemisphere and we have not noted differential survival rates for oocytes injected in opposing hemispheres.

The oocytes to be injected are placed under the 12× magnification of the dissecting microscope and the needle is positioned such that the tip will enter the oocyte approximately 45° above the horizontal axis. This angle ensures that the needle punctures the oocyte and does not cause it to roll away. The fine adjustment on the micromanipulator is used to penetrate the oocyte. Before penetration, the needle tip dimples the oocyte; once the tip has broken the membrane, the dimple disappears as the oocyte surface expands around the needle. The injector dispenser is advanced to release the desired amount and the needle is then slowly removed. If excessive leakage occurs, this may indicate that the needle tip is too large. An additional problem is that the injection solution may become clogged, and no solution will be expelled into the oocyte, even though the injector dispenser is still turning. If sufficient volume is being injected into the oocyte (~50 nl), then a successful injection can be verifed by noting a swelling of the oocyte as it incorporates the solution volume. An additional verification can be employed by noting the injector reading at which the solution was first backfilled. If no difficulties have occurred during injection, then only paraffin oil and no injection solution should be present at the needle tip at this reading. Significant amounts of solution remaining in the needle beyond this value indicate a problem. Once the entire row of oocytes is injected, they are placed in ND96-HS and returned to the incubator. With practice, several hundred oocytes can be injected per hour with a single needle filling.

Preparation of Injection Solution

Our lab uses three different methods for expressing an exogenous message in oocytes. These include (i) standard RNA injections into cytoplasm, (ii) DNA injections into the nucleus, and (iii) a novel method involving cytoplasmic injections of vaccinia virus.

Injection of RNA

Three different forms of RNA are routinely used to obtain expression of excitability proteins in oocytes. Total RNA, extracted directly from the tissue of interest, can be used if the message is present in abundance. A number of different extraction methods are available, including phenol chloroform (5, 26), urea/LiCl (1, 12), and guanidinium isothiocyanate (43) or guanidinium hydrochloride (4, 11, 38). The reader is referred to these papers for specific methods. Briefly, the appropriate choice of extraction method is based on the expected molecular weight of the message, contaminating factors, and the desired end product. The phenol chloroform and urea/LiCl methods work well with small amounts of tissue and with RNA species of low molecular weight; however, the RNA end product may contain unwanted proteins and DNA. The guanidinium methods work well with RNA species of higher molecular weight and provide strong RNase protection. Furthermore, the RNA end product is purer than that with other methods; however, it requires larger amounts of tissue.

Almost all (>95%) of the extracted "total" RNA is ribosomal and therefore not of particular interest for expression of excitability proteins. Thus, to obtain suitable expression levels it is often necessary to extract the desired mRNA from the total RNA. This is accomplished by taking advantage of the fact that most mRNA contains a 3′ tail that is polyadenylated (3). The traditional method for isolating poly(A) RNA is affinity chromatography through a column composed of oligo(dT) cellulose (11, 36). More recently, we have successfully extracted poly(A) RNA using oligo(dT) primers that couple to streptavidin-linked magnetic particles ("MagneSphere" and "Poly-ATract," Promega Corp., Madison, WI, Cat No. Z5200).

The highest expression levels in oocytes are typically obtained when RNA is prepared *in vitro* from a cloned cDNA encoding the message of interest. Once again, this procedure has been described in detail in many publications (11, 19, 36). Briefly, this cRNA is made from a cDNA that is subcloned into a plasmid expression vector that contains a bacteriophage polymerase promoter (usually SP6, T3, or T7). The plasmid DNA is then linearized using a restriction enzyme that cuts at a unique restriction site. Transcription is initiated on the linearized DNA by an RNA polymerase that is specific to the given promoter. Recovery of the cRNA independent of DNA, free nucleotides, and proteins is then accomplished using a combination of DNase, phenol extraction, and spin column centrifugation. A number of additions to this general procedure have resulted in enhancement of cRNA expression in oocytes. The most common methods include the addition of a cap analog [5'-(7-methyl)-guanosine-guanosine triphosphate] to the 5' end of the cRNA and the addition of a polyadenylated tail to the 3' end (13). Recently, our laboratory has been successful at increasing expression using a method of expression polymerase chain reaction (PCR) (22). In this method, a 5' primer is constructed in the sense direction that contains the bacteriophage promoter, the untranslated leader sequence from the alfalfa mosaic virus, the Kozak box, and at least 18 bases of the cloned cDNA sequence. A 3' primer is constructed in the antisense direction that contains a poly(T) tail (usually 20 bases) followed by at least 18 bases complementary to the cloned cDNA sequence. In the PCR, the plasmid containing the cDNA is used as the template and the aforementioned oligonucleotides serve as primers for the reaction. cRNA is synthesized *in vitro* from the PCR product. We have found that this method increases expression levels in oocytes for both transporter and channel proteins.

The RNA can be stored for long periods if it is resuspended in sterile water or, preferably, in sterile water containing 0.1% diethyl pyrocarbonate (DEPC, a RNase inhibitor). The RNA should be aliquoted and kept at $-80°C$. We have found that storage of the RNA at higher temperatures, or repeated freezing and thawing, decreases expression. The injection amount and the translation time must be determined empirically for each protein under study. For example, we typically generate 1–2 μA of current using the two-electrode voltage-clamp method for serotonin receptors injected at a concentration of 10 pg/oocyte; to generate the same current with neuronal nicotinic acetylcholine receptors, we typically inject 20 ng/oocyte for each of the two-receptor subunits. Maximal expression usually occurs between 2 and 7 days postinjection. The typical volume injected into the cytoplasm is 50 nl, although 100 nl can be tolerated by a larger oocyte if the solution is dispensed slowly.

Injection of DNA

The major disadvantage in using cRNA derived from either traditional or expression PCR methods is the time needed to perform the *in vitro* synthesis. This is a significant consideration when a large number of messages, or pools of messages, need to be constructed. One alternative is to use the oocyte's nuclear machinery to synthesize the desired product *in vivo* from a cDNA which resides downstream from a constitutive or inducible promoter (7). For these experiments, we use the protocols designed by Bertrand *et al.* (2), with only slight modifications. Compared to cytoplasmic injections, nuclear injections require one additional step; namely, the oocytes must be centrifuged in order to bring the nucleus visibly near the surface. An easy and fast method that we have found is to place an individual oocyte (animal pole up) into a 0.5-ml Eppendorf tube containing approximately 200–300 μl of ND96-PG. If the oocyte is dropped from the surface of the solution, it typically settles at the bottom of the tube with the animal pole facing up. The tube is then placed in a tabletop centrifuge (15,000 rpm, Eppendorf Micro Centrifuge, Brinkmann Instruments, Model 5414) and spun for 5 sec. The position of the nucleus will be marked by a relatively large, white discoloration on the animal hemisphere. This spot often remains for up to 1 hr, so a large number of oocytes can be spun and injected together.

The typical volume that we inject into each oocyte nucleus is 10 nl. If it is not necessary to accurately quantify the injection amount (e.g., to assay for positive expression only), then we routinely use the Drummond 510 injector described above; however, this device is probably not accurate for such small injection volumes. For accurate injection, a calibrated pump (Nanopump, World Precision Instruments, Sarasota, FL, Model A1400) is employed. The same procedures as described above for cytoplasmic injections are used, including the preparation of injection needles and backfilling of injection solution. We have obtained successful expression of neuronal nicotinic acetylcholine receptors using cDNA resuspended in sterile water or OR-Mg. These injection solutions containing DNA are stored at $-20°C$.

Injection of Vaccinia Virus

An alternative to both the labor involved in creating mRNA and the technical difficulties of nuclear injections is a novel method developed in our laboratory for protein expression using vaccinia virus injections into oocyte cytoplasm (45). For complete details on vaccinia virus vectors in ion-channel expression, see Karschin *et al.* (24). The impetus for this idea comes from the use of recombinant vaccinia virus (VV) for high-level expression in mammalian cells (23, 28). Two different strategies involving VV have been shown to work in oocytes. The first is to subclone the cDNA of interest into a vaccinia recombinant plasmid (34) downstream from the end of the 5′ flanking region containing the vaccinia P7.5 (7.5 kDa) early promoter. This plasmid containing viral DNA and cDNA is then transfected along with genomic wild-type VV into thymidine kinase-negative L cells. Homologous recombination between the wild-type VV and the insert plasmid results in the production of a recombinant VV containing the cDNA inserted at the site of the thymidine kinase gene. Disruption of this site by the foreign DNA provides a criteria for selection of appropriate viral particles (34). This recombinant is then injected into oocytes.

The second method is to co-inject a mixture of recombinant VV and the desired cDNA in a circular plasmid. The recombinant VV (vTF7-3) contains the bacteriophage T7 RNA polymerase gene inserted downstream from the VV P7.5 promoter (17, 32). The recombinant vTF7-3 is available from B. Moss (Laboratory of Viral Diseases, National Institute of Allergy and Infectious Diseases, National Institutes of Health, Bethesda, MD 20892). The cDNA is inserted in a plasmid downstream from a T7 promoter. Insertion of the untranslated leader sequence from the encephalomyocarditis virus between the T7 promoter and the exogenous gene might be expected to enhance expression (16); however, in the case tried, expression was poor and the oocytes died (45). The virus and plasmid are then co-injected into oocytes where the VV T7 promoter drives the expression of the encoded cDNA.

Yang *et al.* (45) used the first method to express the Shaker H4 potassium channels in oocytes. A total of 200–2000 viral particles of VV:H4 (28) were injected per oocyte in 50 nl of injection solution (PBS-MB: Ca^{2+}- and Mg^{2+}-free Dulbecco's phosphate-buffered saline, 1 mM $MgCl_2$, 0.1% bovine serum albumin). The second method was used to express the rat brain IIA sodium channel [inserted in either pBluescript-SK (Strategene, La Jolla, CA, Cat. No 212205) or pTM1 (32)], and the four subunits of the mouse muscle acetylcholine receptor (in pBluescript-SK). In these experiments, each oocyte was coinjected with 5–25 ng of circular cDNA and 3800 vTF7-3 viral particles. Functonal expression was obtained in approximately 80% of the oocytes for both channels and receptors in a time course similar to that of mRNA-injected oocytes; however, the expression level of VV co-injected ACh receptors was about one third of that obtained in oocytes injected with 5 ng mRNA. Note that extra care should be taken when working with vaccinia virus. We use an injector set aside specifically for virus work, and we incubate the oocytes enclosed in plastic containers. Bleach is used to clean all surfaces that may have contacted virus. Additional VV safety procedures may be obtained from us or from B. Moss.

Methods for Examining Oocyte Expression

Our methods to investigate heterologous expression of excitability proteins in oocytes routinely employ four techniques. These include (1) two-electrode voltage-clamp electrophysiology, (ii) patch-clamp electrophysiology, (iii) flux assays, and (iv) binding assays on oocyte membranes. The electronics necessary for the electrophysiological measurements have been treated in depth previously (21, 30, 35). Therefore, for these sections, we instead discuss the perfusion, electrode, and oocyte-handling techniques that are used. We then detail both the flux methods that have been used in the lab to study transporters and the preparation of oocyte membranes for biochemical assays. Last, we briefly discuss methods for internal perfusion of oocytes.

Two-Electrode Voltage Clamping

In the simplest case, one electrode is used to monitor the transmembrane potential of the oocyte. An amplifier compares the resting potential recorded by the voltage electrode to the desired clamping potential. Current is then injected into the oocyte from the second electrode to minimize this difference. We use several different commercially available clamp systems (e.g., Axoclamp 2A, Axon Instruments, Foster City, CA; Model 8500, Dagan Corp., Minneapolis, MN) to achieve this circuit. Clamp current is monitored by oscilloscope and chart recorder. Voltage protocols, data storage, and analysis are done using pCLAMP software and T1-1-Labmaster acquisition hardware (Axon Instruments), running on an MS-DOS computer with an industry standard bus.

The electrodes are pulled from capillary tubes (Kimax-51 glass, Kimble Products, Vineland, NJ) to a resistance of approximately 1 MΩ and filled with 3 M KCl. The low resistance of the electrodes reduces clamp noise and allows for the injection of large amounts of current into the oocyte. With electrodes of even lower resistance, faster clamp speeds can be obtained; however, it is necessary to fill the ("leaky") electrodes with a saline solution that will not damage the oocyte (30). We attach the electrodes to the voltage-clamp circuitry using holders that contain a AgCl wire (E. W. Wright, Guilford, CT). These are held in place by the micromanipulators described earlier. The proximity of the two electrodes can produce capacitive coupling which can be reduced by placing a grounded foil shield between them. The reference electrode is a virtual ground amplifier from the oocyte bath chamber (see below). The electrical coupling is done through a AgCl pellet either embedded in the chamber or connected to the chamber through an agar bridge (3% agarose mixed with 3 M KCl). Agar bridges must be used when Cl$^-$ substitution experiments are performed in order to eliminate junction potentials.

The oocyte recording chamber is made from a 40 $\times$ 40 $\times$ 4-mm piece of Plexiglas. In the center of the Plexiglas, a 10 $\times$ 5-mm rectangular chamber is cut into the surface to a depth of 3 mm. Sylgard (184 Silicone, Dow Corning Corp., Midland, MI) is used to fill the chamber; a V-shaped wedge is cut out of the Sylgard to create a valley in which the oocyte will rest. The final volume of the chamber should be as small as possible to minimize the time required for solution changes. Our chamber volumes are approximately 150–300 μl. A tunnel is drilled through the Plexiglas parallel to its horizontal surface and enters the oocyte chamber at its far edge, perpendicular to the length of the chamber. A 2-mm inner-diameter plastic tube is inserted through this tunnel and serves as the solution entry point. The diameter of the tube should be large enough to permit rapid solution flow, but small enough not to disrupt the stability of the oocyte or recording electrodes. Orienting the tube entry perpendicular to the chamber length also helps to increase stability. A second tunnel is cut into the Plexiglas for the insertion of a AgCl pellet and wire for the reference electrode.

The superfusion system uses gravity flow and begins with 100-ml syringes suspended approximately 30 cm above the chamber height. One end of intravenous drip tubing (Venoset 72, Abbott Laboratories, North Chicago, IL) is connected to each syringe; the other end is connected to a rotary valve (type 50, Rheodyne, Inc., Cotati, CA). The inner diameter of the tubing is 3 mm, and each tube has an adjustable clamp to control flow rate. The rotary valve contains six input ports and a central output port that is connected to the chamber tubing. The dead time for solution changes is determined by the volume of solution between the valve and chamber; thus, this distance should be minimized. Using this setup, the maximum flow rate through the chamber is approximately 10 ml/min, resulting in an exchange time in the chamber of about 1 sec. The dead time is about 2 sec. The solution is removed from the far end of the chamber by suction through a 2-mm inner-diameter plastic tube suspended above the chamber. The height of the suction tip can be adjusted to control the height of the solution flowing through the chamber. The output of the suction tube is a simple vacuum trap.

Patch Clamping

For patch clamping, changes from the two-electrode clamp are necessary in the electronics, the electrodes, and the treatment of the oocytes. Briefly, we collect oocyte patch data using standard patch-clamp amplifiers (Model 8900, Dagan Corp.; Axopatch 1D, Axon Instruments). For some applications, signals are sampled by a pulse-code modulator and stored on videocassette tapes. These signals are digitized and analyzed using pCLAMP software (Axon Instruments).

The patch electrodes are pulled from borosilicate glass (1.5 mm o.d., 0.86 mm i.d., Sutter Instrument Co., Novato, CA, cat No. BF 150-86-10) to a tip diameter of approximately 1 μm using a horizontal puller (Sutter Instrument Co., Model P-87). This produces an electrode resistance between 10 and 40

MΩ (in 150 mM salt) and permits seals of around 100 GΩ that can routinely be maintained for ~1 hr. Another advantage of the oocyte system is that the amount of expression can be titrated by the quantity of the message injected in order to obtain the desired density of channels per patch.

To obtain the high-resistance seals necessary for detecting single channels, it is necessary to remove the vitelline membrane that surrounds the oocyte's plasma membrane. The oocyte is placed in a solution containing ND96 in which an extra 100 mM NaCl is dissolved or in a high-potassium solution (200 mM monopotassium aspartate, 20 mM KCl, 10 mM HEPES, 10 mM EGTA, 1 mM MgCl$_2$, pH 7.2). This causes the oocyte to release water and shrink after 5–15 min. Under a dissecting microscope, it is possible to see the vitelline membrane isolated from the plasma membrane. Using a pair of fine forceps, one can peel away the vitelline membrane. If it is difficult to resolve the two membranes, it is often helpful to spin the oocyte in a 35-mm petri dish containing the hypertonic solution for approximately 10 min on a horizontal shaker prior to manual dissection. The oocyte is extremely fragile and sticky following this procedure, so care should be taken when handling. In particular, the oocyte ruptures if it contacts the air–water interface. We place the devitellinized oocytes in ND96-PG at 18°C in a petri dish coated with agar to prevent sticking. These oocytes usually remain viable for several hours following vitelline membrane removal.

Flux Measurements

Studies of neurotransmitter transporters in our laboratory (18, 19) are performed by measuring the uptake of [^{3}H] neurotransmitters in transporter mRNA-injected oocytes. The oocytes are preincubated at room temperature in ND96 for 15 min. A standard pipetter (in which the disposable yellow tip has been cut to a sufficient diameter for easy withdrawal of an oocyte) is used to transfer the oocytes along with 40 μl of ND96 to individual wells of a 96-well microtiter plate. To each well, 10 μl of the reaction buffer is added. For a typical γ-aminobutyric acid (GABA) uptake assay, the reaction buffer for one oocyte contains 5 μl 2$\times$ ND96, 4 μl distilled water, and 1 μl [^{3}H]GABA (1 μCi/μl, NEN DuPont, Boston MA). The reaction buffer should be added in timed intervals to maintain constant uptake time. After the desired uptake time, the reaction is terminated by (i) the rapid addition of 200 μl ND96 to the microtiter well and (ii) the immediate removal of the oocyte. Each oocyte is passed through four washes of ND96 and then placed in a 7-ml copolymer plastic scintillation vial (Research Products Intl., Mount Prospect, IL, Model 125508) containing 500 μl 10% SDS. The oocyte is solubilized (approximately 45 min) at 42°C in a rotating water bath (New

Brunswick Scientific, Edison, NJ, Model G76). Approximately 3 ml of water-compatible scintillation cocktail is then added to the vial and the amount of uptake is determined by scintillation counter.

Preparation of Membranes

The methods we use are based on those of Colman (6). Twenty or fewer oocytes are placed in a small glass homogenizer ($\frac{3}{8}''$ diameter) along with 2 ml of sucrose buffer [0.32 M sucrose, 1 mM EDTA, 10 mM potassium phosphate (pH 7.0), 1 mM phenylmethanesulfonyl fluoride (PMSF, Boehr-inger-Mannheim, Cat No. 236 608), 10 μg/ml aprotinin (Boehringer-Mann-heim, Cat No. 236 624)] and homogenized using five strokes. The homogenate is placed into a 15-ml plastic test tube and centrifuged (4°C, 1000 g) for 10 min. The supernatant is placed in an ultracentrifuge tube and kept on ice. The oocyte pellet is resuspended in 2 ml sucrose buffer and the procedure is repeated.

The ultracentrifuge tubes containing the oocyte supernatant are filled to the top with sucrose buffer. The membranes are centrifuged at 4°C and 100,000 g for 30 min. The supernatant is discarded and the membrane pellet is resuspended in Tris buffer (50 mM Tris–HCl, 10 mM MgCl$_2$, PMSF, and aprotinin as above). The membranes are stored at -80°C. These crude membranes are often sufficient for many assays. If purer membrane is required, sucrose gradient fractionation can also be performed to remove nuclear and cytosolic proteins (6). Protein assays reveal that 0.5–2 μg of membrane protein are obtained per oocyte.

Internal Perfusion

For some applications, it is desirable to control cytoplasmic solutions in the oocyte. If there is sufficient time for the oocyte to recover from the procedure, then the solution can be injected as described above for RNA injections. For example, EGTA injection to chelate endogenous calcium is typically performed 4–24 hr prior to measurement. If 50 nl is injected, then a 20$\times$ concentration of the solution should be used, since the volume of a typical stage V or VI oocyte is 1 μl. OR-Mg or HEPES (pH 7) is used as the solution buffer.

Internal perfusion can also be performed during oocyte measurement if iontophoretic or pressure injection is utilized (9, 10, 40). In iontophoresis, a glass electrode is pulled to a resistance of 50–100 MΩ and filled with the desired injection solution. The electrode is connected to a constant current source (Grass, Inc., Quincy, MA, Model CCU1) and the current is monitored by oscilloscope. In pressure injection, glass micropipettes are pulled and broken to a diameter of 2–4 μm. The micropipette is filled with injection solution and the pipette base is connected to a pneumatic pump (Nanopump, World Precision Instruments). Calibration of the injection amount per timed pump pulse can be determined by immersing the tip of the pipette in paraffin oil, pulsing the pump, and measuring the diameter of the resulting drop of solution present in the oil. Recently, two additional methods for internal perfusion have been described. One uses an oocyte wedged into the bottom of a Pasteur pipette and broken at the lower surface (9); a second uses an open-surface oocyte separated from the recording apparatus by a Vaseline gap (41).

Conclusions

The oocyte expression system has proven to be a valuable tool for the study of excitability proteins for many reasons. For example, (i) the oocyte size (approximately 0.5–1 mm diameter) makes it convenient for both injection of exogenous messages and electrophysiological measurement, (ii) the oocyte is capable of undertaking many of the posttranslational modifications necessary for protein function (e.g., glycosylation, phosphorylation), (iii) the oocyte can correctly assemble and orient multiple subunits into functioning entities, and (iv) the oocyte, with some exceptions (8, 42), is capable of translating most exogenous messages. However, this system does have some limitations: (i) there can be substantial variation in expression within and especially across batches of oocytes, (ii) expression is transient, and (iii) the oocyte contains a number of endogenous proteins that can interfere and distort study of the protein of interest.

Acknowledgments

The authors thank the following people for sharing their protocols and expertise: Janis Corey, Nathan Dascal, Norman Davidson, Joe Farley, Andreas Karschin, and Nael McCarty. This work was supported by NS-11756, GM-29836, the Muscular Dystrophy Association, the California TRDRP, and the Cystic Fibrosis Foundation.

References

1. C. Aufrey and F. Rougeon, *Eur. J. Biochem.* **107,** 303 (1980).
2. D. Bertrand, E. Cooper, S. Valera, D. Rungger, and M. Ballivet, *in* "Methods in Neurosciences" (P. M. Conn, ed.), Vol. 4, p. 174. Academic Press, San Diego, 1991.
3. G. Brawerman, *Prog. Nucleic Acid Res. Mol. Biol.* **17,** 117 (1976).
4. J. J. Chirgwin, A. E. Przbyla, R. J. MacDonald, and W. J. Rutter, *Biochemistry* **18,** 5294 (1979).
5. P. Chomczynski and N. Sacchi, *Anal. Biochem.* **162,** 156 (1987).
6. A. Colman, *in* "Transcription and Translation: A Practical Approach" (B. D. Hames and S. J. Higgins, eds.), p. 271. IRL Press, Oxford, 1984.
7. G. Dahl, T. Miller, D. Paul, R. Voellmy, and R. Werner, *Science (Washington, D.C.)* **236,** 1290 (1987).
8. N. Dascal, *CRC Crit. Rev. Biochem.* **22,** 317 (1987).
9. N. Dascal, G. Chilcott, and H. A. Lester, *J. Neurosci. Methods* **39,** 29 (1991).
10. N. Dascal, E. M. Landau, and Y. Lass, *J. Physiol. (London)* **352,** 551 (1984).
11. N. Dascal and I. Lotan I, *in* "Methods in Molecular Neurobiology" (A. Longstaff, ed.), Vol. 13, Ch. 8. Humana Press, England, 1993.
12. P. Dierks, A. van Ooyen, N. Mantei, and C. Weissman, *Proc. Natl. Acad. Sci. U.S.A.* **78,** 1411 (1981).
13. D. R. Drummond, J. Armstrong, and A. Colman, *Nucleic Acids Res.* **13,** 7375 (1985).
14. J. N. Dumont, *J. Morphol.* **136,** 153 (1972).
15. J. N. Dumont and A. R. Brummett, *J. Morphol.* **155,** 73 (1978).
16. O. Elroy-Stein, T. R. Fuerst, and B. Moss, *Proc. Natl. Acad. Sci. U.S.A.* **86,** 6126 (1989).
17. T. R. Fuerst, E. G. Niles, W. Studier, and B. Moss, *Proc. Natl. Acad. Sci. U.S.A.* **83,** 8122 (1986).
18. J. Guastella, N. Brecha, C. Weigmann, H. A. Lester, and N. Davidson, *Proc. Natl. Acad. Sci. U.S.A.* **89,** 7189 (1992).
19. J. Guastella, N. Nelson, H. Nelson, L. Czyzyk, S. Keynan, M. C. Miedel, N. Davidson, H. A. Lester, and B. Kanner, *Science (Washington, D.C.)* **249,** 1303 (1990).
20. J. B. Gurdon, C. D. Lane, H. R. Woodland, and G. Marbaix, *Nature (London)* **233,** 177 (1971).
21. A. Hodgkin and A. Huxley, *J. Physiol. (London)* **17,** 550 (1952).
22. K. C. Kain, P. A. Orlandi, and D. E. Lanar, *BioTechniques* **10,** 366 (1991).
23. A. Karschin, J. S. Aiyar, A. Gouin, N. Davidson, and H. A. Lester, *FEBS Lett.* **278,** 229 (1991).
24. A. Karchin, B. A. Thorne, G. Thomas, and H. A. Lester, *in* "Methods in Enzymology" (M. Simon and J. Abelson, eds.), Vol. 27, p. 408. Academic Press, San Diego, 1992.
25. D. S. Krafte and H. A. Lester, *J. Neurosci. Methods* **26,** 211 (1989).

26. L. Kushner, J. Lerma, M. V. L. Bennett, and R. S. Zukin, *in* ''Methods in Neurosciences'' (P. M. Conn, ed.), Vol. 1, p. 3. Academic Press, San Diego, 1989.

27. J. P. Leonard, *in* ''Methods in Neurosciences'' (P. M. Conn, ed.), Vol. 1, p. 62. Academic Press, San Diego, 1989.

28. R. J. Leonard, A. Karschin, S. Jayashree-Aiyar, N. Davidson, M. Tanouye, L. Thomas, G. Thomas, and H. A. Lester, *Proc. Natl. Acad. Sci. U.S.A.* **86,** 7629 (1989).

29. H. A. Lester, *Science (Washington, D.C.)* **241,** 1057 (1988).

30. C. Methfessel, V. Witzemann, T. Takahashi, M. Mishima, S. Numa, and B. Sakmann, *Pfluegers Arch.* **407,** 577 (1986).

31. R. Miledi and I. Parker, *J. Physiol. (London)* **415,** 189 (1989).

32. B. Moss, O. Elroy-Stein, T. Mizukami, W. A. Alexander, and T. R. Fuerst, *Nature (London)* **348,** 91 (1990).

33. M. W. Quick, J. Naeve, N. Davidson, and H. A. Lester, *Biotechniques* **13,** 357 (1992).

34. C. M. Rice, C. A. Franke, J. H. Strauss, and D. E. Hruby, *J. Virol.* **56,** 227 (1988).

35. B. Sakmann and E. Neher, *in* ''Single Channel Recording'' (B. Sakmann and E. Neher, eds.), p. 37. Plenum, New York, 1983.

36. J. Sambrook, E. F. Fritsch, and T. Maniatis, ''Molecular Cloning: A Laboratory Manual,'' 2nd Ed. Cold Spring Harbor Laboratory Press, Cold Spring Harbor, NY, 1989.

37. A. A. Smith, T. Brooker, and J. Brooker, *FASEB J.* **1,** 380 (1987).

38. T. P. Snutch, J. P. Leonard, J. Nargeot, H. Lubbert, N. Davidson, and H. H. A. Lester, *in* ''Soc. Gen. Physiol. Ser.'' (L. J. Mandel and D. C. Eaton, eds.), Vol. 42, p. 154. Rockefeller University Press, New York, 1987.

39. R. E. Straub, Y. Oron, and M. C. Gershengorn, *in* ''Methods in Neurosciences'' (P. M. Conn, ed.), Vol. 1, p. 46. Academic Press, Orlando, 1989.

40. K. Sumikawa, I. Parker, and R. Miledi, *in* ''Methods in Neurosciences'' (P. M. Conn, ed.), Vol. 1, p. 30. Academic Press, San Diego, 1989.

41. M. Taglialatela, L. Toro, and E. Stefani, *Biophys. J.* **61,** 78 (1992).

42. W. B. Thornhill and S. R. Levinson, *Biochemistry* **26,** 4381 (1987).

43. A. Ullrich, J. Shine, J. Chirgwin, R. Pictet, E. Tischer, W. J. Rutter, and H. M. Goodman, *Science (Washington, D.C.)* **196,** 1313 (1977).

44. G. Valle, J. Besley, and A. Colman, *Nature (London)* **291,** 338 (1981).

45. X. C. Yang, A. Karschin, C. Labarca, O. Elroy-Stein, B. Moss, N. Davidson, and H. A. Lester, *FASEB J.* **5,** 2209 (1991).

Section V

Imaging and Electrophysiology

[15] Characterization of Glutamate Receptor Function Using Calcium Photometry and Imaging

Amy B. MacDermott, Victor Henzi, and David B. Reichling

Introduction

Glutamate is a ubiquitous amino acid neurotransmitter in the mammalian central nervous system. Activation of glutamate receptors can cause changes in the intracellular concentration of free Ca^{2+} ions ($[Ca^{2+}]_i$) by direct permeation of Ca^{2+} through the receptor–channel complex, by modulation of voltage-gated Ca^{2+} channel activation, or by causing release of Ca^{2+} from intracellular stores. Thus a relatively simple, rapid, nonintrusive method for addressing some aspects of glutamate receptor diversity, function, and localization is to record changes in the $[Ca^{2+}]_i$ using fluorescent indicator dyes (1).

Due to the increasing numbers of glutamate receptors being identified by physiological and molecular biological means, our view of glutamate receptor function over the past few years has become more complex. There are two broad families of "ionotropic" glutamate receptors called the N-methyl-D-aspartate (NMDA) and the non-NMDA receptors. The ionotropic glutamate receptor subunits gate cation channels and have molecular structural characteristics of ligand-gated channels. When functionally expressed in oocytes or cell lines, ionotropic subunits display diverse properties. Further complexities are revealed when the receptor subunits are permitted to form hetero-oligomers (2). Recent evidence has revealed the presence of an entirely different family of "metabotropic" glutamate receptors. These receptors activate second messenger systems rather than functioning as receptor–channel complexes. Metabotropic glutamate receptors have seven presumed membrane-spanning domains typical of G protein-coupled receptors (3).

In this chapter, we discuss why Ca^{2+}-sensitive dye recording is well suited for the study of glutamate receptor function. Next, we present two different approaches to $[Ca^{2+}]_i$ measurement using two dyes, indo-1 and fura-2. We intend to provide sufficient detail to assist novices in the field of $[Ca^{2+}]_i$ measurement in choosing the appropriate technique for their experimental needs and to aid them in implementing their decision. Other

Methods in Neurosciences, Volume 19

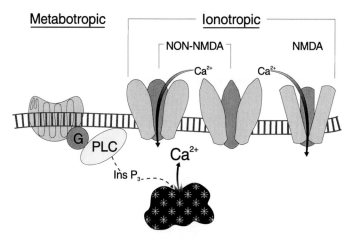

FIG. 1   Schematic illustrates how the different families of glutamate receptors expressed in mammalian neurons influence $[Ca^{2+}]_i$. Metabotropic glutamate receptors are coupled to G proteins (G) and at least two of the identified metabotropic receptors activate phospholipase C (PLC) which may in turn result in release of Ca^{2+} from intracellular stores. The NMDA receptors are all Ca^{2+} permeable and some of the non-NMDA ionotropic receptors are Ca^{2+} permeable. Not shown are interactions of the different receptors with voltage-gated Ca^{2+} channels.

reports should be consulted for more detailed considerations of dye properties and the technical difficulties involved in fluorescence measurements (e.g., 4–6). Finally, we consider some of the specific applications of these two techniques for the study of glutamate receptors in neurons and in cells transfected with cDNA clones encoding glutamate receptors.

$[Ca^{2+}]_i$ as an Indicator of Glutamate Receptor Function

Some Glutamate Receptors Are Permeable to Ca^{2+}

The NMDA and non-NMDA ionotropic glutamate receptors are receptor–channel complexes that open following ligand binding and allow cation flow across the membrane (Fig. 1). Typically, the monovalent cations Na^+ and K^+ permeate these channels. In the case of the NMDA receptor and some configurations of the non-NMDA receptors, the divalent cation Ca^{2+} also permeates these channels.

There are several subfamilies within the non-NMDA ionotropic receptor family. These subfamilies, defined by the sequence similarity and functional expression of their composite subunits, are named by the non-NMDA receptor ligand to which they are the most sensitive. The precise nomenclature for the non-NMDA receptor subunits has yet to be agreed upon. The first set of subunits to be identified are currently referred to as the ''AMPA receptors'' since they show relatively high affinity for the non-NMDA agonist α-amino-3-hydroxy-5-methyl-4-isoxazole propionic acid (AMPA) relative to kainate (7–9). These four subunits and their splice variants are called either GluR1–4, GluRA–D, or α1–4. AMPA receptors appear to occur in the brain as pentamers of GluR1–4 subunits (10). When the adult brain form of GluR2 is included as part of a hetero-oligomer with the other AMPA receptor subunits the functional receptors are Ca^{2+} impermeant. Site-directed mutagenesis studies indicate that these permeability characteristics are controlled by a single amino acid substitution that occurs in the putative second transmembrane domain of GluR2. Further work has shown that this substitution occurs *in vivo* by RNA editing (2). GluR2 in the mature nervous system has an arginine in place of a glutamine. GluR1, 3, and 4 have glutamine in the equivalent location. When a receptor lacks GluR2, or has an unedited GluR2 with glutamine rather than arginine, the activated receptor is permeable to Na^+, K^+, Ca^{2+}, and Mg^{2+}. The functional role for Ca^{2+}-permeable AMPA receptors is currently under study, but an important clue may be that the unedited GluR2 is expressed during nervous system development [reviewed in (2)].

Another subfamily of non-NMDA receptors, referred to as ''kainate receptors,'' appears to be composed of two groups of subunits. The three subunits in the first group are called GluR5–7 or β1–3. These may form hetero-oligomeric receptors by combining with a second group of subunits made up of the molecules called KA1 and 2 or γ1 and 2. These kainate receptors show high affinity for the non-NMDA agonist kainate relative to AMPA in both homomeric and heteromeric configuration. GluR5 and 6 are RNA edited at the glutamine/arginine position in the second transmembrane segment, suggesting the possibility of Ca^{2+}-permeant forms. However, little is known about the functional occurrence of these receptors in neurons or their permeability characteristics *in situ*. Individual subunits have been localized within the brain, but little information is available about their normal configuration or role in the function of the nervous system [reviewed in (2)].

Some NMDA receptor subunits have been cloned (11–13), and again it appears that these receptors are composed of heteromeric oligomers. Two groups of subunits, referred to as NMDAR 1 or ζ1 and NMDAR 2 A–D or ε1–4, contribute to functional receptor formation. Homomeric NMDAR 1 or heteromeric NMDAR 2 receptors form functional channels that are permeable to Na^+, K^+, and Ca^{2+}. Subunit configuration can determine the sensitivity to channel block by Mg^{2+} as well as the sensitivity to glycine and modulation following application of phorbol esters, compounds which activate protein kinase C (12).

Some Receptors Stimulate Ca^{2+} Release from Intracellular Stores

Currently there are five known forms of the metabotropic glutamate receptor, named mGluR1–5. Two of these receptors, mGluR1 and 5, are coupled to the phospholipase C-inositol triphosphate signaling pathway (3, 14) as illustrated in Fig. 1. Inositol triphosphate formation can stimulate release of Ca^{2+} from intracellular stores bearing receptors for this molecule. Such Ca^{2+}-release events may be discriminated from those mediated by Ca^{2+}-permeable receptors by demonstrating that the agonist-induced $[Ca^{2+}]_i$ rise does not depend on extracellular $[Ca^{2+}]_i$.

Fluorescent Ca^{2+} Indicators

Functional Aspects of Ca^{2+}-Sensitive Dyes

Roger Tsien and his colleagues designed the first members of the rapidly expanding group of fluorescent Ca^{2+}-indicator dyes that are derivatives of the Ca^{2+}-selective chelator BAPTA (1). The two characteristics of these dyes that are most important in determining their utility are their fluorescent properties and their affinities for Ca^{2+}. The two techniques for $[Ca^{2+}]_i$ detection described in this chapter use different dyes: the photometry experiments use indo-1 and the video experiments use fura-2 (1). Both dyes undergo wavelength shifts when they bind Ca^{2+}. Indo-1 changes its emission spectrum and fura-2 changes its excitation spectrum. These shifts in wavelength spectra provide two fluorescence values that change in opposite directions with changes in $[Ca^{2+}]_i$. For indo-1, $[Ca^{2+}]_i$ is nonlinearly proportional to the ratio of the fluorescence detected at 405 and 480 nm following excitation of the dye at 350 nm. As $[Ca^{2+}]_i$ increases, the fluorescence emission at 405 nm increases while that at 480 nm decreases. Conversely, for fura-2, the wavelength shift is in the excitation spectrum. With increasing $[Ca^{2+}]_i$, fluorescence at 500 nm increases when excited at 340 nm and decreases when excited at 380 nm. The ability to calculate a ratio of fluorescence at two wavelengths for both indo-1 and fura-2 allows direct estimation of Ca^{2+} concentration without the need to measure the path length or concentration of the dye (1).

The affinities of indo-1 and fura-2 for Ca^{2+} are similar, with K_d's between

200 and 300 nM in salt solution designed to mimic mammalian intracellular ionic conditions. While these values are little affected by pH and Mg^{2+} within physiological ranges, they are sensitive to viscosity and ionic strength (4). Thus, for example, the K_d's for Ca^{2+} can be different in cells with relatively lower (e.g., amphibian) or higher (e.g., molluscan) intracellular ionic strength. The K_d's for Ca^{2+} are optimal for measuring many of the physiological fluctuations in $[Ca^{2+}]_i$ that occur in mammalian neurons in the nanomolar range. If changes in $[Ca^{2+}]_i$ reach to the micromolar range, however, dyes with higher K_d's such as furaptra may be more suitable, although the ionic selectivity of furaptra is problematic when it is used as a Ca^{2+} indicator (15, 16).

Details of Dye Usage

In our laboratory, we have used indo-1 in association with dual photometry to measure small or fast changes in $[Ca^{2+}]_i$, while we have used fura-2 with video imaging to detect localized changes in $[Ca^{2+}]_i$ or to monitor many cells simultaneously. The following sections describe specific experimental arrangements used in these experiments.

Loading Cells with Dye

Both indo-1 and fura-2 are available in a charged or acid form and an esterified form. The charged, membrane-impermeant form is typically injected directly into cells, while the esterified form can diffuse into cells and, in most cases, be de-esterified and trapped inside. Most cells dissociated from central or peripheral nervous tissue and either prepared acutely or grown in tissue culture can be readily loaded with the ester forms of the dyes. In addition, a large variety of transfectable cell lines are easily loaded with the esterified dye whether the cells are attached to a substrate or are in suspension. In contrast, it is difficult to load esterified dyes into neurons in acutely prepared slices of mature brain tissue or in organotypic cultures of neonatal brain tissue, perhaps because nonneuronal cells limit access of dye to neurons. In some cases, this problem can be overcome by focally applying a stream of esterified dye onto the slice of tissue (17). It is more common in these preparations, however, for the charged dye to be introduced into the cytoplasm by injection through a patch pipette or sharp microelectrode [see below and (18)].

Cells may be loaded with esterified indo-1 or fura-2 by exposing them to the acetoxymethyl ester forms of the dyes for 10–45 min at room temperature in the dark. A 10 μM solution of indo-1 ester is prepared by dissolving 10 μl of a stock solution of 1 mM indo-1 ester (in dimethyl sulfoxide) into a pH-buffered, balanced salt solution. To improve solubilization of the dye, 0.025% pluronic acid (Molecular Probes) is added to the solution which can then be sonicated for 1–3 sec and/or vigorously vortexed for up to 1 min.

A major potential artifact related to dye loading is compartmentalization of dye within the cell. Thus, fluorescence emitted by dye enclosed in the nucleus, mitochondria, and other organelles will not accurately reflect cytoplasmic $[Ca^{2+}]_i$ levels, thereby contaminating photometrically measured signals. Compartmentalization can sometimes be reduced by shortening loading times, loading with lower concentrations of dye, and loading at lower temperatures (4). A second possible source of contaminating dye signal is incompletely deesterified dye that remains in the cytoplasm. This pool of dye is insensitive to $[Ca^{2+}]_i$ and, in contrast to the fully acidic forms of the dyes, cannot be quenched by manganese.

To inject cells with the charged dye, the dye is dissolved in the electrode solution appropriate for the recording condition. For patch recording, a relatively low concentration of dye, ~150 μM, is included in the electrode solution. For sharp electrode dye injection, higher concentrations (in the millimolar range) are necessary to pass sufficient dye through the fine pipette tip.

Limitations on Intracellular Dye Concentration

Although the dyes will faithfully report $[Ca^{2+}]_i$ regardless of dye concentration, they can at the same time significantly modify $[Ca^{2+}]_i$. This is because the dyes, like BAPTA itself, are Ca^{2+} buffers and always affect Ca^{2+} dynamics to some extent. Therefore it should be emphasized that the practice of estimating the intracellular concentration of dye is important for most experiments. It is always preferable to use the minimum concentration of dye that yields measurements with an acceptable signal-to-noise ratio. Dye concentrations less than 150 μM are generally considered to minimally affect Ca^{2+} dynamics. Higher dye concentrations improve signal detection but have been found to introduce prominent changes in the apparent amplitude and kinetics of Ca^{2+} signals, especially when rapid Ca^{2+} changes are being considered (6, 16). Increasing the intensity of the exciting light will improve signal detection but when raised too far it will cause dye bleaching and phototoxicity. In experiments where Ca^{2+} detection rather than Ca^{2+} quantitation and kinetics is the primary objective, modest overloading with indicator dye may be an acceptable way to improve the signal-to-noise ratio.

Optical Considerations

The Microscope

When using a microscope to determine changes in $[Ca^{2+}]_i$, the microscope objective serves both as a condenser for the exciting light and a collector lens for the emitted light. In practice, the main considerations in choosing an objective are its magnification, which should suit the experiment, and the objective's efficiency in collecting emitted light, a characteristic that is quantified by the numerical aperture (NA). A higher NA objective is preferred because it typically yields greater signal-to-noise ratios with less bleaching and phototoxicity than lower NA objectives, assuming exciting light is the same. An upright or inverted microscope may be used for these experiments. However, inverted microscopes are generally more flexible when electrophysiological access to the preparation is required, particularly if a higher NA, lower working distance, oil-immersion objective is in use. Thus, inverted microscopes are a good choice if the cell preparation is relatively thin and sufficiently transparent to permit the visibility necessary for electrophysiological maneuvering. Dissociated neurons grown in culture, thin organotypic cultures of central nervous system tissue, or monocultures of cell lines are all appropriate preparations for recording with an inverted microscope.

The Optical Path

The optical path arrangement for using indo-1 is simpler than that for fura-2 because, with indo-1, neither the excitation nor emission wavelengths need to be changed mechanically during an experiment (see Fig. 2). Rather, indo-1 fluorescence emission is optically split to allow simultaneous and continuous measurement of fluorescence at each of the two wavelengths. Using fura-2, in contrast, requires a mechanism for rapidly switching the exciting light between two wavelengths. However, since only one emission wavelength is collected for fura-2 (unlike indo-1), only one detector is required. Therefore, fura-2 is the better-suited dye for video microscopy because using only one video camera is less expensive and does not require alignment of video images.

Exciting Light

Figure 2 illustrates the essential features of the optical path of both the exciting and emitted light in our Ca^{2+} experiments using indo-1 and fura-2. For indo-1, the exciting light is filtered at 350 ± 10 nm by an interference

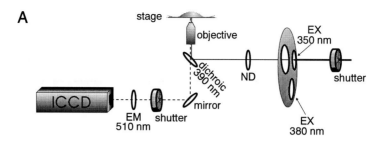

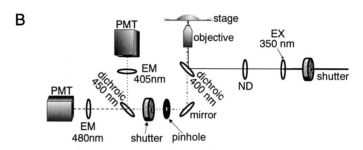

FIG. 2   Two schematics show the basic elements of the optical paths for (A) the fura-2 video-imaging system and (B) the indo-1 photometry system. The exciting light comes in from the right, goes through the shutter to the microscope objective and the emitted light goes out to the fluorescence detectors. EX and EM indicate excitation and emission filters. The appropriate peak transmitted wavelengths are given below each filter. ND, neutral-density filter; PMT, photomultiplier tube and the cutoff wavelengths are specified for the dichroic filters.

filter. Neutral-density filters should be used to adjust light intensity to the minimum required for sufficient activation of the dye. To minimize bleaching and phototoxicity, a remote-controlled shutter is useful to allow light to the preparation only when required. For fura-2 recording, the exciting path is similar except that there must be a provision for switching the wavelength of the exciting light in a manner coordinated with data acquisition. Although there is some flexibility, typical excitation filters used for fura-2 are 380 ± 10 nm and either 340 or 350 ± 10 nm. If the exciting light path is primarily quartz optics, 340 nm may be used. However, with glass optics, 350 nm is more appropriate since glass will strongly filter light at wavelengths below 350 nm. Finally, after the proper wavelengths are selected by the excitation filters, the exciting light is reflected toward the specimen by a dichroic filter (see Fig. 2).

Emitted Light

Unlike the excitation light, which is reflected up to the specimen by the dichroic filter, the higher wavelength emitted fluorescence passes through the dichroic filter toward a final filter or set of filters. For indo-1, the dichroic filter passes light at wavelengths above 400 nm and, for fura-2, the dichroic filter passes wavelengths above 390 nm.

During dual-photometer recordings using indo-1, an appropriately sized aperture restricts the collection of emitted light to that of the single cell under study. The emitted light is again split by a mirror that reflects wavelengths less than 450 nm. The two beams are passed through emission interference filters of 405 or 480 ± 10 nm before being detected by high-sensitivity photomultiplier tubes.

After traversing the dichroic filter and before reaching the photodetector, the emitted fura-2 fluorescence passes through a barrier filter. Depending on the need to limit the background fluorescence, the barrier filter may either be a long-pass filter with a cutoff around 490 nm or a wide band-pass filter. For example, we use a filter which passes 510 ± 20 nm.

Indo-1 Photometry

Appropriate Applications—Fast Kinetics and Small Changes in $[Ca^{2+}]_i$

Indo-1 together with dual photomultipliers may be used to follow changes in $[Ca^{2+}]_i$ in cell bodies and in thicker cell processes. Because the fluorescence emission from a significant portion of the cell is spatially averaged by sensitive photomultipliers, a favorable signal-to-noise ratio can usually be achieved with minimal dye loading. Thus, this system is useful for reliable detection of small changes in $[Ca^{2+}]_i$. For example, we have used indo-1 photometry as a sensitive means to detect Ca^{2+} entry following activation of NMDA receptors in the presence of the channel-blocking ion, Mg^{2+} (Fig. 3). At resting membrane potential, it is frequently assumed that little or no current flows through NMDA receptors in the presence of 1 mM Mg^{2+} due to the voltage-dependent channel block produced by this ion. However, we are often able to detect small but distinct $[Ca^{2+}]_i$ signals under such conditions.

Another important application of the indo-1 photometry system is in resolving rapid changes in $[Ca^{2+}]_i$ when detailed knowledge of the kinetics

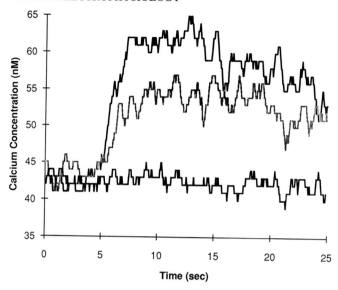

FIG. 3   Calcium responses are shown following three consecutive applications of 5×10^{-7} M glutamate to an embryonic rat dorsal horn neuron grown in tissue culture. These signals were recorded using the indo-1 photometry system. The responses were evoked in the presence of Mg^{2+} at concentrations of 10^{-3} (bottom trace), 10^{-4} (middle trace), and 0 mM (top trace). This concentration of glutamate is low enough to be selective for NMDA receptor activation.

of the $[Ca^{2+}]_i$ transient is desired. Such information may reveal differences in mechanisms of $[Ca^{2+}]_i$ elevation. For example, we have found that there are two components of the $[Ca^{2+}]_i$ response induced by the non-NMDA agonists, kainate and AMPA. One component is due to indirect activation of voltage-gated Ca^{2+} channels and is completely blocked by moderate concentrations of the trivalent cation, La^{3+}. The second component is resistant to La^{3+} and apparently results from Ca^{2+} entry directly through AMPA and kainate-gained non-NMDA receptors (19). The kinetics of the $[Ca^{2+}]_i$ transients evoked at the cell bodies of embryonic dorsal horn neurons by kainate in the presence and absence of the voltage-gated Ca^{2+} channel (VGCC) blocker, La^{3+}, differ substantially. These experiments were performed with agonist applications of several seconds duration. The rise times of the two responses were readily distinguishable illustrating the ability of this system to readily and accurately track rapid signals. Thus, the dual-photometer system provides a relatively simple way to continuously track fluorescence changes on the subsecond time scale because it does not require changing the filter or light path.

Technical Aspects

The Photomultiplier Tube

Either photodiodes or photomultipliers may be used as light detectors in $[Ca^{2+}]_i$ photometry systems. Photodiodes are inexpensive and have better quantum efficiency. This means that for a given number of photons, a large percentage will be sensed by the detector. However, while photomultiplers have a lower quantum efficiency, they have high quantum gain. Thus for every photon detected by a photomultiplier, the output signal is boosted many fold higher, making the signal much higher than the thermal or dark current noise generated by the photomultipler tube itself. Thermal noise generated by photomultipler tubes is sometimes reduced by cooling the tubes. However, this is usually not necessary in $[Ca^{2+}]_i$ photometry systems where the level of thermal noise is small compared to the biological and instrument noise. In our indo-1 experiments, we use two photomultipliers to record fluorescence at the two different emission wavelengths (Fig. 2B). It is worthwhile noting that once the dual photomultiplier system is calibrated for $[Ca^{2+}]_i$, it is important not to change the gain on the photomultipliers by changing the voltage applied to each tube since this will change the calibration.

Data Acquisition and Analysis

Transformation of the recorded fluorescence signal into a value that is proportional to or equal to $[Ca^{2+}]_i$ requires a number of arithmetic operations. Therefore, it is generally convenient to digitize and store the signal output from the photomultipliers for analysis after sampling has been completed. At the beginning or end of each experiment, a background measurement at each emission wavelength is made by recording fluorescence from an indo-1-free area of the preparation near the cell under study. These background values are then subtracted from the digitized fluorescence signal recorded from the cell at each of the two emission wavelengths. A ratio is calculated from these subtracted values for each time point. The $[Ca^{2+}]_i$ can then be calculated using the equation,

$$[Ca^{2+}]_i = K_d \times F_{min}/F_{max} \times [(ratio - ratio_{min})/(ratio_{max} - ratio)],$$

where K_d is the dissociation constant of indo-1 for Ca^{2+} determined *in vitro* or in the cell (4). This determination is performed *in vitro* using a series of accurately calibrated Ca^{2+}-buffered solutions (19, 20). F_{min}/F_{max} represents the ratio of fluoresence at 485 nm in zero nominal $[Ca^{2+}]_i$ and in saturating $[Ca^{2+}]_i$; ratio refers to the experimentally measured ratio of fluorescence at

405 and 485 nm; $ratio_{max}$ and $ratio_{min}$ are the ratios of fluorescence at 405 and 485 nm determined in neurons in conditions of saturating and minimal $[Ca^{2+}]_i$, respectively.

It is important to be aware that when a $[Ca^{2+}]_i$ transient is recorded, the calculated signal may display an associated increase in noise that could be misinterpreted as actual biological oscillations of $[Ca^{2+}]_i$. This artifact arises when fixed-amplitude, instrument-derived noise is amplified by different amounts as the intensity of the fluorescent signal changes with $[Ca^{2+}]_i$. This occurs for two reasons. First, because the signal-to-noise ratio decreases when increasing $[Ca^{2+}]_i$ causes the intensity of the 480-nm emitted fluorescence signal to decrease toward a minimum value. That decrease in signal-to-noise ratio is then reflected by increased noise in the calculated 405/480 ratio. Second, during the calculation that converts the 405/480 ratio into $[Ca^{2+}]_i$ (equation given above), noise is further amplified whenever the $[Ca^{2+}]_i$ reaches levels that begin to saturate the dye. In this way, an unchanging level of nonbiological noise in the recorded signal can appear deceptively similar to stimulus-induced $[Ca^{2+}]_i$ oscillations of biological origin.

Fura-2 Video Imaging

Appropriate Applications—Spatial Resolution and Population Screening

Video imaging is especially well suited for screening populations of cells since it is possible to record changes of $[Ca^{2+}]_i$ from many cells at the same time. This may be used to identify individual cells in a population that express different types of glutamate receptors. The Ca^{2+}-permeable receptors and metabotropic receptors are examples (19).

Video imaging may be used to indicate localized changes in $[Ca^{2+}]_i$. Figure 4 shows data recorded from six different locations on a fura-2-filled neuron. This neuron was located in a hippocampal slice growing in organotypic culture for 2 weeks and was filled with dye by injection through a sharp-tipped microelectrode. The same electrode was then used to apply two, 100-msec depolarizing pulses to the soma (location 1). It is apparent that the amplitude and kinetics of the $[Ca^{2+}]_i$ responses varied with location on the cell from which the measurements were made.

Video miscrosopy experiments similar to that shown in Fig. 4 should be helpful in identifying where higher densities of Ca^{2+}-permeable glutamate receptors are located on the surfaces of neurons. For example, it is possible to use video microscopy in combination with fura-2 to investigate the localization of the Ca^{2+}-permeable AMPA or kainate-activated receptors in embry-

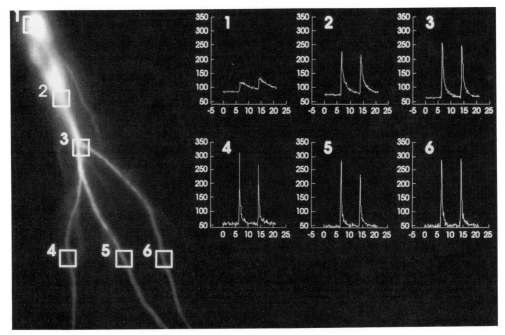

FIG. 4   Subcellular measurements of $[Ca^{2+}]_i$ were made from a hippocampal neuron filled with fura-2. Rapid measurements of $[Ca^{2+}]_i$ were made by time zero ratio generation at the six sites selected by the white boxes drawn over the image of the neuron. In this case, the cell was stimulated by 2 100-msec depolarizing pulses of current through the recording electrode. It is clear from this figure that different regions of the dendrites show different changes in $[Ca^{2+}]_i$. These data also demonstrate the level of signal detection in long dendritic processes that can be achieved with the video-imaging system. Hippocampal slices grown for 2 weeks in cultures were used in the experiment illustrated here. The neuron was impaled with a sharp microelectrode containing 14 mM fura-2 at the tip.

onic dorsal horn neurons. By blocking Ca^{2+} entry through VGCCs, we can assume that the Ca^{2+} transients we record following agonist applications are due to Ca^{2+} entry through glutamate receptors. Although localized differences in Ca^{2+} buffering can influence such observations, both the spatial and the temporal aspects of the $[Ca^{2+}]_i$ transient can give insight about receptor localization.

Photomultipliers have high quantum gain and are readily used with indo-1, the dye that does not require filter changes to obtain ratio measurements. Thus, Ca^{2+} measurement by photometry may be faster and in some cases more sensitive than those with video microscopy. However, video cameras offer strong advantages over photomultiplier tubes or even arrays of photodiodes, when detailed spatial information about Ca^{2+} changes is desired or when many individual neurons are to be monitored simultaneously. However, as spatial resolution increases, the signal-to-noise ratio decreases because there is less spatial averaging. Thus, spatial resolution should be sought only to the level that is required to address the experimental question. In cases where the signal-to-noise ratio is sufficiently poor, temporal averaging becomes a necessity. In the special case of confocal microscopy, the ability to spatially localize Ca^{2+} fluxes can be brought to its theoretical maximum, but at the cost of a low signal-to-noise ratio.

Technical Aspects

The basic elements of a video-imaging system designed to follow changes in $[Ca^{2+}]_i$ include a video camera sensitive to low light levels, an image acquisition and processing device, data storage hardware, and a computer that can control the peripheral optics and shutters while interacting with the image acquisition hardware. When building or buying such a system, investigators should establish their own minimum requirements of convenience and speed, but some general guidelines apply.

The master control computer should permit coordination between the shutters controlling access to the exciting light, the camera, the excitation filters, and image acquisition. The image processor should have the capability to average successive images, to allow acquisition and subtraction of background images, and to calculate a ratio image from these averaged and background-subtracted images. It should allow measurements from specific regions of the image, preferably both during and following data collection. Finally, the system should be able to use the ratio image to calculate $[Ca^{2+}]_i$ based on appropriate calibration procedures. The data storage device must have a large capacity, with the capability to store hundreds of megabytes of digitized data.

The Camera

Currently, three types of cameras are most commonly used for low-light, quantitative applications. The silicon-intensified target (SIT), the intensified charge-coupled device (ICCD) and the cooled charge-coupled device (cooled CCD). The advantages and disadvantages of each camera define which is best suited for individual experimental requirements.

The SIT camera produces a standard video output (30 frames/sec). Because it is an analog photosensor rather than an array of discrete photosensitive elements, a SIT camera has the best spatial resolution of the three cameras. The SIT cameras are being utilized less frequently for $[Ca^{2+}]_i$ measurement, however, because of their slow response time. For example, delays on the order of 500 msec may be necessary whenever the excitation filter is switched to avoid gathering erroneous and temporally smeared data from a SIT camera.

Like the SIT camera, the ICCD gives a standard video output. This is an important feature since it allows interfacing with image processors and video monitors that are designed for this format. Incoming photons are intensified to overcome the thermal noise of the ICCD's solid-state photosensors. The response speed is excellent, but the limitations of the high-voltage intensifier contribute to its poor spatial resolution and quantum efficiency. At present, the ICCD is less costly than the cooled CCD and is the camera of choice for image processors that require a standard video input.

The cooled CCD differs fundamentally from the SIT and the ICCD in that its output is not standard, but user defined. While this aspect adds flexibility, it also adds considerable complexity. Thus, image-processing hardware and software designed to optimize the flexibility of the cooled CCD must differ intrinsically from those used for the SIT or the ICCD. For example, the cooled CCD allows the camera's output to be customized by the user, defining which pixels are read or averaged at the level of the camera itself. Largely because there is no intensification of the incoming photons, cooled CCDs offer the best quantum efficiency of the three cameras. However, they are inherently slow in gathering enough photons to achieve good signal-to-noise ratio and they are slow to read out the data.

Speed of Measurement

A disadvantage of video imaging compared to photometry is the relative slow speed of video acquisition. This is due in part to the use of fura-2, which typically requires sequential measurement at two discrete excitation wavelengths. In addition, image acquisition requires more sophisticated and expensive hardware to process efficiently. Finally, video measurements often require temporal averaging to compensate for the increased noise arising from the reduced spatial averaging.

Of these factors, the one that can be minimized most easily, yet for most investigators is the rate-limiting step, is the speed of the image-processing hardware and data storage devices. Therefore, before purchasing image-processing and data storage hardware, it is worthwhile to consider hardware capabilities and limitations. For example, frame averaging, background subtraction, and ratio calculation are essential features of an image processor. Depending on the hardware, each of these operations can take up to hundreds of milliseconds to perform. In practice, it is advantageous for the processor to display the image for experimenter feedback during the experiment. This operation may also take up to hundreds of milliseconds. These additional delays are substantial compared to the 33 msec needed to acquire a single video frame.

Speed and efficiency of image acquisition are also limited by the time required to store the image data. When digitized, a single frame of data occupies about 250 kilobytes. Digital storage to even a relatively fast hard disk drive is slow (on the order of 200 msec), thereby considerably delaying data acquisition. However, as the amount of computer RAM commercially available on computers increases, RAM will increasingly provide a fast, high-capacity alternative for temporary storage of images during an experiment. Furthermore, computer bus rates and disk drives are also being designed with increased speed and efficiency. Some image processors allow the storage of only small portions of an image, a property that greatly speeds data storage rates. A more expensive alternative is to save the frame data to an optical disk recorder. These devices can store image data at video rates but may introduce image distortion if the digitized ratio image is stored.

Other aspects of the video imaging system that can potentially slow down data acquisition include the speeds of the camera, filter switcher, and shutter. The relative speeds of different cameras have been discussed in the previous section. The time required to switch the exciting light between two wavelengths varies considerably among the available devices, from on the order of 1 to 100 msec or more.

As a means of increasing data acquisition speed, approximate measurement of $[Ca^{2+}]_i$ is sometimes performed without repetitive switching between two excitation wavelengths. In this practice, the dual-wavelength fluorescence is measured at the first time point or "time zero" in the usual fashion for ratio measurements [see, for example (21)]. Subsequently, measurements are made only at one wavelength, 380 nm, and $[Ca^{2+}]_i$ is estimated based on comparison to the $[Ca^{2+}]_i$ at time zero. This approach allows rapid data acquisition with fura-2 by bypassing the mechanism for switching exciting light, which is mechanically slow. However, it opens the door to artifacts arising from changes in fluorescence not due to changes in $[Ca^{2+}]_i$, such as dye bleaching, changes in the intensity of the exciting light, or changes in dye distribution. With standard ratio imaging, these artifacts are compensated for by the calculation of a dual-wavelength ratio for every image.

$[Ca^{2+}]$ Quantitation and Calibration

The $[Ca^{2+}]_i$ detected by fura-2 may be estimated using the same equation as that used for indo-1 (see above) where the K_d must be determined, ratio is fluorescence excited by 350/380 nm, and F_{min}/F_{max} is the fluorescence excited by 380 nm in zero and saturating $[Ca^{2+}]$. Since the K_d is sensitive to viscosity and the ionic strength of the solution, intracellular calibrations are preferable to in vitro calibrations, although often difficult to achieve due to the active participation of the cell in Ca^{2+} buffering (4, 6).

Prior to calculating the ratio image, background images must be subtracted from the images acquired using the 350- and 380-nm excitation. Background images at each wavelength allow nonuniformities in the image generated by the hardware and the biological preparation to be subtracted such that the ratio is calculated from fura-2 fluorescence alone. The ideal background image would be one obtained of the exact field under study prior to any addition of fura-2. However, this is often impractical and compromises must be made that allow one to approximate the ideal condition. [See Refs (5) and (16) for further discussion.]

After background subtraction, the fluorescence value of some of the pixels in an image may be near zero. This is particularly true for the pixels representing small diameter regions of cells with low intracellular volume and consequently few dye molecules, or extracellular areas with small amounts of autofluorescence. These regions of the image may be removed from the ratio image by thresholding prior to calculating the ratio. Indeed, investigators frequently choose an arbitrary threshold level of fluorescence which must be exceeded for each pixel to be accepted as reliable data.

Applications with Neurons

Since Ca^{2+} permeates through some glutamate ionotropic receptors and not through others, $[Ca^{2+}]_i$ transients can be used to identify the presence of the Ca^{2+}-permeant receptor species. In this way, $[Ca^{2+}]_i$ measurements are useful to approach some of the questions of immediate interest in the field of glutamate receptors, such as identifying the configurations of receptor subunits that occur in specific neuronal populations and determining where in the complex neuronal morphology these receptor species are expressed. However, neuronal membranes also bear VGCCs that could confuse interpretation of the source of Ca^{2+} following receptor activation.

A Problem with Voltage-Gated Calcium Channels

Because glutamate receptors are cation-selective, channel opening leads to membrane depolarization under physiological conditions. Except when a neuron is voltage clamped (with good space clamp) during the experiment, glutamate receptor activation may stimulate opening of voltage-gated channels. Thus, when utilizing $[Ca^{2+}]_i$ measurement as a tool to probe glutamate receptors, one must be aware of the potential contribution of Ca^{2+} entry through VGCCs to the measured $[Ca^{2+}]_i$ signals. This problem is further complicated by the fact that the divalent cations typically used to block VGCCs are sometimes permeable themselves and can interact with the indicator dye. Furthermore, some of these divalent cations also interact with the glutamate receptors themselves, potentially changing channel permeability and conductance. Since specific toxins for all VGCC types are not yet available, the complication of VGCCs makes experimental design a nontrivial endeavor.

Our approach to the problem has been to use La^{3+}. Lanthanum is extremely effective in blocking VGCCs, without itself being particularly permeable through these channels. We have characterized the effects of La^{3+} on NMDA and non-NMDA receptor-mediated currents in cultured and acutely dissociated spinal cord dorsal horn neurons (22). Our results indicate that La^{3+} can be a useful tool for isolating Ca^{2+} entry through some glutamate receptors.

Receptor Identification

Calcium measurements can be used in conjunction with other techniques to tackle the problem of which receptor subunits are expressed in which neurons. For example, the demonstration of the presence or lack of Ca^{2+} entry through non-NMDA receptors can be used as a criterion to soundly establish the presence of particular glutamate receptor subunit species in different neuronal cell populations when used in conjunction with *in situ* hybridization analysis, electrophysiological measurements, and immunocytochemical observations. The La^{3+}-resistant AMPA and kainate-evoked changes in $[Ca^{2+}]_i$ recorded in embryonic dorsal horn neurons (19) are specific examples of this type of application.

Identification of Subpopulations of Neurons

Video imaging of $[Ca^{2+}]_i$ can be used to rapidly screen a population of cells for a particular property or set of properties. For example, using dissociated hippocampal neurons, Iino *et al.* (23) have used electrophysiological techniques to describe a small and discrete subpopulation of neurons that express Ca^{2+}-permeable non-NMDA receptors. With the appropriate experimental design, Ca^{2+} measurements could be used to determine the distribution of the subpopulation of cells and to target cells for subsequent electrophysiological characterization.

Localization of Receptors on Single Neurons

Just as video imaging of $[Ca^{2+}]_i$ can be used to identify disparities in Ca^{2+} signals between cells, it may also yield information about variations in the spatial profile of Ca^{2+} increases within a single cell. For example, Ca^{2+} signals arising from Ca^{2+} release from intracellular stores are likely to differ from those mediated by Ca^{2+} entry in both their spatial and temporal characteristics. Furthermore, particular glutamate receptor subtypes may be confined to different regions of the dendritic arbor. While it is possible to obtain information from electrophysiological measurements indicating spatial distribution of glutamate receptors (24), $[Ca^{2+}]_i$ imaging techniques are also suited for those types of questions.

Applications with Transfected Cells

One approach to studying the relationship between glutamate receptor structure and function is to transfect cell lines with specific receptor subunits (7). When the functional receptors are expressed they can then be studied using physiological, biophysical, and immunocytochemical techniques. An aspect of cell transfection which makes it a particularly powerful technique is that it allows the receptors to be placed in a new non-neuronal environment allowing individual subunits or subunit combinations to be studied in isolation. Finally, with recent technical advances in molecular biology, transfection experiments allow expression of receptors following site-directed mutagenesis of the cloned receptor subunits in order to facilitate the study of structure/function aspects of glutamate receptors.

Calcium measurements can be a useful tool for identifying and studying transfected cells for several reasons. For example, since transfection efficiency is never 100%, video imaging of $[Ca^{2+}]_i$ can be used to rapidly and efficiently identify the cells that have been successfully transfected. [An additional tool to identify transfected subpopulations is by the combined use of indo-1 and flow cytometry (18)]. The simplicity and noninvasiveness of Ca^{2+} measurements make it a convenient assay to detect mutagenetic modification or pharmacological modulation of glutamate receptor properties, such as their Ca^{2+} permeability. Finally, since many of the cell lines most commonly used for transfection experiments do not express VGCCs, transfection experiments with glutamate receptors and Ca^{2+}-indicator dyes are simplified.

Conclusion

It has been our purpose in this chapter to draw attention to the emergence of an exciting new group of questions about glutamate receptor function and to describe the theoretical and practical aspects of how $[Ca^{2+}]_i$ measurement can be useful in addressing some of these questions. Calcium measurements are well suited for identifying subpopulations of cells containing particular glutamate receptors as well as for deducing information about their localization within the cells. Additionally, analysis of the kinetics of $[Ca^{2+}]_i$ changes can reveal properties that bear on the functional role of the glutamate receptors. Calcium measurement alone will not be sufficient to answer all of these questions. However, because it is a simple and rather noninvasive means to probe cells, it can be used in conjunction with electrophysiological, molecular biological, and immunocytochemical approaches to unravel the functional heterogeneity of glutamate receptors and the impact this diversity has on neuronal physiology.

Acknowledgments

We thank Ira Schieren and Mark Heath for helpful discussions on the material covered in this chapter. We are also very grateful to Ira Schieren for lending his skill and artistry in computer graphics to the development of some of our figures.

References

1. M. Grynkiewicz, M. Poenie, and R. Y. Tsien, *J. Biol. Chem.* **260**, 3440 (1985).
2. B. Sommer and P. H. Seeburg, *TIPS* **13**, 291 (1992).
3. Y. Tanabe, M. Masu, T. Ishii, R. Shigemoto, and S. Nakanishi, *Neuron* **8**, 169 (1992).
4. M. W. Roe, J. J. Lemasters, and B. Herman, *Cell Calcium* **11**, 63 (1990).
5. R. Y. Tsien and A. T. Harootunian, *Cell Calcium* **11**, 93 (1990).
6. M. Wahl, M. J. Lucherini, and E. Gruenstein, *Cell Calcium* **11**, 487 (1990).
7. K. Keinanen, W. Wisden, B. Sommer, P. Werner, A. Herb, T. A. Verdoorn, B. Sakmann, and P. H. Seeburg, *Science (Washington, D.C.)* **249**, 556 (1990).
8. J. Boulter, M. Hollmann, A. O'Shea-Greenfield, M. Hartley, E. Deneris, C. Maron, and S. Heinemann, *Science (Washington, D.C.)* **249**, 1033 (1990).
9. N. Nakanishi, N. A. Schnieder, and R. Axel, *Neuron* **5**, 569 (1990).

10. R. J. Wenthold, N. Yokotani, K. Doi, and K. Wada, *J. Biol. Chem.* **267,** 501 (1992).

11. H. Monyer, R. Sprengel, R. Schoepfer, A. Herv, M. Higuchi, H. Lomeli, N. Burnashev, F. Sakmann, and P. H. Seeburg, *Science* (*Washington, D.C.*) **256,** 1217 (1992).

12. T. Kutsuwada, N. Kashiwabuchi, H. Mori, K. Sakimura, E. Kushiya, K. Araki, H. Meguro, H. Masaki, T. Kumanishi, M. Arakawa, and M. Mishina, *Nature* (*London*) **358,** 36 (1992).

13. N. Nakanishi, R. Axel, and N. A. Schneider, *Proc. Natl. Acad. Sci. U.S.A.* **89,** 8552 (1992).

14. T. Abe, H. Sugihara, H. Nawa, R. Shigemoto, N. Mizuno, and S. Nakanishi, *J. Biol. Chem.* **267,** 13361 (1992).

15. B. Raju, E. Murphy, L. A. Levy, R. D. Hall, and R. E. London, *Amer. J. Physiol.* **256,** C540 (1989).

16. W. Regehr and D. W. Tank, *J. Neurosci.* **12,** 4202 (1992).

17. W. Regehr and D. W. Tank, *J. Neurosci. Methods* **37,** 111 (1991).

18. N. T. Slater, S. R. Glaum, S. T. Alford, D. J. Rossi, and G. L. Collingridge, this volume [18].

19. D. B. Reichling and A. B. MacDermott, *J. Physiol.* (*London*) **469,** 67 (1993).

20. I. Schieren and A. B. MacDermott, "Cell Lines for Neurobiology: A Practical Approach (John Wood, ed.), Vol. 161. Oxford University Press, 1992.

21. H. Miyakawa, W. N. Ross, D. Jaffe, J. C. Callaway, N. Lasser-Ross, J. E. Lisman, and D. Johnston, *Neuron* **9,** 1163 (1992).

22. D. B. Reichling and A. B. MacDermott, *J. Physiol.* (*London*) **441,** 199 (1991).

23. M. Iino, S. Ozawa, and K. Tsuzuki, *J. Physiol.* (*London*) **424,** 151 (1990).

24. O. Arancio and A. B. MacDermott, *J. Neurophysiol.* **65,** 899 (1991).

[16] Phosphorylation of Glutamate Receptors Using Electrophysiology Techniques

John F. MacDonald, Lu-Yang Wang, and Michael W. Salter

Introduction

It is generally accepted that postsynaptic glutamate receptors which incorporate an ion channel fall into at least three major subclasses: (a) N-methyl-D-aspartate (NMDA), (b) α-amino-3-hydroxy-5-methyl-4-isoxazoleproprionic acid (AMPA), and (c) high-affinity kainate (relative to kainate's affinity for AMPA receptors). The recent cloning of excitatory amino acid (EAA) receptor subunits from each of these major categories has confirmed this classification (1). Many of these glutamate receptor clones possess a variety of sites of potential phosphorylation by seronine–threonine kinases and recent evidence suggests that phosphorylation of some subunits by protein kinases can upregulate their function (2–4). Consistent with the hypothesis that glutamate receptors are regulated through phosphorylation by protein kinases are the reports that intracellular applications of protein kinase C and/or activators of this kinase can potentiate NMDA currents in a variety of preparations (5, 6). Similarly, AMPA/kainate receptors in cultured hippocampal neurons can be regulated by cAMP-dependent protein kinase (7, 8).

We have been interested in using patch-clamp techniques to help answer a general question about glutamate receptors: Does direct phosphorylation of these receptors (or at least a closely associated protein) functionally regulate glutamatergic transmission in the mammalian central nervous system?

Whole-Cell Patch-Clamp Recordings

Whole-cell patch clamping of mammalian neurons is now routinely performed by many laboratories working on preparations ranging from cultured neurons to *in vitro* neurons in the slice. At first glance this technique has significantly advanced electrophysiological studies of central neuronal excitabilty. Most of the advantages of this technique over conventional sharp electrode recordings relate to an increased ease of making the recordings and an improvement in the quality of the recordings. The high resistance of the seal of the patch in combination with the low-resistance pathway the electrode provides into the cell interior permits substantial improvements in the electrical resolu-

Methods in Neurosciences, Volume 19

tion of the recording. This can be particularly advantageous if one is attempting to resolve small events such as spontaneous miniature synaptic potentials. It can also be argued that the temporal and spatial properties of a voltage clamp should also be greatly improved. However, central mammalian neurons (e.g., in culture or slice) are seldom adequately space clamped regardless of the recording technique employed. Also attention must be paid to the difficulties which may arise as a consequence of the diffusional exchange of solutes between the electrode and the interior of the cell (9). This disruption of the neuron's intracellular milieu can raise questions about the physiological significance of observed events. Furthermore, in the case of neurons with anything more than a very simple geometry (e.g., small-diameter acutely isolated neurons) there is absolutely no guarantee that the concentrations of solutes in the electrode are going to be attained by the cell nor for that matter that they will reach the same steady-state concentration in all regions of the same cell (e.g., the soma versus fine processes). As with the problems of assuring an adequate space clamp it may also be difficult to ensure a steady state and uniform spatial distribution of solutes applied through the patch electrode.

During whole-cell voltage clamp the interior of the cell is exchanged by diffusion with the solution contained within the recording electrode. If channels are regulated intracellularly their activity may decrease or "wash out" as second messengers (e.g., cAMP) or cofactors (GTP, ATP, etc.) are lost from the cell's interior. For example, intracellular regulation of calcium (Ca^{2+}) channels has been demonstrated using this technique. Ca^{2+} currents wash out, but this can be reversed by including ATP and the catalytic subunit of cAMP-dependent protein kinase in the dialysis solution (10). Furthermore, nucleotides, activated kinases, and protease inhibitors (e.g., calpastatin) may be required to sustain channel activity in membrane patches (11). This suggests that some Ca^{2+} channels might be directly phosphorylated before they can demonstrate full activity. However, the washout or rundown of channel activity is likely to be much more complex than the previous statements imply. In some cases, a transitory "runup" of channel activity has been observed (12). Furthermore, a number of processes which may be completely independent of dephosphorylation can contribute to channel rundown (13–15). In particular, whole-cell recordings may fail to maintain low and stable concentrations of intracellular calcium concentration ($[Ca^{2+}]_i$) even though high concentrations of buffers are employed. Elevated intracellular Ca^{2+} may then contribute to Ca^{2+} channel inactivation and/or trigger a number of calcium-dependent process.

Glutamate currents also wash out or run down in cultured hippocampal neurons (8, 9). In some cases the rate of washout was uninfluenced by wide variations in the Ca^{2+} buffer capacity of the pipette solutions. In addition

NMDA channels recorded from outside-out macropatches demonstrate a time-dependent rundown (15). Such evidence suggests that intracellular factors may regulate both NMDA and AMPA/kainate channels in cultured neurons. When an "ATP regenerating solution" (16) was employed in the patch pipette glutamate currents did not wash out as rapidly and in some cases the washout could be at least partially reversed (8, 9) suggesting that the maintenance of phosphorylation reactions may be important for retarding but not necessarily preventing washout. There are of course other interpretations of such data. Some channels are directly modulated by nucleotides. However, the ATP-regenerating solution rapidly lost its ability to prevent washout if it was permitted to degrade and nonhydrolyzable ATP analogs such as β, γ-methylene ATP did not retard rundown while the thiophosphate analog of ATP (ATPγS) was effective (17). In contrast ATP solutions did not prevent the rundown of NMDA single-channel currents recorded in outside-out patches (15). It has been known for some time that elevations in intracellular Ca^{2+} concentration are associated with a depression of NMDA currents (18). Thus, uncontrolled increases in intracellular Ca^{2+} might contribute to the rundown of glutamate currents during whole-cell recordings. The ATP-regenerating solutions may have simply helped to maintain low $[Ca^{2+}]_i$ perhaps as a consequence of the upregulation of Ca^{2+} pumps. Despite this potential problem it has been possible to demonstrate that when techniques are employed to monitor $[Ca^{2+}]_i$ (see below) there clearly is a Ca^{2+}-independent component of the washout of NMDA currents (18a).

The rundown of glutamate receptor-channel activity (and other channels for that matter) either in whole-cell recordings or in recordings from patches likely results from a wide variety of cellular-, membrane-, and receptor-dependent changes. Knowledge about these alterations can potentially provide information about pathological processes in the cells. However, it should be remembered that in most cases the objective of such experiments is not to fully characterize washout but rather to determine the mechanisms that are functionally important for regulating glutamate receptors.

In an attempt to determine if glutamate receptors can be functionally regulated by protein kinases we are employing several different approaches. The most direct is to introduce an activated kinase directly into the recorded cell using the whole-cell patch electrode itself. In some cases cultured hippocampal neurons have been employed and the activities of endogenous and presumably native receptors are assayed using conventional voltage-clamp measurements. In addition, the same techniques can be used to examine the effects of the kinase on glutamate receptor clones expressed in cell lines (e.g., HEK293 cells). Appropriate peptide inhibitors of these kinases can also be perfused in order to test the specificity of the kinase's actions. In addition, potential sites of phosphorylation found in the amino acid sequence of a given glutamate clone can be selectively altered using site-directed mutagenesis techniques. It would be predicted that mutants lacking the appropriate site(s) would fail to respond to the application of the kinase. The success of this approach depends on having a method for applying high-molecular-weight substances to the cell interior. We have developed a simple and inexpensive technique for perfusing the inside of the patch electrode during whole-cell recording that is described below.

Other patch techniques can offer viable alternatives to whole-cell recordings. In particular, the perforated patch technique developed by Horn and Marty (19, 20) can overcome the disadvantage of diffusional exchange of the interior. In this technique a pore-forming substance such as Nystatin is placed in the patch electrode. The presence of a high density of Nystatin pores provides a low-resistance pathway into the cell but prevents the diffusion of substances any larger than the simple monovalent cations (e.g., Na^+, K^+, and Cs^+) and anions (Cl^-). This technique is far less disruptive to the internal milieu and presumably will overcome interior diffusional exchange problems. We have adopted this technique for recording from cultured hippocampal neurons for the purpose of examining the functional modulation of glutamate receptors by endogenous protein kinases and phosphatases.

The Intracellular Perfusion Technique

A variety of different types of internal perfusion systems (21) have been successfully applied to study both whole-cell and single-channel currents in various preparations. Some of these techniques can be elaborate and require considerable effort, time, and equipment in the preparation of the electrode assembly. Our objective was to develop an easy and simple method of perfusing substances into neurons during whole-cell recordings. We also wanted to record from a neuron for a considerable period prior to applying substances and in this way cells could serve as their own controls. Considering that routine long-term (e.g., 1–2 hr) whole-cell recordings can be difficult to make our procedure was developed with the objective of keeping preparation time and complexity to a minimum. In addition, we did not want to rely on expensive and/or complex manipulators for locating an internal perfusion device within the patch electrode. The easiest solution to this problem is to insert a second, but smaller diameter, glass pipette into the patch electrode and connect it to a perfusion system. This inner pipette can be constructed using whatever electrode puller is currently available in the laboratory. While such an assembly can introduce some electrical noise into the recording

electrode, and therefore is not particularly useful for single-channel recordings, it is more than adequate for the delivery of high-molecular-weight
substances such as protein kinases into cultured hippocampal neurons. We
also developed a particularly simple procedure for placing the inner electrode
tip close to the opening of the patch electrode, thereby reducing the time
required to change the solution in the pipette and thus facilitating entry into
the cell soma. The entry of substances into the cell is thus dependent on
other factors such as the access resistance into the cell and the mobility of
the substance in solution (22). This configuration of electrodes allows us to
make repeated changes of the solution entering the neuron. Using this system
we are able to characterize the time course of the actions of protein kinases
as well as construct dose–response and current (I)–voltage (E) relationships
for excitatory amino acids before and after the injection of a given protein
kinase. In addition, having such a routine system permits us to collect large
numbers of miniature excitatory synaptic currents (mEPSCs) from cultured
hippocampal neurons both prior to and following the intracellular perfusion
of protein kinases. Thus, the effects on synaptic transmission that a protein
kinase delivers exclusively to the postsynaptic neuron can be explored in
some detail. We believe that this particular technique has dramatically expanded the flexibility of our experiments.

Design of the Electrode Holder

In keeping with our plan for simplicity we designed a series of modifications
of an existing commercial electrode holder. The holder (Fig. 1) is a modified
E. W. Wright (Guilford, CT, 06437) holder (A037) which is chosen to take
a patch electrode of 1.5 mm o.d. A rubber gasket with a Plexiglas plug and
screw cap arrangement secures the pipette. This holder also has a pin and
silver–silver chloride pellet assembly which connects to the head stage (Axo-
patch-1B, Axon Instruments, Foster City, CA). A standard port also comes
off to the right of this axis and provides the connection for the suction tubing
used to apply a vacuum during seal formation. In addition, this port provides
an escape for solutions displaced during perfusion of test solutions into the
patch pipette. Our contribution to the holder was simply to request that a
second pipette port together with screw cap and rubber gasket be placed at
the opposite end of the holder relative to the first (A026, 0.04″ pin). This
pipette port is chosen to hold a pipette of 0.75 mm o.d. High-quality glass
of this diameter (6250) is readily available from A-M systems, Inc. (Everett,
WA 98203) and it fits easily within the inner diameter of our patch electrode glass.

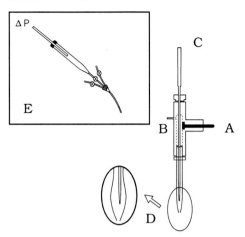

FIG. 1 A schematic of the electrode holder used for intracellular applications of various solutions. (A) The electrical connection from the silver–silver chloride pin to the head stage of the amplifier; (B) the suction port; (C) the internal pipette port. The enlargement at the bottom (D) shows the bubble formed in the tip of patch pipette with the tip of the internal pipette placed inside. (E) The simple manifold and a single 1-ml syringe reservoir for intracellular solutions.

Internal Perfusion Pipette

The internal pipette (0.75 mm o.d.) is pulled on a conventional electrode puller (in our case a David Kopf Instruments, Tujunga, CA, Model 700C) under relatively high heat. The pipette which is formed has a long taper (approximately 1 cm), a total length of about 10 cm, and a much smaller tip than would normally be used for a patch electrode. Tip diameter is not important, however, as the perfusion pipette is broken back to give an outer diameter of about 10 μm. As anticipated the diameter of this pipette determines the rate of flow of solution into the outer patch electrode. Furthermore, the distance that the tip of the inner electrode is placed relative to that of the outer pipette is restricted by this diameter. The tip of the inner pipette obviously must not be jammed into the patch pipette and we describe below an alteration of the patch electrode which eliminates this potential problem.

Patch Electrodes

Conventional patch-clamp electrodes are constructed using 1.5-mm-o.d. thin-walled glass tubing (TW150F-4, WPI) which contains a thin glass filament to facilitate filling. Patch electrodes are pulled on a vertical puller (Narishige

Scientific Instruments Laboratory, Tokyo, Japan; Model PP-83) and fire polished to a final tip diameter of approximately 2 μm. The inner pipette is then placed inside the patch electrode as described below. Recently we have routinely modified the shape of the patch electrode in order to facilitate placement of the inner pipette tip and also to enhance the exchange rate of pipette solution with the cell interior. This modification is based on a similar and previously described technique (23).

A piece of polyethylene tubing and a syringe are attached to the patch pipette and it is placed in a microforge (Narishige, Scientific Instruments Laboratory, Tokyo, Japan; Model MF-83) used for fire polishing electrode tips. This forge allows vertical placement of the electrode in apposition with a perpendicularly oriented platinum wire filament, while under visual observation with a built-in compound microscope. The patch electrode is lowered laterally to the plane of the wire so that its tip can be placed well below the level of the wire. This permits the tapered portion of the pipette just above the tip to be positioned next to the wire filament. The glass is then heated while pressure is applied to the interior of the pipette using the syringe. An expansion of the pipette is formed near the tip. This expansion will then accommodate the relatively large tip of the internal pipette. The expansion itself can readily be located less than 50 μm from the tip of the patch electrode.

Assembly

The perfusion system is assembled simply by feeding the solution-filled inner pipette backward through the screw cap of the port which will eventually hold the patch pipette and out of the opposite or inner pipette port and screw cap. The position of the inner pipette can be adjusted so that any desired length of the inner pipette protrudes from the patch pipette port. The inner pipette is held snugly by tightening the screw cap and gasket arrangement of its own port. The chamber of the holder is then filled with solution in order to ensure electrical contact with the silver–silver chloride pellet. The filled patch pipette is then carefully threaded by hand over the inner pipette and through its screw cap and then clamped in position. The suction port is clamped off so that the solution in the holder is prevented from leaving. Some fluid will leak from the gasket of the patch pipette port when the electrode is inserted. At this stage the inner pipette tip does not need to be close to the patch tip. Final adjustment of the inner tip is performed under a dissecting microscope. The upper screw cap is loosened and the inner pipette is placed into the patch pipette's bubble by hand. The holder and screw cap hole are themselves sufficient to guide the inner pipette. The protruding rear end of the internal pipette is connected to thin plastic tubing (Nalge Co., Rochester, NY, 14602, No. 8000-0020) which is fed into a tiny manifold. The manifold is constructed from a plastic pipette tip threaded on the end of the tubing supplying the inner pipette. Three pieces of similar tubing are then inserted and glued into the opening of the plastic pipette tip. Various intracellular solutions that are to be perfused into the patch pipette are stored in 1-ml syringes and forced into this manifold through one of the three pieces of tubing. The perfusion solution is forced into the electrode under constant pressure (maximum 10–15 psi) provided by a connection to a picospritzer II (General Valve Corp., E. Hanover, NJ). Small clamps are placed on each of the individual tubes in order to control which of the solutions will be injected.

Both the patch and internal pipettes are first filled with control solution in order to prevent contamination with any of the perfusion solutions. The internal pipette is then inserted into the recording pipette and positioned within the patch electrode's bubble. The patch is then formed in a conventional manner while all of the tubes are clamped off to prevent the suction from draining the solution out of the inner pipette. Once a high-resistance seal has been achieved, and the patch ruptured, the suction is immediately released. Internal perfusion is then initiated by unclamping the appropriate tubes and applying pressure to the interior of its reservoir syringe. The pressure differential is applied gradually rather than turned on suddenly in order to avoid any vibration of the electrode assembly. Solution forced out of the internal pipette displaces that in the patch electrode and the overflow leaves via the suction tubing which is left open to atmospheric pressure. The flow rate is kept at about 5–10 μl/min by adjustment of the pressure applied to the pipette. If the flow rate appears to be lower than desired then a slight vacuum can be applied to the suction tubing. The patch electrode tip is rapidly exchanged with the perfusion solution. For example, when a fluorescent dye is applied the solution in the patch tip is completely exchanged in much less than 5 min and some cases it is fully exchanged in about 1 min.

Perforated Patch Technique

We have successfully applied this technique to cultured hippocampal neurons by making some small modifications to the original technique described by Horn and Marty (19). As illustrated in Fig. 2 we have chosen to monitor patch patency by observing voltage-dependent Ca^{2+} currents (suggestion of Dr. Phillipe Ascher) although the use of fluorescent dyes is equally effective (24). These currents wash out within several minutes following rupture of the patch and therefore loss of these currents quickly

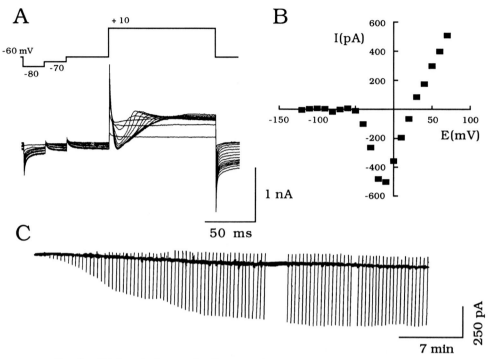

FIG. 2   (A) A typical example of progressive perforation of a Nystatin patch monitored by three voltage steps. Depolarizations from -60 to $+10$ mV activate voltage-activated Ca^{2+} currents and are preceded by two hyperpolarizing steps used for calculating ongoing changes in leak conductance. Inward Ca^{2+} currents are detected 10 min after the formation of seal and increase in amplitude over time. These currents reached a steady state after 20 min and remain stable at least for 1 hr. Representative current traces taken every 2 min (2 to 62 min) are superimposed. Note the gradual increase in the capacitive transients and shift in leak conductance. (B) Current–voltage relationship constructed about 1 hr after the formation of seal reveals the stability of voltage-activated Ca^{2+} currents. (C) The patch perforation can also be monitored by repeatedly applying agonist (kainate, 250 μM) to cell soma. Kainate currents gradually increase in amplitude as Nystatin enters the membrane and remain unaltered for up to 2 hr once stable. The transient capacitive currents above the baseline resulted from hyperpolarizing steps from the holding potentials -60 to -70 mV which were applied before each application of kainate in order to monitor capacitance and leak conductance.

signals rupture of the patch. Using the techniques described below we have been able to record stable excitatory amino acid currents for extended periods of time. In addition, we have been able to obtain high-resolution recordings of mEPSCs.

Recordings are made with patch electrodes with tip diameters of 2–3 μm filled with (in mM) 140 CsCl/KCl, 35 CsOH/KOH, 10 HEPES, 11 EGTA, 2 MgCl$_2$, 2 TEA, 1 CaCl$_2$ (pH 7.3), 320 to 335 mOsm. The presence of Cs$^+$ ions in the electrode, together with their permeation through the perforated patch, permits the blockade of various potassium currents. In some cases 50 mM CsCl/KCl and 90 mM Cs$_2$SO$_4$/K$_2$SO$_4$ are used to replace 140 mM CsCl/KCl with the objective of achieving a more rapid Cl$^-$ equilibrium. The extracellular solution contains (in mM) 140 NaCl, 1.3 or 13 CaCl$_2$, 5.4 KCl, 25 HEPES, 25 glucose, and 0.001 tetrodotoxin (TTX) (pH 7.4), 320 to 335 mOsm. Elevated concentrations of extracellular Ca^{2+} can be employed in order to accentuate the Ca^{2+} currents used to monitor patch patency.

Preparation of Nystatin Solutions

A stock solution of 50 mg/ml Nystatin (Squibb/Sigma) is made in dimethyl sulfoxide (DMSO) and stored in a −20°C freezer. Solutions are protected from light as much as possible. This stock solution is thawed immediately before use and refrozen for storage. This stock will remain effective for about 1 week.

The stock solution is diluted to a final concentration of 125 μg/ml in the recording solution and vortexed vigorously for 1–2 min just prior to filling of an electrode. We have found that no sonication is required. In fact excessive sonication may result in a loss of potency (20). This final recording solution is refrigerated and remains effective for at least 2 hr.

Perforated Patch Electrodes

The recording electrodes used for the perforated patch are constructed from borosilicate thin-walled glass (1.5 mm o.d., TW150F-4, WPI). This type of glass contains a filament which is absolutely necessary for controlling the volume of solution reaching the tip of the pipette during filling (see below). The electrodes used for perforated patch recordings typically have resistances of about 2 MΩ and as little taper at the tip as possible. Long tapered electrode tips are not recommended for perforated patch recordings.

The most critical element in making a successful perforated patch recording is the filling of the patch electrode. The tip of the recording pipette must first be filled with a Nystatin-free solution because the presence of Nystatin at the electrode tip dramatically reduces the success of making a high-resistance seal. On the other hand, if Nystatin reaches the tip too slowly then a stable perforation of the patch will not be achieved during the period of the recording. One answer is to perfuse Nystatin into the patch (e.g., as above); however, we have developed a simpler method for routinely making perforated patches. This method owes its success to a procedure which allows us to control the volume of the Nystatin-free solution initially placed in the tip.

First, with a small piece of tissue paper in one hand a drop of Nystatin-free solution is touched to the rear end of the patch pipette. The presence of the glass filament rapidly transports a small volume of the solution to the tip. Immediately following application of the drop, the tissue is placed on the rear end of the pipette in order to blot excess solution. If the volume in the tip is greater than desired, the tissue is simply kept a few minutes longer on the electrode and excess solution is drawn out of the electrode by capillary action along the glass filament. The volume required in the tip will depend on the shape and size of the patch electrode tip. These factors will determine the time required to achieve a steady-state perforation of the patch.

Second, the electrode is quickly backfilled with a Nystatin-containing solution and a seal to the cell made (usually within 2 min) before Nystatin is able to diffuse to the very tip of the electrode. In our hippocampal cultures, a seal can be obtained without applying any positive pressure to the pipette as its goes through the air–water interface. The quality of the seal usually improves over the time even if the initial seal resistance is not that high. The cell is then held -60 mV and repeated voltage steps (-80 to 0 mV) are made in order to monitor the appearance of the perforation and the detection of Ca^{2+} currents. As Nystatin starts to enter the membrane (usually within 5 min), capacitive transients and voltage-gated Ca^{2+} currents increase in amplitude reaching a steady-state in about 15 to 30 min. Stable responses can be maintained for at least 1 hr following the initial stage of patch perforation.

Simultaneous Patch-Clamp Recordings and Fluorescent Measurements of Intracellular Calcium

The pivotal role of intracellular Ca^{2+} as a key regulator of numerous intracellular biochemical pathways is well known. Glutamate receptors both affect $[Ca^{2+}]_i$ and are affected by it. For example, activation of NMDA receptors increases $[Ca^{2+}]_i$, via influx of Ca^{2+} through NMDA channels (25) and release of Ca^{2+} from intracellular stores (26). On the other hand, NMDA channels themselves appear to be modulated by $[Ca^{2+}]_i$ such that increases in Ca^{2+} depress NMDA-channel activity (18). Studies on the role of phosphorylation in regulating glutamate channels must therefore consider whether phosphorylation affects the EAA-evoked increases in $[Ca^{2+}]_i$, and, as well, whether Ca^{2+} itself might be a mediator of the effects of phosphorylation. In order to directly measure $[Ca^{2+}]_i$, we have implemented a system for simultaneous patch-clamp recording and Fura-2 measurement of $[Ca^{2+}]_i$. Voltage clamp in concert with the measurement of $[Ca^{2+}]_i$ is crucial because holding the membrane potential at a constant level prevents the activation of voltage-dependent Ca^{2+} channels which is concomitant with activation of glutamate receptors in unclamped cells. The simultaneous recording approach also provides the opportunity to study the second messenger pathways causing changes in $[Ca^{2+}]_i$ when the whole-cell mode is utilized. Fluorescent dyes for the measurement of a number of ion and other molecules are becoming increasingly available and thus simultaneous optical and electrical measurement will undoubtedly become important in studying other important intracellular regulators such as H^+, Na^+, Mg^{2+}, and cyclic AMP.

Optical–Electrical Recording System

A number of systems for combined recording of optical and electrical signals have been described (e.g., 27). The setup we use is illustrated schematically in Fig. 3. An important consideration in the development and implementation of this setup is that recording optical and electrical signals simultaneously should be about as easy as recording electrical signals alone. Our setup was designed for dual excitation dyes in order to use the ratiometric method to calculate directly intracellular ion concentration. We chose to develop the system for Fura-2 because, relative to the other principally used dye, indo-1, Fura-2 is more readily loaded into cells in the membrane-permeant form, has a higher quantum efficiency, and is less susceptible to photobleaching. The use of a dual excitation dye requires that the cells be alternatively illuminated at two excitation wavelengths. In the present setup this is accomplished by a chopper blade which alternates the excitation light through either a 340 or a 380 narrow band pass filter. The light is then recombined in a quartz fiber optic cable which is coupled to the epifluorescence adapter of a Nikon Diaphot microscope. The fiber optic cable transmits essentially no vibration and hence serves to mechanically isolate the chopper from the recording setup. The chopper can run at rates up to 300 Hz which allows sampling of the optical signals at rates similar to those for the electrical signals.

Individual cells are visualized using Hoffman modulation contrast (Modulation Optics, East Hills, NY 11548). In order to maximize the throughput of the fluorescence signal the modulator, which is normally located in the back of the objective attenuating light transmission by about 50%, is repositioned after the Fura-2 signal is split off. Maximum sensitivity for detecting low light levels is achieved by using a photomultiplier tube in single-photon counting mode. This mode has the added advantage of being linear over

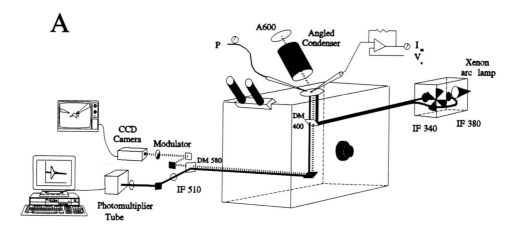

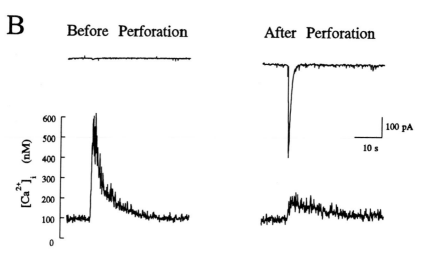

FIG. 3   (A) Schematic diagram of optical–electrical recording system. (B) Examples of simultaneous current (upper traces) and $[Ca^{2+}]_i$ (lower traces) measurements during perforated patch recording from a dorsal horn neuron in culture. The traces on the left were recorded before the patch had begun to perforate. These traces show the increase in $[Ca^{2+}]_i$ caused by applying L-aspartate (250 μM, 100 msec) to an unclamped cell. On the right are traces obtained 10 min after the perforation had stabilized. At this time, the inward current caused by L-aspartate is now observed as is the increase in $[Ca^{2+}]_i$.

more than four orders of magnitude. A rectangular diaphragm just before the photomultiplier tube is used to limit the emitted light to that from soma of the cell under study. The hardware and software we use for measuring $[Ca^{2+}]_i$ are from Photon Technologies Inc., Canada (London, Ontario).

Whole-Cell Recording and Fura-2 Measurements

For whole-cell recordings and $[Ca^{2+}]_i$ measurements the neurons may be preloaded with dye by incubation with the membrane permeant form, Fura-2 acetoxymethyl ester (Fura-2AM). We incubate for 90 min at 21–23°C with 1 μM Fura-2AM in the extracellular HEPES-buffered Hanks' balanced salt solution described above with 0.5% bovine serum albumin added. Fura-2AM is stored at $-20°C$ as a stock in dry DMSO and is diluted 1/1000 immediately before use. We have found that the working solution should be shaken vigorously and then sonicated for about 30 sec for optimum loading. After being loaded, the cells are washed in the extracellular recording solution. The intracellular recording solution contains the following (in mM): Fura-2 pentapotassium salt, 0.05; KCl/CsCl, 140; KOH/CsOH, 35; HEPES, 10; EGTA/BAPTA, 0.1–30; MgATP, 0 or 4; pH 7.25. A small amount of EGTA or BAPTA is included to prevent grossly overloading the cells with Ca^{2+} while still being able to observe EAA-evoked increases in $[Ca^{2+}]_i$. Higher levels of EGTA or BAPTA blunt the response to EAAs but may be used to maintain low baseline levels of $[Ca^{2+}]_i$ when required (18a).

When the cells are preloaded with dye, Ca^{2+} signals can be recorded as soon as the whole-cell configuration is attained and this is crucial when, for example, washout of EAA responses is being studied. Alternatively, when the cells are not preloaded with Fura-2AM time is required for the Fura-2 pentapotassium to diffuse before reliable, low-noise Ca^{2+} signals can be recorded. The latter approach has the advantage of limiting compartmentalization of Fura-2 into mitochondria and other noncytoplasmic areas which is quite important if cells sequester the dye. We have found that the loading strategy described above produces limited compartmentalization as the vast majority of the fluorescence washes out of the cells if Fura-2 is not included in the intracellular solution. We have also used washin/out of the fluorescence signal to estimate the intracellular concentration of the dye following preloading with Fura-2AM as 25–100 μM in our cells.

Preloading cells also allows control responses to be recorded before the intracellular milieu is altered by dialysis. This is an additional advantage when the dialysis produced by whole-cell recording is being used to deliver agents intracellularly for studying second messenger pathways and/or releasable pools of Ca^{2+} (28, 29).

Perforated Patch Recordings and Fura-2 Measurements

As the perforated patch technique produces limited disturbance of the intracellular microenvironment, it is well-suited to studies of $[Ca^{2+}]_i$ in voltage-clamped cells while altering intracellular Ca^{2+} metabolism as little as possible. For such studies the cells must be loaded with Fura-2AM as the Nystatin pores are impermeable to the free acid form of Fura-2. The perforated patch recordings are made just as described above. We do not measure Ca^{2+} routinely until the patch has stabilized so that the cell is exposed to as little UV irradiation as possible in order to minimize damage to the cell and photobleaching of the dye. However, recording one, or more, responses before the perforation of the patch begins establishes a control level of response, in the unclamped cell, which can be used for later comparison (Fig. 3B). We have found that intracellular Ca^{2+} can be measured reliably for more than 30 min after stabilization of the patch providing that it does not rupture. Rupture of the patch is readily detected during the simultaneous recordings by a rapid decrease in baseline levels of $[Ca^{2+}]_i$ as cells are dialyzed with the patch solution containing 10 mM EGTA and 1 mM Ca^{2+}.

Ackknowledgments

We thank the Medical Research Council of Canada Group "Nerve Cells and Synapses" and the Nicole Fealdman Memorial Fund for funding much of the research on which this chapter is based.

References

1. G. P. Gasic and M. Hollmann, *Annu. Rev. Physiol.* **54,** 507 (1992).
2. B. U. Keller, M. Hollmann, S. Heinemann, and A. Konnerth, *EMBO J.* **11,** 891 (1992).
3. T. Kutsuwada, N. Kahiwabuchi, H. Mori, K. Sakimuar, E. Kushiya, K. Araki, H. Meguro, H. Masaki, T. Kumanishi, M. Arakawa, and M. Mishina, *Nature (London)* **358,** 36 (1992).
4. M. Yamazaki, H. Mori, K. Araki, K. J. Mori, and M. Mishina, *FEBS Lett.* **300,** 39 (1992).
5. L. Chen and L. Y. M. Huang, *Nature (London)* **356,** 521 (1992).
6. S. R. Kelso, T. E. Nelson and J. P. Leonard, *J. Physiol. (London)* **449,** 705 (1991).
7. P. Greengard, J. Jen, A. C. Nairn, and C. F. Stevens, *Science (Washington, D.C.)* **253,** 1135 (1991).
8. J. F. MacDonald, I. Mody, and M. W. Salter, *J. Physiol. (London)* **414,** 17 (1991).
9. L.-Y. Wang, M. W. Salter, and J. F. MacDonald, *Science (Washington, D.C.)* **253,** 1132 (1991).
10. J. E. Chad and R. Eckert, *J. Physiol. (London)* **378,** 31 (1986).
11. C. Romanin, P. Grosswagen, and H. Schindler, *Pfluegers Arch.* **418,** 86 (1991).
12. S. L. Mironov and H. D. Lux, *Neurosci. Lett.* **133,** 175 (1991).
13. B. Belles, J. Hescheler, W. Trautwein, K. Blomgren, and J. O. Karlsson, *Pfluegers Arch.* **411,** 353 (1988).

14. T. M. Egan, D. Dagan, J. Kupper, and I. B. Levitan, *J. Neurosci.* **12**(5), 1964 (1992).
15. W. Sather, S. Dieudonne, J. F. MacDonald, and P. Ascher, *J. Physiol.* (*London*) **450**, 643 (1992).
16. P. Forscher and G. S. Oxford, *J. Gen. Physiol.* **85**, 743 (1985).
17. M. C. Bartlett, M. W. Salter, and J. F. MacDonald, *Soc. Neurosci. Abstr.* **218.13**, 107 (1989).
18. M. L. Mayer, A. B. MacDermott, G. L. Westbrook, S. L. Smith, and J. L. Barker, *J. Neurosci.* **7**, 3230 (1987).
18a. Y. T. Wang, Y. S. Pak, and M. W. Salter, *Neurosci. Lett.* **157**, 183 (1993).
19. R. Horn and A. Marty, *J. Gen. Physiol.* **92**, 148 (1988).
20. S. J. Korn, A. Marty, J. A. Connor, and R. Horn, *in* "Methods in Neuroscience" (P. M. Conn, ed.), Vol. 4, p. 364. Academic Press, San Diego, 1991.
21. J. Hescheler, M. Kameyama, and R. Speicher, *in* "Practical Electrophysiological Methods" (H. Kettenmann and R. Grantyn, eds.), p. 241. Wiley–Liss, New York, 1992.
22. M. Pusch and E. Neher, *Pfluegers Arch.* **411**, 204 (1988).
23. J. Siara, J. P. Ruppersberg, and R. Rudel, *Pfluegers Arch.* **415**, 701 (1990).
24. R. A. Silver, S. F. Traynelis, and S. G. Cull-Candy, *Nature* (*London*) **355**, 163 (1992).
25. M. L. Mayer and G. L. Westbrook, *J. Physiol.* (*London*) **361**, 65 (1985).
26. I. Mody, K. G. Baimbridge, J. A. Shacklock, and J. F. MacDonald, *Exp. Brain Res. Ser.* **20**, 75 (1991).
27. S. A. Thayer, M. Sturek, and R. J. Miller, *Pfluegers Arch.* **412**, 216 (1988).
28. E. Neher, *Neuroscience* **26**, 727 (1988).
29. M. W. Salter and J. L. Hicks, *Soc. Neurosci. Abstr.* **18**, 1000 (1992).

[17] Imaging Ion Channel Dynamics in Living Neurons by Fluorescence Microscopy

Barry W. Hicks and Kimon J. Angelides

Introduction

The distribution of ion channels and receptors over the neuronal surface is important for the receipt of incoming synaptic inputs and for the integration of these inputs. Most voltage-gated and ligand-gated ion channels have non-homogeneous distributions in the neuronal membrane, many being restricted to dendritic, axonal, or somatic domains and further localized within these domains to regions such as dendritic spines, nodes of Ranvier, or synaptic junctions (Poo and Young, 1985; Augustin *et al.*, 1987; Black *et al.*, 1990). Determining where and how ion channels are distributed and maintained is important for a variety of reasons. Ion channels in growth cones have a role in neurite outgrowth mechanisms (Saffell *et al.*, 1992): they are obligatory for synaptic transmission and they are required for amplification of neurotransmitter signals in the postsynaptic membrane (Beam *et al.*, 1985). Changes in ion channel distributions are an important aspect in development and plasticity (Stya and Axelrod, 1983; Joe and Angelides, 1991). Finally, several neurological disorders, including multiple sclerosis, are characterized by a change in ion channel distribution which is partially responsible for the clinical manifestations of the disease (Black *et al.*, 1991). A restricted distribution of ion channels can be initiated and maintained by a variety of mechanisms including interactions with components of the extracellular matrix (Koppel *et al.*, 1981; McCloskey and Poo, 1984), indirect or direct interactions with elements of the cytoskeleton (Stya and Axelrod, 1983; Srinivasan *et al.*, 1988), and formation of tight junctions (Diamond, 1977).

Fluorescence techniques are invaluable in the study of ion channel and receptor distributions and dynamics. Fluorescence microscopy has extremely high sensitivity, the lower limit being the ability to detect a single fluorophore under the optimal conditions (Mathies and Stryer, 1986). Fluorescence measurements are specified by the wavelengths of emission and excitation. The specificity in most applications is not limited by the fluorescent molecule but by the probe from which the fluorescent analog is made. The spatial resolution, limited by the emission wavelength and the microscope optics, can be as small as 200 nm. However, objects as small as 30 nm can

Methods in Neurosciences, Volume 19

be visualized and distances less than the 100 nm can be determined using resonance energy transfer.

Methods

General Considerations

The study of the distribution of nicotinic acetylcholine receptor in muscle has been facilitated by its very high density ($\sim$5000 receptors/μm^2). Unfortunately, this high density has been the exception rather than the rule. One of the difficulties encountered in the study of other ion channel and receptor distributions is detection of the very low densities of some of these proteins. For example, in developing nerve voltage-dependent sodium channel (NaCh) densities have been estimated at 5 channels/μm^2. Detection of proteins at this density by fluorescence microscopy can be further complicated by cellular autofluorescence in the 350–500 nm range due to NADH and flavins (Taylor and Salmon, 1989) that can often conceal faint specific signals. Several strategies can be used to enhance the detectability of fluorescent probes.

Developments in low-light level video cameras like the silicon-intensified target (SIT), intensified silicon target (ISIT), and cooled charge-coupled device (CCD) cameras have extended the limits of signal detection (Spring and Lowly, 1989; Aikens *et al.*, 1989). If the outputs from these sources are interfaced to a digitizer and monitor the images are visualized immediately, saving considerable time and minimizing frustration often associated with developing film. Additionally, through digital image enhancement, fluorescent signals that are not detectable by eye can be produced with adequate contrast. Each type of camera has a characteristic spectral sensitivity that should be considered when deciding what type of fluorescent probe to use if sensitivity is a problem. In most cases we are limited in the choice of camera to what is available or to what is affordable, but there are other ways to increase the detectability of a given protein.

It is important to choose carefully the fluorescent molecule used to prepare the analog. Fluorophores are characterized by three parameters, the extinction coefficient, the quantum yield, and the fluorescence lifetime. The extinction coefficient (ε) is a measure of the molecule's ability to reach an excited state. Typical values for extinction coefficients are in the range of 10,000–100,000 M^{-1} cm^{-1}. The quantum yield (Q) is a measure of the efficiency of an excited molecule to fluoresce rather than relax through other mechanisms. Values for the quantum yield range from 10 to 100%. The fluorescent intensity of a single fluorophore is proportional to the product of the extinction coefficient and the quantum yield (εQ). For most fluorophores, bleaching can be problematic if the analysis is conducted over a long time. Although the exact mechanism by which bleaching occurs is not known, it is a process of photooxidation. In fixed samples it can be minimized by removal of molecular oxygen from mounting solutions and addition of anti-oxidants. In the study of living cells these options are largely eliminated. For this reason, the fluorescence lifetime should also be considered as long intrinsic lifetimes render the excited fluorophore more susceptible to beaching. Thus, given the choice of fluorescein, BODIPY, and NBD (all of which are excited at ~480 nm and emit at ~520), BODIPY and fluorescein (ε ~80,000 M^{-1} cm^{-1}) would be the preferred over NBD (ε ~20,000 M^{-1} cm^{-1}). BODIPY would be preferred to fluorescein as its emission spectra suggests a shorter fluorescence lifetime. However, NBD is a smaller molecule than fluorescein or BODIPY and might be less likely to perturb biological activity by preparation of a fluorescent analog.

In addition to selecting a fluorescent probe with the best possible spectral qualities for the detection system being used, a variety of strategies can be used to amplify the signal from the chosen probe (Taylor and Salmon, 1989). Phycobiliproteins have fluorescent intensities equivalent to more than 100 individual fluorophores. One of the best fluorescent probes that has emerged for amplifying signals are fluorescent microspheres. Fluorescent latex microspheres are available in a variety of sizes, with a variety of emission wavelengths, and with a variety of surface functional groups. Figure 1 shows the images of single 30-, 93-, and 280-nm fluorescent carboxylate modified latex beads obtained with both cooled CCD and SIT cameras. In all three of the SIT images, pixels are saturated near the center or the bead, whereas only the 280-nm beads saturated the CCD camera under the conditions used demonstrating the better linear response of this camera over a greater range of intensities. A single 93-nm bead can have the fluorescent intensity equivalent to several thousand fluorophores and individual beads can be visualized by eye. The beads are resistant to bleaching because the fluorophores are trapped within the latex and are thus minimally accessible to oxygen. The wide range of commercially available products allow production of individual probes to be tailored to the user's needs. The beads also have the advantage of being detectable by electron microscopy if information at the ultrastructural level is needed (Egensperger and Hollander, 1988).

Tissue Culture

Currently we are examining ion channel distributions and dynamics in several primary neuronal types including dorsal root ganglion neurons, superior cervical ganglion neurons, and hippocampal neurons. Procedures for isolat-

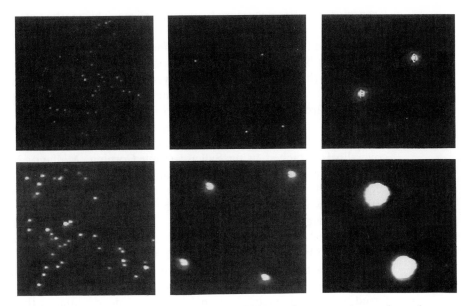

FIG. 1   Images of (left to right) 30-, 93-, and 280-nm fluorescent latex microspheres. The top images were obtained with a cooled CCD camera using a 4-sec exposure. The lower images were obtained with a SIT camera by averaging 128 frames (~4 sec). The gain, offset, and sensitivity settings for each camera were the same for all three images.

ing and maintaining these neuronal cultures are described elsewhere (Sharar *et al.*, 1989; Joe and Angelides, 1991).

We use three basic types of cultures. In some experiments, neurons are enzymatically and mechanically dissociated. In other experiments, tissue explants are used if neurite outgrowth is the primary domain of interest. Cultured explants have the advantage of not suffering from large losses in the neuronal populations or synaptic connections due to the harsh treatments necessary for complete dissociation. Explants or dissociated cells plated on coverslips are available to study by phase, differential interference contrast (DIC), and fluorescence microscopy measurements. These systems also have the advantage of exposing neuronal surfaces so that they are accessible to analyses with antibodies and other large probes for fluorescence microscopy. Last, organotypic tissue slices are examined. This technique, while being limited to smaller molecular probes because of the limited access of large protein probes to the cells, has the advantage that neuronal interactions with each other and with glia are preserved within the slice. Generally, confocal laser microscopy is necessary to examine slices. This method can, however, provide information that is not available in dissociated cell systems.

Information obtained from ion channel and receptor dynamics of dying neurons can be extremely misleading. Ideally, the cells should be cultured in a manner that allows them to be moved directly from the incubator to the microscope stage. For fluorescence microscopy, this means that the thick plastic culture dishes normally used for cell culture are inadequate. We culture cells in chambers that are made by cutting a 20-mm circle from the center of a 35-mm culture dish. In place of the plastic, a number 1 glass coverslip is attached using Sylgard (Dow Corning, Midland, MI). These chambers are inexpensive, thus disposable, and the thin glass surface allows access to cells for epifluorescence microscopy using oil-immersion objectives with short working distances on an inverted microscope. Prior to plating cells, the coverslips are coated with a mixture of poly-D-lysine (100 μg/ml) and 100 μg/ml of either collagen or laminin to promote attachment and neurite outgrowth. Maintaining these cultures on the microscope stage over long time periods requires that the pH, temperature, osmolarity, and nutrient concentration in media all be controlled. Bicarbonate buffering is the preferred method for maintaining pH as it most nearly mimics the system *in vivo*, and other buffers can be toxic at relatively low concentrations. We maintain pH by perfusing the chamber holder on the stage with a constant stream of 95% air, 5% CO_2. The chamber is heated to 36 $\pm$ 1°C and a 40-mm number 1 coverslip is placed on top of the dish or a layer of mineral oil is placed across the media to prevent loss of osmotic balance due to evaporation. In the cases where cells are monitored for more than 24 hr, additional medium is pipetted directly into the chamber. To ensure that the cells are examined in a manner that is minimally destructive, the cells should remain healthy under the same conditions for at least 24 hr after the experiment is terminated.

Preparation of Labeled Toxins and Antibodies

A number of low-molecular-weight polypeptides and heterocyclic toxins have ion channels and neurotransmitter receptors as targets. While antibodies prepared against the purified protein or against peptides from site-directed sequences can be used in fluorescence studies, the low-molecular-weight toxins are ideally suited as probes of these channels. Most toxins bind to extracellular domains of the target protein which is obligatory for studying dynamics on living cells as cytoplasmic domains are generally inaccessible. They can demonstrate extremely high affinities and selectivities for individual ion channels or receptors that have been engineered by evolution, in some instances to be isoform specific (Deneris *et al.*, 1991). In addition to specificity and high affinity, the toxin must be amenable to chemical modification with fluorescent molecules to be valuable. This means that amines, carboxylates, sulfhydryls, hydroxyls, or other active groups must be present within the toxin. The first step in preparing a toxin analog is the chemical modification of the chosen reactive group. After terminating the reaction with an excess of the reactive group, the reaction mixture is separated by high-performance liquid chromatography (HPLC). A complete separation of the analog from the unmodified toxin is essential as the unmodified toxin can compete for binding sites. Purity from starting materials and verification of modification can be performed by mass spectroscopy. It is necessary to characterize the toxin analog to ensure that chemical modification did not destroy biological activity. Competition for binding sites with a radiolabeled analog of the toxin is one useful method for ensuring that binding activity has not been destroyed. In addition, electrophysiological measurement of the toxin activity as an agonist or antagonist is necessary to ensure that the bound analog exerts effects similar to those of the unmodified toxin. All of these properties of the modified toxin should be demonstrated to ensure that the analog faithfully represents the unmodified toxin. Detailed considerations and procedures for preparing and using fluorescent toxin analogs have been discussed (Angelides, 1989).

Here we give the details used to prepare the voltage-dependent NaCh probe, rhodamine-labeled tityus-γ toxin. Tityus-γ toxin is a 62 amino acid polypeptide with six lysine residues that recognizes the voltage-dependent sodium channel from a variety of species (Barhanin, 1982). The toxin is treated with the succinamidyl ester of tetramethylrhodamine in a 1:1 mol ratio by adding 7.2 μg of activated rhodamine from a stock solution in DMSO to 100 μg of toxin in 50 μl of 0.1 M sodium bicarbonate buffer at pH 8.5. After 30 min at room temperature, the reaction is terminated by the addition of 10 μl of 1 M glycine. The labeled toxin is purified from starting materials by reverse-phase HPLC on a C-18 column using a gradient from 0.1% ammonium acetate in water to 0.1% trifluoroacetic acid in acetonitrile. Fractions are collected and dried under vacuum, reconstituted in 50 μl of distilled water, and tested for their ability to inhibit binding of [^{125}I]tityus-γ toxin to rat brain synaptosomal membranes.

When preparation of labeled antibodies is necessary, periodate oxidized antibodies are labeled using biotin hydrazide or the hydrazide of the chosen fluorophore. This method is preferable as it does not interfere with the antigen recognition site causing loss of activity as can often occur when labeling with amine or carboxylate reactive fluorophores. In addition, the major contaminants in isolation of antibodies from serum, hybridoma supernatant, or ascites fluid are albumin and transferrin. Some of these proteins are glycosylated so background labeling is usually reduced in comparison with other methods. Production of Fab fragments of antibodies is done using papain immobilized on beaded agarose (Harlow and Lane, 1988).

Preparation of Labeled Microspheres

Fluorescent microspheres offer a signal enhancement that is unequaled by other methods. The fluorescent microspheres are available from several manufacturers (Sisken, 1989) with emission wavelengths ranging from 400 to 650 nm, in sizes ranging from 0.01 to 10 μm, and with different functional groups for modification including amines, carboxylates, and aldehydes. To minimize cross-linking on the cell surface, use of the smallest possible probe is desirable; however, as the bead size decreases the ability to identify individual particles and monitor displacements accurately over long time periods on the surface of living cells becomes more difficult. To minimize the damage to the cell from excitation, use of wavelengths greater than 500 nm is preferable. Attachment of protein to the beads can be accomplished via hydrophobic interactions by simply incubating the protein of interest with the beads or covalently by taking advantage of the various functional groups that are commercially available. Although noncovalently prepared beads have shorter lifetimes, they have the obvious advantage of being easier to prepare.

Microsphere probes that incorporate lectins, antibodies, or avidins covalently can be produced. To make NaCh probes we use monoclonal antibodies directed against the purified channel from rat brain (Elmer *et al.*, 1990). These monoclonal antibodies are obtained from serum-free medium and partially purified and concentrated in Amicon Centricon concentrators with a 100,000 molecular-weight cutoff (MWCO). To make avidin probes, we use Neutralite avidin because its lower p*I* value and deglycosylation render it less susceptible to nonspecific binding. Neutralite avidin and fluospheres were obtained from Molecular Probes. We attach our antibody or avidin covalently to the carboxylate beads using procedures modified from the manufacturer's recommendations. In a disposable 75-mm glass test tube, 0.5 ml of 0.1-μm carboxylate-modified orange latex microspheres and 0.5 ml of protein solution containing 1.0 mg/ml protein in 15 m*M* MES buffer, pH 5.5, are mixed and allowed to stand for 30 min at room temperature. This period allows proteins to adhere passively to the surface so that they will be more likely to react with the modified carboxylate groups. This mixture is then transferred to a separate glass test tube containing 4.0 mg of [1-ethyl-3-(3-dimethylaminopropyl)carbodiimide]. Dicyclohexylcarbodiimide is not used because the solubility limits the concentration in aqueous solution to much lower values and because the activated groups cause agglutination which is difficult to reverse. As the carboxylate groups become activated with the positively charged carbodiimide, agglutination becomes readily visible. After 5 min, 0.01 N NaOH is added to adjust the pH to 6–7, the approximate value determined by spotting a few microliters of the reaction mixture on pH paper. The beads are dispersed in a bath sonicator several times for 10 sec until the large aggregates are no longer present. The reaction of the protein with the activated carboxylate groups is allowed to proceed for 4 hr at room temperature. The reaction is stopped by adding an excess of a primary amine. Several amines can be used, the choice depending partially upon the intended use of the probe. Typically, we add either glycine or ethanolamine at a final concentration of 0.1 M at pH 8.5. Use of glycine returns a terminal carboxylate to the beads and helps minimize nonspecific interactions with negatively charged cell surfaces, but can cause problems with beads attaching to polylysine-coated coverslips. Ethanolamine results in an uncharged terminating group and might leave the beads more susceptible to react through hydrophobic interactions, but helps prevent labeling of polylysine surfaces. The modified beads are added to an equal volume of PBS containing 5% BSA to block hydrophobic sites. The entire mixture is dialyzed in tubing with a 300,000 MWCO to remove unattached protein from the beads. Dialysis is performed against $0.1\times$ PBS for 72 hr with three changes of buffer. The beads are recovered and diluted to 20 ml with PBS containing 1.0% BSA and 5 mM sodium azide. The mixture is concentrated in an Amicon Centricon concentrator with a 100,000 MWCO until the final volume is about 1–2 ml. The modified beads are aliquotted and stored at 4°C. The lifetimes for the covalently modified beads are at least several months at 4°C.

Any protein or peptide can be attached to the fluorescent microspheres. Toxin probes can be prepared by direct conjugation of the toxin to the microspheres or by creation of a biotinylated toxin to examine distributions with avidin-labeled microspheres. In cases where small molecules are conjugated to the beads, a sufficient linker should be included to ensure that binding is not prohibited by steric interactions.

Labeling Cells with Toxins, Antibodies, or Modified Fluorescent Microspheres

To study channel dynamics by spot fluorescence recovery after photobleaching (FRAP) cells are labeled with the fluorescent toxin analog or Fab fragments to the cells in chemically defined medium. Media containing serum are not preferred because serum proteases might inactivate the probe. Whole antibodies or other multivalent probes are generally not used as cross-linking can cause artifically depressed values for diffusion coefficients and mobile fractions. After 20 min at room temperature, the cells are rinsed three times with medium lacking phenol red. Near neutral pH, phenol red can absorb the excitation light from the 514.5-nm band of the argon laser and increase the amount of photooxidation that occurs and cause an increase in background fluorescence. As our spot FRAP analyses are done at room temperature, all labeling and washing procedures are done at room temperature.

For video FRAP or distribution imaging, cells in chambers are labeled with the toxin analog or the primary antibody in chemically defined medium for 20 min at 37°C. After three 5-min washings, secondary fluorescent probes such as antibodies or avidins, diluted into medium containing 10% fetal bovine serum, are added and allowed to bind for 20 min at 37°C. Cells are then transferred to the microscope stage for analysis. Imaging and video FRAP are generally done at 37°C, so all labeling and incubations are done at this temperature.

To label neurons for single-particle tracking, the modified beads are first diluted into medium containing 10% fetal bovine serum and sonicated briefly in a bath sonicator. The dilution factor is determined previously by examining serial dilutions on microscope slides to provide a concentration that is high enough to label all the cells while not providing an excess which makes identification of individual beads difficult. After dilution, the beads are filtered through a nylon filter with a pore size about two times the diameter of the beads (e.g., 0.1-μm beads are filtered through a 0.22-μm filter). This ensures that aggregates including most dimers and trimers are not used to label the cells. The medium in the chamber is removed and replaced with fresh medium containing the microspheres. Labeling is typically accomplished by incubating the cells with the medium containing the microspheres for 10 min at room temperature followed by two rapid washings with fresh medium.

Fluorescent Imaging of Living Cells

To determine qualitative and quantitative distributions of ion channels and receptors, we perform digitally enhanced fluorescence microscopy. Cells in the culture chambers are placed in a heated, CO_2-equilibrated stage manufactured by Precision Assemblies (Somerville, MA) and custom modified in our machine shop. The stage rests on a Zeiss Axiovert inverted microscope equipped with Hamamatsu (Hamamatsu City, Japan) SIT and Newvicon cameras and a cooled CCD camera with a 14-bit 512 × 384-pixel memory plane (Photometrics, Tucson, AZ). The video camera outputs are digitized by a Biovision image processor (Perceptics, Nashville, TN) into 8-bit

512×512 memory planes and the digital CCD output goes directly into the image processor memory. The microscope is equipped with a 75-W xenon arc lamp for epifluorescence and a 100-W mercury lamp for phase and DIC microscopy. The images, usually obtained through a Zeiss 63X Plan-apochromat objective with a 1.4 NA or a 100X Plan-neofluar objective with a 1.3 NA, are stored either in digital form on an Optical Access optical drive, in analog form on a Panasonic Optical Disk Recorder TQ 2028F, or on standard VHS videotape using a Panasonic NV8200 VCR. The image processor and microscope are run interactively through a Macintosh IIx computer. The digitized images can be enhanced with a variety of mapping equations at video frame rates. In addition, for sharpening images a number of spatial convolution masks are available for images from stored media or images produced during time-lapse microscopy. The enhanced images can be printed immediately with a Sony UP-5000 color video printer.

For most applications where quantitative information is desired, we use the CCD camera. The CCD camera offers several advantages including a better linear response, a better signal-to-noise ratio, and greater sensitivity. Only in situations where images have to be obtained rapidly do we use the SIT camera because it is capable of producing images at video frame rates. To ensure that the images obtained represent the true distributions, artifacts due to nonuniform excitation and to camera aberrations are corrected for using flat-field and bias images. Several correction algorithms are available, but we typically apply the algorithm,

$$\text{Corrected image} = (\text{image} - \text{bias})/(\text{flat field} - \text{bias}),$$

where the flat-field image is obtained with a uniformly fluorescent slide in the same optical plane as that used for the specimen, and the bias image is obtained in the dark without opening shutters. Figure 2 shows the DIC image of superior cervical ganglion neurons labeled with rhodamine tityus-γ and the corresponding NaCh distributions using the CCD and SIT cameras. The ability to expose the CCD camera for longer periods increases its sensitivity relative to the SIT camera and cooling the camera minimizes thermal noise. Both fluorescent images were enhanced by linear mapping.

Fluorescence Recovery after Photobleaching

For the analysis of ion channel dynamics by spot FRAP we use an 8-W argon ion laser (Laser Ionics, Orlando, FL) with a 1.8-mm beam width that is tuned to 514.5 nm for rhodamine analogs, a Zeiss microscope modified to allow the beam to enter through a side port, a cooled photomultiplier tube

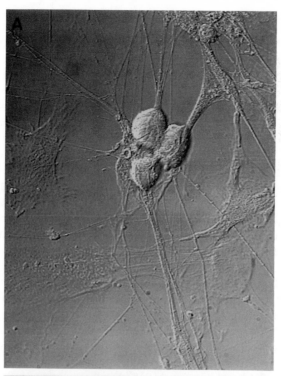

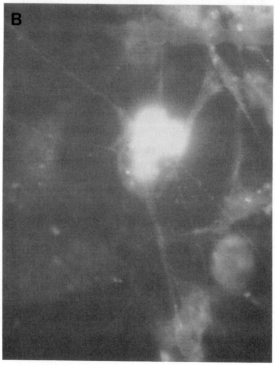

FIG. 2   Images of superior cervical ganglion neurons. (A) DIC image taken with the CCD camera. (B) CCD image of fluorescence from NaChs labeled with rhodamine tityus-γ toxin using a 90-sec exposure time. (C) SIT camera image of NaCh distributions made by averaging 256 video frames.

(Products for Research, Danvers, MS), and an IBM PC to control shutters and receive output from the photomultiplier tube. In a typical experiment, cells are labeled as described with a rhodamine-modified probe and placed on the microscope stage. After focusing on the cell surface with phase optics through a 100X, 1.3 NA oil-immersion lens, the fluorescence signal from the attenuated beam of the laser is monitored to produce a stable baseline. A program is then executed which monitors the excitation for 10 sec before bleaching with the unattenuated beam for about 5–50 msec which causes a decrease of about 60–70% in fluorescence intensity as determined by the photon counts. Recovery is monitored for periods ranging from 30 sec to several min. Diffusion coefficients and mobile fractions are obtained from curve-fitting routines.

Figure 3 shows the results of a typical analysis of the diffusion of voltage-dependent NaChs on the soma of a superior cervical ganglion neuron using rhodamine-conjugated tityus-γ toxin. In this case the majority of the channels are immobile in the membrane. The small fraction that did recover did so with a diffusion coefficient of about 10^{-10} cm^2/sec.

Video FRAP

For most sodium-channel FRAP videos, cells are labeled with MAb 3 (Elmer *et al.*, 1990) or Fab fragments followed by rhodamine-conjugated goat anti-mouse Fab fragments. To make the video recordings, the Zeiss Axiovert microscope described above is used. The xenon lamp is attached to an adapter that allows importation of a laser into the optical path for excitation so that photobleaching and recovery can be filmed.

Video FRAP of sodium-channel recoveries is filmed in a time-lapse mode, using either the SIT or the CCD camera, depending on the acquisition speed

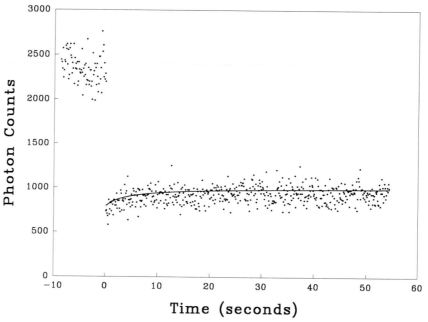

FIG. 3   Typical spot FRAP curve from rhodamine tityus-γ-labeled superior cervical ganglion neurons. The mobile fraction is 12% and the diffusion coefficient of the mobile fraction is 1.3 × 10^{-10} cm^2/sec.

that is required. If the SIT camera is used, an individual image normally is the arithmetic average of multiple frames to reduce noise. An image histogram is analyzed after setting the camera gain, offset, and sensitivity to ensure saturation is not occurring. Control of the bleaching shutter is done interactively between image acquisition steps to ensure that the shutter protecting the camera is not exposed to laser light. Figure 4 shows a region of dorsal root ganglion (DRG) axons that were labeled with NaCh antibodies, and rhodamine-labeled secondary antibodies, less than 1 min after bleaching and after nearly 90 min. Similar images obtained with rhodamine-labeled phosphatidyl ethanolamine over a 6-min period are shown below. In contrast to the lipids, the channels appear to be essentially immobile in the region studied.

Both spot FRAP and video FRAP rely on describing the behavior of ensembles. For a variety of reasons, including answering some mechanistic questions, it would be useful to study the movements of individual ion channels in the neuronal membrane. Video-enhanced techniques developed in the past decade have made the study of structures considerably smaller than the diffraction-limited resolution of light possible (Allen, 1985) using transmitted light. It is equally possible to visualize particles smaller than 200 nm using fluorescence microscopy.

Video Records Using Modified Latex Beads: Single-Particle Tracking

Chambers containing labeled neurons are placed in the preheated, CO_2-equilibrated holder on the microscope stage. Images are collected using the SIT or cooled CCD cameras mounted on a $4\times$ adapter. The additional magnification reduces the pixel size so that more accurate tracking is possible. Exposure times range from 0.03 to 2.0 sec, depending on the size of the microspheres, the camera being used, and the image storage medium. The camera and exposure time used depend on the rate of diffusion which we are attempting to measure. Minimizing the exposure time in time-lapse recordings minimizes photo damage to the cells when records are obtained for long periods. The resulting images are stored in digital form to allow quantitative information on the movement to be obtained and onto the optical disk in analog form for sequence viewing. We have monitored the movements of beads in time-lapse mode for more than 12 hr.

An image of 93-nm anti-sodium-channel antibody-labeled fluorescent microspheres on DRG axons is shown in Fig. 5. This image is made by simultaneously exposing the CCD camera to both transmitted light and fluorescent light emitted by the beads. The image is enhanced using a sharpening mask and a linear map and zoomed to the area of interest. Beads bound to the

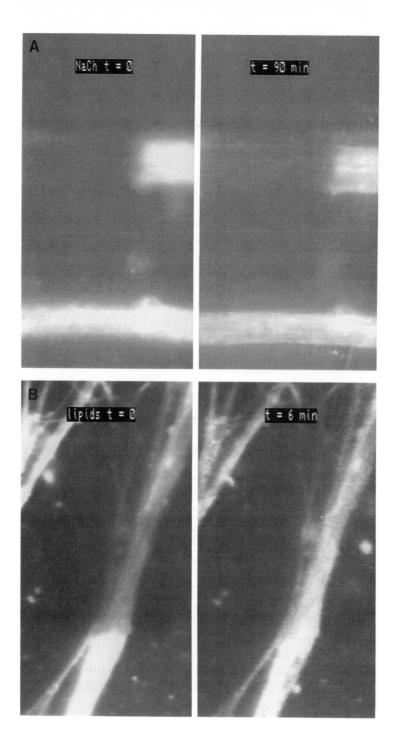

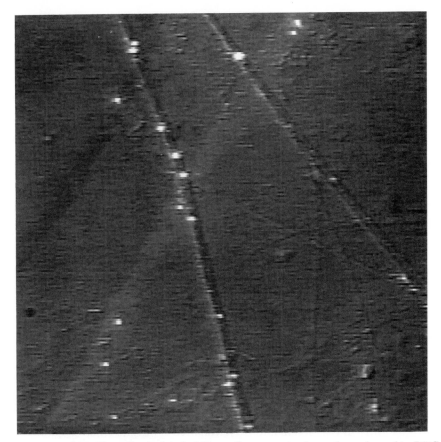

FIG. 5   Anti-NaCh antibody-labeled 93-nm fluorescent microspheres bound to DRG axons on a collagen-coated gridded coverslip.

substratum are useful reference markers. The movements of individual beads can be analyzed with high precision using a similar strategy to that of Gelles *et al.* (1988). Briefly, a region of 30 × 30 pixels containing a given bead is selected. Figure 6 is a graphical representation of the pixel intensities from a 30 × 30 array of the CCD image of the 93-nm bead shown in Fig. 1. The

FIG. 4   Video FRAP micrographs obtained with the SIT camera. (A) DRG axons labeled with NaCh antibodies and rhodamine-labeled secondary antibodies immediately after bleaching and 90 min after bleaching. (B) DRG axons labeled with rhodamine-labeled phosphatidylethanolamine immediately after and 6 min after bleaching.

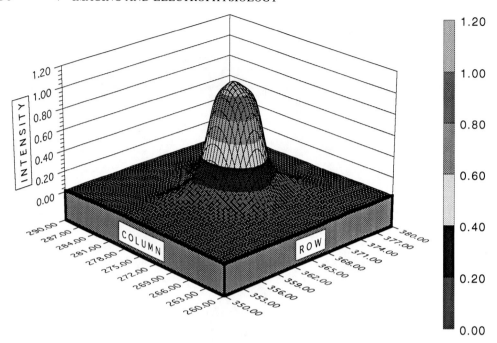

Fig. 6   Intensity distributions of a 30 × 30 array of pixels from the CCD image of the 93-nm bead shown in Fig. 1.

actual pixels which form the bead image are isolated by selecting only those pixels with intensity values above a defined threshold intensity value (e.g., 0.20 in Fig. 6). An intensity-weighted centroid is then calculated. Movements can be traced by analyzing the position of the centroid in sequential images and estimating diffusion coefficents by plotting the mean square displacements against time. The tracking accuracy depends on the pixel size and thus on the magnification used to obtain the image.

Conclusions and Difficulties

A number of artifacts can lead to improper interpretations in spot FRAP experiments (Wolf, 1989), and this technique is limited in its ability to provide spatial information. Although it is theoretically possible to obtain different results depending on the mode of recovery, in practice this is seldom realized. Thus, it is difficult to distinguish between recovery due to directed motion or flow and recovery due to isotropic diffusion. In practice, it is also limited in its ability to determine mobilities of proteins with diffusion coefficients smaller than about 1×10^{-12} cm^2/sec. Although most proteins are prohibited from diffusing freely, this does not exclude the possibility that they are able to diffuse over large distances given sufficient time. One alternative to spot FRAP is video FRAP. The advances in low light level cameras allow the use of video FRAP (Kapitza *et al.*, 1985) to provide information on the spatial aspects of recovery and this technique can be used in time-lapse mode to extend the practical time over which information can be obtained.

There are advantages of using a bead-based strategy to determine diffusion coefficients over FRAP analysis. First, the laser is eliminated, reducing cost and preventing photooxidation and cell damage often caused by misuse of these techniques. Second, the practical time period over which processes can be monitored is extended to make diffusion coefficients smaller than 1×10^{-13} cm^2/sec accessible to study by this technique. As it is becoming apparent that many proteins have at least partially immobile populations when analyzed by FRAP, this becomes quite important. Most neurons will survive for the entire life of an animal. Therefore, very slow diffusion in the membrane might play an important role in cellular processes that occur over days or weeks. A statistical analysis of bead movements (Qian *et al.*, 1991) in different parts of the cell might provide information on the mechanisms used to distribute and maintain channels in the neuronal membrane over long time periods.

The biggest problem with labeling cells using these fluorescent microspheres is the amount of nonspecific binding that occurs due to the surface properties of the latex. This binding, while nonspecific with respect to the protein of interest, is extremely tight. Carboxylate beads on poly-D-lysine-coated coverslips can bind to the surface through ionic and hydrophobic interactions so tightly that they cannot be washed off and will not detach from the bound site even after many hours. Using controls that have beads that are either unmodified or beads that have been modified and brought to 95°C for 10 min to denature proteins on the surface, we find that the labeling index for the active probes is generally several times higher than that for controls. A second difficulty in using beads is the possibility of cross-linking proteins on the cell surface with larger beads. Sheetz and co-workers have found that ~500 nm concanavalin A beads either diffuse randomly or move rearward in keratocytes but never toward the leading edge, whereas 50-nm beads are capable of random diffusion and rapid movement toward the leading edge (Kucik *et al.*, 1991). For controls we have used avidin beads on the surface of cells that have been labeled with biotin-phosphatidylethanolamine. Lipid probes as large as 40 nm are capable of diffusing in the membrane with diffusion coefficients essentially identical to the values measured by FRAP (Lee *et al.*, 1991).

There are several difficulties encountered in attempting to extend the time frame for determining mobilities of very slow-moving components in the membrane. First, it is necessary that the probe remains bound for the duration of the experiment. In most receptor–ligand assays, ligands that have dissociation half-lives beyond several hours are usually considered to be irreversible because of the time scale of the experiment. In extending the experimental window to days, even probes with first-order dissociation half-lives of the order of 20 hr ($\sim K_d = 10^{-12}$ M) must be used with caution. High-affinity toxins, which can have much slower dissociation rates than antibodies, are perhaps better suited for experiments performed over long time periods. A biotin–avidin system is capable of being used for periods of many hours as the half-life of dissociation is many days. We are currently pursuing the production of covalent probes capable of incorporating biotin into the protein of interest. Second, no frame of reference is available on motile cells.

The use of fluorescence microscopy to study ion channel dynamics has provided a great deal of information on how ion channels and receptors are maintained within specific neuronal domains. The combination of dynamic measurements with effectors of cytoskeletal elements such as colchicine or cytocalasin can provide information on the cellular mechanisms used to maintain nonhomogeneous distributions of these proteins which is necessary for proper neuronal function.

Acknowledgments

Thanks to Richard Rodriguez for tissue isolation and culture, to Dr. Michael Sheetz for image analysis software, and to Dr. Eun-hye Joe for preparation of the rhodamine-labeled tityus-γ toxin. This work was supported by the National Institutes of Health (NS 28072) and Muscular Dystrophy Association. B. Hicks is a NIH Postdoctoral Fellow (NS 09196).

References

1. R. S. Aikens, D. A. Agard, and J. W. Sedat, *Methods Cell Biol.* **29** part A, 292 (1989).
2. R. D. Allen, *Annu. Rev. Biophys. Chem.* **14**, 265 (1985).
3. K. Angelides, *Methods Cell Biol.* **29** part A, 13 (1989).
4. G. J. Augustin, M. P. Charlton, and S. S. Smith, *Annu. Rev. Neurosci.* **10**, 633 (1987).
5. J. Barhanin, J. Giglio, P. Leopold, A. Schmid, S. Sampaio, and M. Lazdunski, *J. Biol. Chem.* **257**, 12553 (1982).
6. K. G. Beam and J. T. Campbell, *Nature (London)* **313**, 588 (1985).
7. J. A. Black, J. D. Kocsis, and S. G. Waxman, *Trends Neurosci.* **13**, 48 (1990).

8. J. A. Black, P. Felts, K. J. Smith, J. D. Kocsis, and S. G. Waxman, *Brain Res.* **544,** 59 (1991).
9. E. S. Deneris, J. Connolly, S. W. Rogers, and R. Duvoisin, *TIPS* **12,** 34 (1991).
10. J. M. Diamond, *Physiologist* **20,** 10 (1977).
11. R. Egensperger and H. Hollander, *J. Neurosci. Methods* **23,** 181 (1988).
12. L. W. Elmer, J. A. Black, S. G. Waxman, and K. J. Angelides, *Brain Res.* **532,** 222 (1990).
13. J. Gelles, B. J. Schnapp, and M. Sheetz, *Nature (London)* **331,** 450 (1988).
14. E. Harlow and D. Lane, *in* "Antibodies: A Laboratory Manual," pp. 626–632. Cold Spring Harbor Laboratory, Cold Spring Harbor, NY, 1988.
15. E.-h. Joe and K. Angelides, *Nature (London)* **356,** 333 (1991).
16. H. G. Kapitza, G. McGregor, and K. A. Jacobson, *Proc. Natl. Acad. Sci. U.S.A.* **82,** 4122 (1985).
17. D. E. Koppel, M. P. Sheetz, and M. Schindler, *Proc. Natl. Acad. Sci. U.S.A.* **78,** 3576 (1981).
18. D. F. Kucik, S. C. Kuo, E. L. Elson, and M. P. Sheetz, *J. Cell Biol.* **114,** 1029 (1991).
19. G. M. Lee, A. Ishihara, and K. A. Jacobson, *Proc. Natl. Acad. Sci. U.S.A.* **88,** 6274 (1991).
20. R. Mathies and L. Stryer "Applications of Fluorescence in the Biomedical Sciences" (D. L. Taylor, A. Waggoner, R. Murphy, F. Lanni, and R. Birdge, eds.), pp. 129–140, Liss, New York, 1986.
21. M. McCloskey and M. Poo, *Int. Rev. Cytol.* **87,** 19 (1984).
22. M. Poo and S. Young, *Annu. Rev. Neurosci.* **8,** 369 (1985).
23. H. Qian, M. P. Sheetz, and E. L. Elson, *Biophys. J.* **60,** 910 (1991).
24. J. L. Saffell, F. S. Walsh, and P. Doherty, *J. Cell. Biol.* **118,** 663 (1992).
25. A. Sharar, J. Vellis, A. Vernadakis, and B. Haber, eds., "A Dissection and Tissue Culture Manual of the Nervous System." A. R. Liss, New York, 1989.
26. J. E. Sisken, *Methods Cell Biol.* **29B,** 113 (1989).
27. K. R. Spring and R. J. Lowly, *Methods Cell Biol.* **29A,** 270 (1989).
28. Y. Srinivasan, L. Elmer, J. Davis, V. Bennett, and K. Angelides, *Nature (London)* **333,** 177 (1988).
29. M. Stya and D. A. Axelrod, *Proc. Natl. Acad. Sci. U.S.A.* **80,** 449 (1983).
30. D. L. Taylor and E. D. Salmon, *Methods Cell Biol.* **29A,** 207 (1989).
31. D. E. Wolf, *Methods Cell Biol.* **29B,** 271 (1989).

[18] Whole-Cell Patch Recording with Simultaneous Measurement of Intracellular Calcium Concentration in Mammalian Brain Slices *in Vitro*

Steven R. Glaum, Simon Alford, David J. Rossi,
Graham L. Collingridge, and N. Traverse Slater

Introduction

Slices of brain tissue have been widely employed by neuroscientists to investigate the biochemical and biophysical properties of neurons and glia. With the advent of fluorescent intracellular ion-sensitive dyes (1), many investigations have focused on the interactions between neurochemical transmitters and membrane properties and the role of the intracellular ionic microenvironment in signal transduction. The most widely employed of these dyes in recent years is the Ca^{2+}-sensitive dye Fura-2 (2). Some of the earlier neurobiological applications for Fura-2 were in the monitoring of intracellular Ca^{2+} concentration ($[Ca^{2+}]_i$) within cells maintained in culture, such as dorsal root ganglion cells, astrocytes, and pyramidal neurons (1). More recently, the calcium-indicator dyes Fura-2 and fluo-3 have been employed in the monitoring of $[Ca^{2+}]_i$ in brain slices.

In addition to advances in the development of Ca^{2+}-indicator dyes, the patch-clamp recording technique developed for whole-cell and single-channel recording in isolated or enzymatically pretreated cells (3) has recently been adapted for use in brain slices (4, 5). By using thin (100–300 μm thick) slices of brain tissue, it is now possible to employ whole-cell recording techniques and simultaneously measure $[Ca^{2+}]_i$ in single neurons in brain slices, an approach that offers great promise in correlating biophysical measurements of excitability and synaptic transmission with changes in intracellular ionic composition. Future developments, such as the use of fluorescent enzyme analogs (6), coupled with improved temporal acquisition of images, should provide a plethora of new information on cellular events in the central nervous system (CNS).

We examine some of the methodological considerations for the use of intracellular Ca^{2+}-sensitive dyes in brain slices and discuss the advantages and potential pitfalls associated with the technique. In particular, we describe

Methods in Neurosciences, Volume 19

the use of patch-clamp recording methods in brain slices combined with the photometric method to monitor neurons dialyzed with Fura-2 and imaging methods using fluo-3. The primary advantage of examining $[Ca^{2+}]_i$ in brain slices by the photometric method is that it provides a rapid temporal means of resolving changes in $[Ca^{2+}]_i$. In addition, the method maintains the advantages of the slice preparation, such as the avoidance of culture or dissociation artifacts, the preservation of a degree of endogenous connectivity, and control of the extracellular milieu. The key disadvantages are twofold. Accurate assessment of emitted fluorescence is only achieved from cells located at or near the surface of the slice. Intact neurons in this location are likely to be few in number and to have a relatively higher degree of cutting damage (i.e., fewer processes). Second, the photometric method cannot provide detailed spatial resolution of $[Ca^{2+}]_i$ distribution, which may be quite heterogeneous in some neuronal populations. Nevertheless, the method can give an accurate measure of $[Ca^{2+}]_i$ for a region of the cell that is simultaneously subject to voltage control in whole-cell patch-clamp recordings. Imaging methods are the option of choice where a low temporal resolution is acceptable, and spatial gradients of $[Ca^{2+}]_i$ are of primary interest.

Preparation of Slices

The special considerations for slice preparation are highly dependent on the region of the CNS to be examined and the particular synaptic connectivity one wishes to maintain. However, some practical points related to optical recordings are in order. Slice thickness is of particular importance in preparing slices for simultaneous whole-cell recording and $[Ca^{2+}]_i$ measurements. Where specific types of neurons in a brain region need to be visually identified prior to recording, it is advantageous to approach the cell under visual guidance at high ($200\times$ or greater) magnification. This may be achieved using transmitted illumination through the slice and optics modified for increased contrast and depth perception. Both Hoffman Modulation Contrast-modified Leitz nonimmersion objectives (25X) and Nomarski-modified Zeiss water-immersion objectives (40X) are well suited to this task, although the performance of these objectives deteriorates rapidly when thick slices are employed due to light scattering. In poorly organized structures, such as those seen in transverse sections of the ventral horn of the spinal cord or brain stem, slices of 200 μm may be the maximum thickness permitting visualization of motor neurons by transmitted light. In more organized structures where specific cells types are densely packed, individual cells need not be directly visualized (e.g., hippocampus or cerebellum), and satisfactory levels of illumination are transmitted through slices of 300 μm or greater. The greater the density of cell somata, which are relatively transparent, the thicker the slice containing the area of interest may be. Also, slices from younger animals are more transparent, and yield better seals for recording purposes.

The actual cutting process will greatly affect the number of intact cells at the surface of the slice. Although adequate results may be obtained in some areas using a tissue chopper, in most cases a vibrating microtome yields superior results. Indeed, for brain stem or spinal cord preparations, a vibrotome is a virtual necessity. We have prepared transverse and longitudinal sections of spinal cord and brain stem, as well as parasaggital sections of cerebellum and transverse sections of hippocampus. In each case, the greatest degree of preservation of neurons at the cut edge of the slices was achieved by cooling of the tissue prior to slicing, using an extremely slow speed for the advancement of the blade through the tissue, and using the highest degree of amplitude (side-to-side motion) of the blade. At low temperature, the osmolarity, rather than the ionic composition of the extracellular solution, appears of greater import to the success of slicing. Thus either 0.32 M sucrose or chilled external medium may be employed with similar results. To achieve sufficiently slow forward motion of the cutting blade, a higher range varistor may need to be substituted for the one supplied by the manufacturer. Finally, the amplitude of the blade motion can be increased by inserting a washer in line with the vibrating solenoid to increase the blade excursion.

Patch Recording in Brain Slices

The general methods for intracellular recording with patch electrodes in brain slices have been given elsewhere (4, 5). Because relatively clean cell somata can be visually identified on the surface of the slice, for photometric measurements of $[Ca^{2+}]_i$ we have not found it necessary to clean the cell surface by a stream of external medium (4), nor is it necessary to apply positive pressure to the electrode prior to obtaining a gigaseal (5), when cells at the slice surface are recorded. However, in imaging experiments where cells near the lower surface of the slice are recorded, the methods of Blanton *et al.* (5) are used, and dye loading of the pipette is achieved after the whole-cell recording is obtained (see below).

Dye Loading Cells

Two basic methods for the loading of intracellular ion-sensitive dyes are generally employed, based on the chemical nature of these compounds. Both intact free acid and cell-permeant ester forms of most dyes are available.

The former provides the advantage of precise control of intracellular dye concentration and rapid equilibrium with the cell cytosol. The most typical configuration employing the free acid form of Fura-2, for example, is in whole-cell patch recording. A known concentration of Fura-2 may be added to the contents of the patch electrode, either the entire contents of the electrode for slice recordings or as a backfill for work with cultured and dissociated cells. On rupturing the gigaseal obtained prior to whole-cell recording, the dye rapidly diffuses into the neuron, reaching an apparent equilibrium in the soma within a matter of minutes. Diffusion to distal processes can require significantly longer. An example of a $[Ca^{2+}]_i$ response to bath-applied acetylcholine in a patch-clamped neuron in the nucleus tractus solitarius measured with this method is shown in Fig. 1A.

The second method for loading cells with ion-sensitive fluorescent dyes is to employ a cell-permeant dye precursor, which is deesterified on entering the cell (1). The chief advantage of this method is the ease with which it can be employed. However, a number of considerations must be taken into account when using this method. The first is that dye which is not completely deesterified can provide false indications of intracellular ionic concentration. Second, the final intracellular concentration of dye is unknown. For example, the Fura-2 molecule is based on the widely employed Ca^{2+} buffer EGTA. At high concentrations, Fura-2 itself can buffer significant levels of $[Ca^{2+}]_i$. Finally, depending on the particular method of loading, fluorescence from cells (either neurons or glia) adjacent to the cell being examined can produce false indications of changing $[Ca^{2+}]_i$ by virtue of light scatter within the slice. Three methods have been employed to load cells in slices with the cell-permeant form of Fura-2, Fura-2 acetoxymethyl ester (Fura-2AM).

A. In the earliest use of Fura-2AM in the slice (7) the entire slice is incubated in an external medium containing Fura-2AM. This method has now been largely abandoned, as it provides an unsatisfactory level of nonspecific loading of astroglia, excessive scatter of emitted fluorescence, and, in many cases, is cytotoxic to neurons in the slice. In our own hands we have noted that, whereas satisfactory loading of neurons in culture by Fura-2AM is relatively unchallenging, loading astroglia in culture proves to be quite difficult. Thus, a curious paradox exists, for while astroglia in slices load readily, neurons in slices load poorly by this slice incubation method.

B. A second method used to load neurons in slices with cell-permeant dye precursors has been employed by Regehr and Tank (8), who used a pipette to direct a narrow, steady flow of dye-laden solution onto the cell body of a Purkinje cell in the cerebellar slice preparation. The primary advantage of this technique is the preservation of the intracellular contents of the neuron. The chief disadvantage is the unknown intracellular dye concentration, the

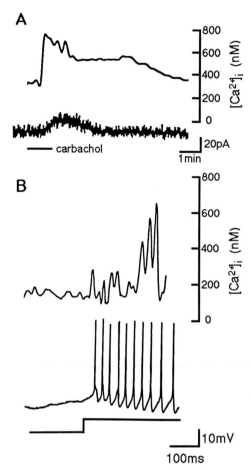

FIG. 1   Examples of data obtained from simultaneous patch-clamp and photometric [Ca^{2+}]$_i$ measurements. (A) Whole-cell voltage-clamp recording from a neuron in the rat nucleus tractus solitarius filled with 100 μM Fura-2 via the internal solution in the patch pipette. Exposure of the slice to 10 μM carbachol resulted in a rapid rise of [Ca^{2+}]$_i$ and a slowly developing outward current. The properties of the carbachol-induced outward current indicated that it reflected the activation of a calcium-dependent potassium current following the mobilization of [Ca^{2+}]$_i$ stores ($V_{hold} = -50$ mV). (B) Current-clamp recording of a rat cerebellar Purkinje cell loaded with Fura-2 using the perforated patch/Fura-2AM method to avoid dialysis of the cell cytoplasm by the patch pipette internal solution. Action potential firing (middle trace) was evoked by direct current injection (200 pA; lower trace), and the [Ca^{2+}]$_i$ response monitored. Conversions of emitted fluorescence ratios to estimates of [Ca^{2+}]$_i$ were made from comparisons with cells loaded by the whole-cell method as in A.

completeness of dye deesterification, and the likelihood that adjacent structures will be loaded with dye and thus contribute to the overall fluorescent signal being received by the detection system.

C. A third method we have employed is a hybrid between the whole-cell method of loading neurons with the free acid form of the dye and the use of cell-permeant dye precursors (1). By using the perforated patch technique for whole-cell recording, where molecules of Nystatin (9) or amphotericin B (10) are permitted to enter the membrane of neurons in the cell-attached configuration, it is possible to satisfactorily load neurons with Fura-2 by including a high concentration (20 μM) of fura-2AM within the patch electrode (Fig. 1B). As no dye is exposed to the surrounding slice, only the electrode itself and the neuron being examined contribute to the emitted fluorescence signal. The primary advantage of this method is the preservation of the intracellular constituents, thus obviating the dialysis of intracellular messengers which may occur with whole-cell recording. The chief disadvantages are similar to those for the AM method in general: there is an unknown intracellular dye concentration and no information regarding the completeness of deesterification of dye precursor. Dye compartmentalization may also be a problem (11), but this can be avoided by working at room temprature. In addition, as the pipette provides a reservoir of Fura-2AM, the cell will be continuously loading with dye. This can produce unacceptably high intracellular concentrations of dye that can significantly buffer $[Ca^{2+}]_i$. Nevertheless, this method could prove useful for the loading of neurons that prove difficult to successfully record from via the whole-cell method, such as cerebellar granule cells. These neurons are very small and electronically compact, thus single neurotransmitter-activated ion channels can readily be resolved using perforated patch recording [Fig. 2 and Ref. (12)], and the combination of this recording method or cell-attached single-channel recording with photometric $[Ca^{2+}]_i$ measures would allow sensitive temporal correlations to be made of single-channel kinetics with changes of $[Ca^{2+}]_i$.

Whatever means is chosen to introduce dye into the cytosol, a balance must be struck between intracellular concentrations sufficient to permit detection of the fluorescence signal and the potential of significant buffering of the ion under investigation. In addition, exposure to longwave uv (330–380 nm) radiation should be limited to the period of data collection to minimize dye photobleaching and potential photolytic damage to the cell. Despite these precautions, photobleaching and some cell damage is inevitable. By monitoring not only the ratio of emitted fluorescence, but also the absolute signal strength at each wavelength, it is possible to monitor for the former concern.

Another important methodological consideration is correction for back-

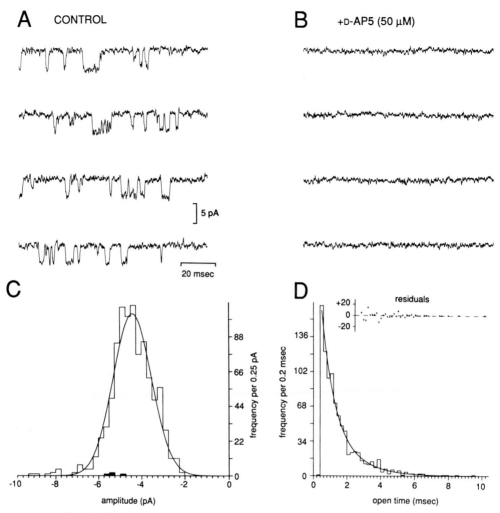

FIG. 2 Single-channel records of spontaneous NMDA receptor-gated channel activity in granule cells of an 18-day-old rat cerebellar slice recorded using the perforated patch-clamp technique for whole-cell recording. (A,B) Representative records of spontaneous single-channel openings before (A) and after (B) the bath application of the NMDA-receptor antagonist D-AP5 in a granule cell voltage-clamped at -80 mV. (C) Amplitude–frequency histogram of channel openings before (open bars, fitted with a single Gaussian function) and after (solid bars) the application of D-AP5. Note that virtually all spontaneous channel openings in this cell were blocked by D-AP5, suggesting that this activity results from activation of NMDA receptors by ambient glutamate levels in the granule cell layer. Such activity is reduced in granule cells prior to migration from the external germinal cell layer and may promote neurite outgrowth during development. (D) Dwell time histogram of the channel openings illustrated in A. Solid line is the fit of a double exponential to the data ($\tau_1 = 0.64$ msec; $\tau_2 = 1.4$ msec); inset shows the residual plot (at a smaller scale) of the fit.

ground fluorescence. It is particularly important to monitor for falsely high ratios resulting from fluorescence measured at 380 nm falling below background. For neurons loaded by the whole-cell patch method, background fluorescence should be determined in the cell-attached mode immediately prior to membrane rupture. For cells in slices loaded by one of the AM methods, satisfactory background subtractions can be made on an adjacent area of the slice, preferably in an area containing a cell of similar dimensions. This method can give false readings, however, as the light-scattering properties of differing areas of the slice are likely dissimilar.

Photometric Methods of $[Ca^{2+}]_i$ Measurement

Perhaps the most difficult question facing the investigator constructing an apparatus for fluorescence detection in slices is the detection system. This is indeed an area of great concern as, in general, it is far easier to get light "in" than to get it "out." Thus, the choice of a detection system must balance the concerns of cost with those of sensitivity and speed. A typical apparatus (see Fig. 3) will consist of a light source, an upright microscope configured for epifluorescence (through the lens) illumination and detection, a photomultiplier tube and power supply, an analog-to-digital converter, and some form of digital storage and analysis (i.e., a computer). Of these, the first priority should be to match the particular slice preparation of interest to the detection system. We have found the Thorn/EMI photomultiplier tube (modified for low light levels) coupled with APED hardware to provide the greatest degree of sensitivity and speed, as well as the flexibility to record from a variety of preparations. Lower cost photomultiplier tubes are readily available from a number of manufacturers; however, serious consideration of future experimental goals should be addressed before reaching a decision on this most important component. On reaching a decision on the detector, the next serious consideration is the illumination system. For Fura-2, two wavelengths of long-wave ultraviolet light are required. Three broad categories of light source are available.

A. *Filter changers*. Light from a high-intensity Xe or Hg lamp is passed through a condenser to a spinning wheel containing two or more narrow band pass filters. A stepper motor controls the position of these filters. The transmitted light (340–360 and 380 nm) is passed to the sample, causing intracellular Fura-2 to fluoresce. The ratio of fluorescence produced by the two wavelengths can indicate the concentration of $[Ca^{2+}]_i$. The primary advantage of the filter changer is the low cost. However, two serious shortcomings make it the least acceptable light source. First, the speed at which

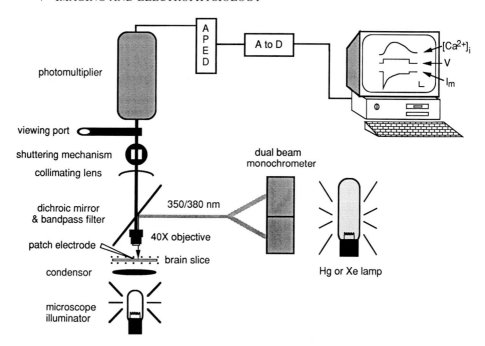

FIG. 3   Schematic illustration of the apparatus for photometric measurement of $[Ca^{2+}]_i$ during whole-cell recording. Light from a mercury or xenon lamp is split by means of a chopper into separate wavelengths by a dual-beam monochrometer. The light is passed to the brain slice via a dichroic mirror and through a 40X water immersion objective. Emitted epifluorescence is passed through a high-pass filter (typically >430 nm) and into a photomultiplier. The raw analog signal is then integrated by hardware (APED), converted to a digital signal with a high sampling rate (>50 kHz), and passed to a computer for storage and analysis. The current and voltage output of the amplifier are also stored on the computer, allowing the simultaneous post hoc analysis of changes in $[Ca^{2+}]_i$ and transmembrane current and voltage.

filters are changed is relatively slow. While adequate for detection of fluorescence by (SIT) or (CCD) cameras (see following sections), this slowness negates the speed advantage of the photomultiplier detection system. A filter changer might be an option, however, should the investigator intend to add or convert the system to imaging at a future date. The second key disadvantage is deterioration of filters with time. This can be partially offset by the relatively low cost of the filters themselves, which are easily replaced (note: manufacturing quality varies widely from the available sources). It is also possible to prolong the life span of the filters by protecting them from the infrared radiation produced by the lamp using a water filter.

B. *Filter disk*. A second approach uses a similar lamp to illuminate a continuously spinning disk that is composed of two or more regions of narrow band pass filter. By varying the speed of rotation, a high degree of temporal resolution can be achieved that is compatible with the photometric detection method. Though more costly than the filter changer, the filter wheel should deteriorate less rapidly. The chief disadvantage of the spinning disk is the limitation in available wavelengths. Different disks can be changed, but the added cost can largely offset the advantage of this method if more than one set of wavelengths are required.

C. *Dual monochrometers*. The dual-monochrometer system provides the greatest flexibility, albeit at the greatest cost. A beam splitter separates the light from the lamp and a variable speed chopper alternately passes it through one of two independent monochrometers. These can be set to a wide variety of wavelengths depending on the experimental needs. The bandwidth of the output of the monochrometers can also be adjusted. The output is recombined and delivered to the sample as with the other light sources described above. Situations calling for differing or narrowly constrained wavelengths are readily achievable with this light source. Again, the chief disadvantage is cost, which is often three times or greater than that of the preceding systems.

Whatever light source is ultimately chosen, compatibility with the photodetection system is an absolute necessity. To achieve an acceptable signal-to-noise ratio, it may be necessary to provide a stronger lamp for a less sensitive photomultiplier. In these circumstances, additional considerations, such as the cost, heat, and ozone output of the lamp, as well as the cost of the lamp housing, should be considered. Small laboratory spaces can quickly become intolerably hot from just one 150-W Xe bulb. Many of the higher wattage bulbs also require magnetic containment fields and water cooling. In general, it is far better to detect emitted fluorescence with high sensitivity than overcome a weak detection system by exposing the slice to excessive, potentially damaging longwave radiation. It should also be noted that a minimum of 100 W is required for transmitted light to penetrate the thickness of the slice and achieve acceptable signal-to-noise ratios.

A fourth method utilizes the whole-cell patch technique of Blanton *et al.* (5) to image $[Ca^{2+}]_i$ in neurons using whole-cell dye-loading methods and confocal microscopy (Fig. 4). To enable a lens of relatively high numerical aperature to be utilized for confocal imaging, it is necessary to limit the working distance between the preparation and lens by the use of an inverted microscope. Slices (200 μm thick) of hippocampus from 2-week-old rats are imaged through a coverslip attached to the bottom of a recording chamber, which is mounted to the microscope stage. Both stage-mounted stimulating and recording electrodes and the tissue are visualized with a conventional

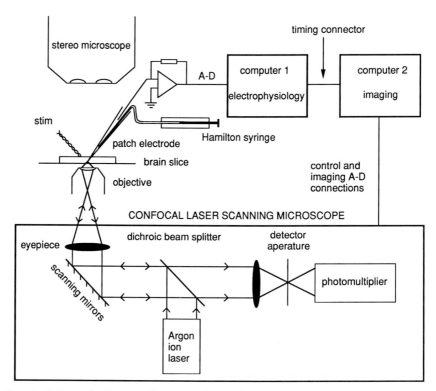

FIG. 4   Schematic diagram of the experimental arrangement for scanning laser confocal microscopic measurement of $[Ca^{2+}]_i$. The slice of brain tissue is mounted on a coverslip in a chamber on the stage of an inverted microscope. Imaging is performed with a confocal laser-scanning head attached to this microscope. Electrophysiological stimulating and recording electrodes are attached to the microscope stage above the slice in a manner similar to the arrangement in conventional brain slice recording setups.

stereomicroscope mounted above the stage. Patch recordings are made by driving the electrode through the tissue to obtain recordings within 20 to 50 μm of the lower surface of the slice. Positive pressure is applied to the interior of the pipette to keep the tip cleaned of debris during the electrode advance. This necessitates filling the electrode with dye after obtaining a whole-cell recording to prevent dye leakage into the surrounding tissue. For this purpose a perfusion system was developed to enable the contents of the electrode to be changed after obtaining a recording. This system is comprised of a fine silica tube (o.d., 170 μm) pulled over a bunsen flame to a tip diameter of 10 to 20 μm. The tube is led through a second port in the electrode holder so that its tip is positioned within 70 to 100 μm of the interior of the pipette tip. Sealed to the second port is an oil-filled hydraulic tube attached to a Hamilton syringe. this tube is preloaded with 5 μl of pipette solution to which 20 μM fluo-3 has been added. After obtaining a whole-cell recording, depressing the syringe fills the electrode tip and neuron with dye.

This technique differs from that discussed above in that neurons are dialyzed with the dye solution. It has the disadvantage that soluble cytoplasmic intracellular components are not well preserved. However, it is advantageous in that the maximal concentration of dye in the neuron will not exceed a known value (the pipette concentration), dye compartmentalization associated with the use of the AM ester of the dye (11) will not occur, and the stable, long-lived recordings make it possible to manipulate the internal contents of the cell when required.

Microscopes and Objectives

Although it is possible to record fluorescence from slices on an inverted stage microscope (see below), for simultaneous electrophysiology and photometric measurement of $[Ca^{2+}]_i$, an upright microscope is advantageous. We employ a fixed stage upright microscope; focusing is achieved by raising or lowering the entire headpiece of the microscope. In some cases, it may also be advantageous to add a hinge modification, which allows the top of the microscope to be quickly raised. This makes the changing of slices in the chamber or objectives far easier. A trinocular headpiece, mounted on the top of the microscope, passes light either to the eyepieces or to the photomultiplier. Between the trinocular and the photodetector, a lens for collimating the emitted fluorescence and a means for shuttering or limiting the field of detection are also required.

The choice of objectives is limited by a number of factors. The water/air interface provides an unacceptable degree of scatter, both of the incident illumination and the emitted fluorescence. Thus, a water-immersion objective is the remaining choice. The primary obstacle to electrophysiological recording under a water-immersion objective is working distance. The best available objective at present, one that balances the factors of optical density and working distance, is the 40X water immersion objective from Zeiss. By providing a 1.3-mm clearance between the focal plane and the bottom of the lens, whole-cell patch electrodes may be positioned along the edge of a neuron for whole-cell recording. This also has the benefit of locating the Fura-2-filled electrode to the side of the cell, where shuttering can largely eliminate stray background fluorescence being emitted from the pipette. At present, no other objective is as widely employed in this type of recording, as working distances become unreasonably small. However, the use of other objectives is clearly possible, should one wish to record fluorescence alone.

Long working distance condensors are also needed to accommodate the depth of slice chambers mounted on the microscope stage. For patch-clamp recording in slices without $[Ca^{2+}]_i$ measurements on an upright microscope, a Hoffman Modulation Contrast-modified Leitz 25X objective (Model NPLFLL) provides excellent optical clarity and a long (13 mm) working distance useful for the placement of multiple electrodes for recording, pathway stimulation, or pressure ejection of drugs.

Photometric Calibration of $[Ca^{2+}]_i$

Dyes like Fura-2 should not be considered to be fully quantitative, but rather semiquantitative, as comparisons between actual $[Ca^{2+}]_i$ values in identical cell types derived between different laboratories can demonstrate. Thus, the rationale for calibration should be to provide a relative idea of intracellular Ca^{2+} and its dynamics in response to various stimuli. Frequent calibration also provides a means to monitor the state of the equipment, as lamps tend to degrade over time.

Two methods of intracellular calibration are widely employed. The first uses ionophores, such as ionomycin, to permeabilize a dye-loaded cell (13). The intracellular concentration of ions like Ca^{2+} can then be directly manipulated by altering the extracellular ionic concentration and allowing time for equilibrium to be established. The primary advantage of this method is that it gives an accurate assessment of dye fluorescence within the microenvironment of the cell. It is particularly important that R_{min} and R_{max} values be determined at some point by this method. The chief disadvantage of this method is the extended length of the calibrations, which can require 30 min or more for equilibrium at a given Ca^{2+} concentration. At the same time, dye tends to leak more readily from permeabilized cells, thus adding an additional variable. Nevertheless, for the most accurate determination of $[Ca^{2+}]_i$, this method is unsurpassed.

The second most commonly employed method is the cell-free calibration. Commercially available kits are available containing buffers of known Ca^{2+} concentration. Dye is added to these and a thin film of each is in turn monitored for fluorescence. Calibration curves of eight or more points can be rapidly constructed by this method. Furthermore, the speed and ease of calibration permits for frequent (daily or weekly) measurements, which are appropriate for an apparatus in daily use. The primary disadvantage of the method is that these solutions clearly do not mimic the intracellular milieu, thus resulting fluorescence ratios are unlikely to precisely follow those of the intracellular calibration method.

Other Hardware and Software Considerations

The decision to purchase a commercially available analysis program for the conversion of raw digital output of the analog-to-digital converter into ratio and ultimately $[Ca^{2+}]_i$ values is largely the choice of the investigator. In our laboratory, we have prepared a custom softare package for the simultaneous photometric monitoring of $[Ca^{2+}]_i$ and electrophysiology. The primary consideration for such a package should be speed of digitization and conversion, storage capacity flexibility, and ease of use. Alternatively, raw data can be stored either on hard disk or tape during an experiment and analyzed off-line. While less desirous, this approach can be mastered by any investigator familiar with one of many commercially available spreadsheet programs. The added advantage of direct graphical output of these programs can also compensate for the loss of on-line monitoring of $[Ca^{2+}]_i$. With monitoring of the output of the photomultiplier system during the course of an experiment, it is often possible to make qualitative judgments about the state of $[Ca^{2+}]_i$ following a given stimulus. One can then return to the data at the conclusion of the experiment and determine actual $[Ca^{2+}]_i$ values.

The choice of recording chamber is also largely dependent on the type of slice and nature of recording. In our laboratory, we have a variety of chambers that suspend the slice between two nets made of nylon. The spacing of the netting must balance the needs of slice stability with access for electrodes. As a cautionary note, some of the chamber and netting materials that may be presently employed in a laboratory, such as bridal veil used as netting, may fluoresce quite readily under longwave uv, thus it is best to avoid making a chamber out of these materials.

Interpretation of Data

The chief advantage of the photometric method for monitoring of $[Ca^{2+}]_i$ is the speed of acquisition. Regardless of the light source used, however, the photomultiplier will have a defined range of linearity, below which background noise is a significant portion of the signal and above which signal saturation occurs. Therefore, adequate time must be allowed to permit switching between wavelengths and data acquisition along the linear region of the emitted signal. This can typically be assessed by examining the raw or integrated signal with an oscilloscope.

While brief $[Ca^{2+}]_i$ transients can be resolved, there are a number of potential pitfalls with such measurements. Because of the light-scattering properties of the slice, changes in $[Ca^{2+}]_i$ occurring within proximal processes, but not the soma, can give false indications of somatic $[Ca^{2+}]_i$

changes. This can be minimized by using the shuttering mechanism to limit the region of detection to the central portion of the soma. This has the added benefit of minimizing false edge effects, which may occur as the result of cells on the surface of the slice being somewhat flattened. As the net Fura-2 signal is an integral of the vertical, two-dimensional column of Fura-2 molecules, the 380-nm emission often falls below background at the relatively thin edge of the cell, giving falsely high $[Ca^{2+}]_i$ values (an identical problem is observed in cell culture, where cells tend to flatten in profile on the coverslip). Finally, as the extracellular perfusate is present between the objective and the sample, care must be taken with certain compounds that might themselves fluoresce.

Imaging of $[Ca^{2+}]_i$ Using Confocal Microscopy

The use of imaging techniques in conjunction with microfluorimetry of individual neurons provides information on the complex spatial anisotropy of ionic movements through neuronal membranes during evoked or physiological activity. Confocal microscopy is ideally suited for imaging of $[Ca^{2+}]_i$ in slices because of its ability to eliminate glare caused by scatter and out-of-focus fluorescence in the slice. Optimizing the image quality in this manner is an important consideration when performing simultaneous intracellular recording and imaging from the same neuron, for it is neither possible to manipulate the slice during the experiment to improve image quality nor possible to select from a number of prefilled neurons.

The details of confocal imaging have been discussed at length elsewhere and are not addressed in depth here. In brief, confocal microscopes eliminate out-of-focus information and scattered light in an image by scanning a pinhole or slit light source at an identical location and depth of field as a detector pinhole or slit. In practice, this is achieved with epifluorescence microscopy by scanning the excitant light through the same lens and with the same mirror system as those of the detected light. Optical sections may then be cut through the tissue to generate sequences of images at selected focal planes. The use of such a system to measure physiological changes in ionic concentrations in three dimensions is not practical, because of the time required to make a series of sections. However, the advantages of such a system for imaging microfluorimetry in tissue slices lies in the ability to eliminate the substantial glare resulting from scattering of light in the tissue and from out-of-focus emitted light generated in more standard epifluorescent microscopes. The resultant image enables greater confidence that measured changes in intensity in the image occur at the particular site of interest and not from contaminating light from elsewhere.

Two different systems are available from a number of manufacturers. The least expensive of these are systems which rely on a scanning slit generating a line of light which is imaged confocally. This has the advantage of low cost and potential scanning speeds: image frames can be scanned at video rates, depending on the detector used and the quantal yield necessary to generate an image. The brightness of the image will be proportional to the dwell time of the light source.

A more expensive alternative is the laser-scanning confocal microscope. The advantage of this system is flexibility and power of the light source. The major technical disadvantage is the difficulty in choice of a suitable laser, and the speed with which image frames may be obtained is inversely proportional to the number of image lines used. With respect to flexibility, the system allows a choice between photometry at a single defined line in an image frame at high speed and imaging at lower speeds at a resolution that is inversely proportional to imaging speed as the number of raster lines is increased. The ability of the microscope to scan in only selected areas of the lens field of view enables a higher resolution to be obtained than a standard microscope and obviates the need to move the sample across the field of view or to change the lens. This is an important consideration in electrophysiological studies where the preparation must be kept mechanically stable.

The choice of laser for this system depends on the absorbance wavelength of the dyes used. Lasers generate lines of light at particular frequencies. This, coupled with the expense, size, and heat generation of uv lasers has limited their use and largely prevents imaging with dual-excitation ratiometric dyes such as Fura-2 (see above). However, the ongoing development of longer wavelength dyes should reduce this problem in the future. Most laser-scanning confocal microscopes commercially available use an argon ion laser generating lines at 488 and 514 nm, useful for exciting rhodamine- and fluorescin-based dyes and dyes with equivalent absorbtion wavelengths, such as fluo-3, Ca^{2+} orange, and Fura red.

We have utilized the nonratiometric Ca^{2+}-sensitive dye fluo-3 for measuring changes in $[Ca^{2+}]_i$ (Figs. 4–6). For this purpose it is not necessary to measure absolute $[Ca^{2+}]_i$, but rather to compare the relative change under various experimental conditions. Following the injection of fluo-3 into the tip of the electrode, the tissue is scanned with the confocal microscope at low resolution. The cell soma fills rapidly with dye and can be imaged immediately after dye loading. Thus, experiments may be performed on Ca^{2+} entry in the soma very soon after whole-cell access is obtained for this reason. Application of depolarizing steps to the cell through the patch pipette activates Ca^{2+} conductances leading to an increase in fluo-3-fluorescence proportional to the amplitude of the Ca^{2+} conductance.

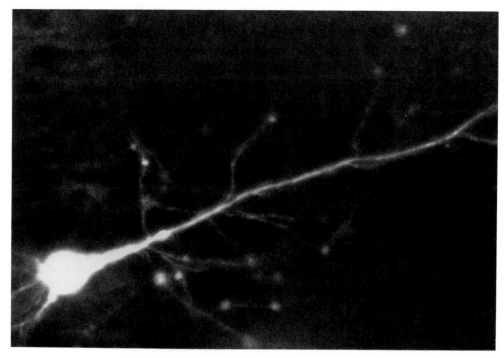

FIG. 5 A low-resolution image of a hippocampal pyramidal neuron filled with fluo-3. This image is of an optical section with a depth of field of approximately 6 μm. Measurements during the course of the experiment were made by scanning only selected portions of dendrite or the soma. For each region the photomultiplier gain was set to maximize the recorded signal without saturation.

Complete dye loading of dendritic processes requires beween 15 and 30 min. It is then possible to measure dendritic changes in $[Ca^{2+}]_i$. The technique was used to measure changes in dendritic $[Ca^{2+}]_i$ induced by stimulation of synaptic inputs to CA1 pyramidal neurons under voltage-clamp conditions (Fig. 6). It has proved possible to demonstrate that rises in $[Ca^{2+}]_i$ in these cells following synaptic stimulation display the voltage dependency predicted from the voltage dependence of the NMDA receptor channel (Fig. 6), with a peak rise in $[Ca^{2+}]_i$ at holding potentials of -35 mV. This rise in $[Ca^{2+}]_i$ could be blocked by the NMDA receptor antagonist D-AP5, but not by the AMPA receptor antagonist CNQX (not shown). The ability to manipulate the membrane potential of a neuron while monitoring fluctuations in $[Ca^{2+}]_i$ thus enables isolation of various components of Ca^{2+} entry not possible when performing measurements of Ca^{2+} flux alone.

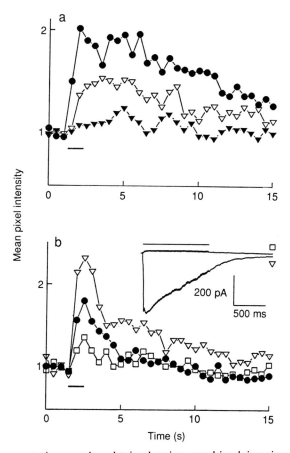

FIG. 6   Representative results obtained using combined imaging and whole-cell patch-clamp techniques in rat CA1 hippocampal pyramidal cells. (a) Data from a neuron in the presence of Mg^{2+}. The amplitude of the fluorescence signal is the average calculated over a 50-μm length of dendrite and thus represents the mean rise in $[Ca^{2+}]_i$ over this region. The rise in $[Ca^{2+}]_i$ is greatest at -35 mV (closed circles), intermediate at -70 mV (open triangles), and smallest at -90 mV (closed triangles). This behavior is typical of Ca^{2+} permeation through the NMDA receptor-gated channel. This conductance has a reversal potential of 0 mV and is blocked by external Mg^{2+} in a voltage-dependent manner, giving a peak inward current at approximately -35 mV in the presence of this ion. (b) The relative rise in fluo-3 fluorescence (plotted with respect to baseline) in a different neuron following high-frequency stimulation of the Schaffer collateral–commisural fiber pathway at three different holding potentials in a Mg^{2+}-free saline. The peak rise in $[Ca^{2+}]_i$ follows stimulation at a holding potential of -70 mV, with a small, but measurable rise at $+5$ mV. The inset traces are the corresponding synaptic currents evoked at holding potentials of -70 mV (open triangles) and $+5$ mV (open squares). The intermediate rise of $[Ca^{2+}]_i$ was recorded at -35 mV (closed circles).

Conclusions

The photometric method provides a high temporal means of simultaneously monitoring the fluorescence of intracellular ion-sensitive dyes and the transmembrane voltage or current in neurons in slices. The key advantage of the method is speed. The chief disadvantage is the limited region of the cell that can be accurately examined and the absence of spatial resolution. Notably, the majority of the components suggested here can be adapted for use in digital imaging. Indeed, hybrid systems containing both imaging, for spatial resolution, and photometry for temporal resolution are gaining popularity. Imaging methods are needed where the spatial resolution of the calcium signal is of interest, and speed can be compromised to some extent.

Acknowledgment

We are grateful to Bob Hughes of E. Nuhsbaum, Inc. for many helpful discussions and comments on the manuscript.

References

1. R. Y. Tsien, *in* "Fluorescence Microscopy of Living Cells in Culture" (D. L. Taylor and Y.-I. Wang, eds.), p. 127. Academic Press, New York, 1989.
2. G. Grynkiewicz, M. Poenie, and R. Y. Tsien, *J. Biol. Chem.* **260,** 3440 (1985).
3. O. P. Hamill, A. Marty, E. Neher, B. Sakmann, and F. J. Sigworth, *Pfluegers Arch.* **39,** 85 (1981).
4. F. A. Edwards, A. Konnerth, B. Sakmann, and T. Takahashi, *Pfluegers Arch.* **414,** 600 (1989).
5. M. G. Blanton, J. J. Lo Turco, and A. R. Kriegstein, *J. Neurosci. Methods* **30,** 203 (1989).
6. K. M. Hahn, A. S. Waggoner, and D. L. Taylor, *J. Biol. Chem.* **265,** 20335 (1990).
7. Y. Kudo, K. Ito, H. Mivakawa, Y. Izumi, A. Ogura, and H. Kato, *Brain Res.* **407,** 168 (1987).
8. W. G. Regehr and D. W. Tank, *J. Neurosci. Methods* **37,** 111 (1991).
9. R. Horn and A. Marty, *J. Gen. Physiol.* **92,** 145 (1988).
10. J. Rae, K. Cooper, P. Gates, and M. Watsky, *J. Neurosci. Methods* **37,** 15 (1991).
11. S. Bolsover and R. A. Silver, *Trends Cell Biol.* **1,** 71 (1991).
12. R. A. Silver, S. F. Traynelis, and S. G. Cull-Candy, *Nature (London)* **355,** 163 (1992).
13. D. A. Williams and F. S. Fay, *Cell Calcium* **11,** 75 (1990).

Section VI

Lipid Bilayers

[19] Reconstitution of Channel Proteins from Excitable Cells in Lipid Bilayers: Authentic and Designed Proteins

Anne Grove, Antonio V. Ferrer-Montiel, and Mauricio Montal

Perspective and Overview

Single-channel recordings of channel proteins reconstituted in planar lipid bilayers have proved of great value for the functional characterization of naturally occurring channel proteins (1–10) and pore-forming peptides [for review, see (11)], as well as *de novo* designed channel-forming peptides and proteins (12–16). As detailed procedures have been described previously (17–21), our objective here is to highlight salient advantages and illustrate potential pitfalls associated with the use of this highly sensitive technique. Two complementary approaches to the characterization of channel proteins exemplify these points, with prototypes of ligand-gated and voltage-gated channels selected for illustrative purposes: the single-channel characterization of a naturally occurring channel protein, the nicotinic acetylcholine receptor (AChR) from *Torpedo californica* (1, 2, 18, 20, 22), and synthetic peptides and oligomeric proteins representing sequences of specific segments of the voltage-gated calcium channel, used in the *de novo* design of functional channel proteins (16).

From Authentic to Designed Channel Proteins

The starting point of this description is the purified channel protein—what information may be secured following reconstitution in lipid bilayers? Single-channel recordings, from protein reconstituted in lipid bilayers (1–10) or from patches of cell membranes expressing native or mutant channel proteins (23, 24), provide a sensitive assay for characterization of the channel at the level of single molecular events. Estimates of pore size, profiles of ionic selectivity, and the sites of action of specific ligands or channel modulators may guide the conjectured assignment of local protein structure and provide indications of molecular determinants of function (23–25).

The study of purified protein incorporated into a lipid bilayer offers the distinctive advantage that the contents of solutions separated by the bilayer membrane, as well as membrane lipid composition and symmetry, are specified. The system is suitable for investigating modulation of single-channel properties: the effect of covalent modifications such as phosphorylation or the mode of action of specific drugs and toxins may be studied under well-defined conditions, and in absence of cellular cofactors or modifiers. In addition, the transmembrane voltage may be controlled and, therefore, the effect of applied electrical field on single-channel characteristics determined.

Voltage-Gated and Ligand-Regulated Channel Proteins

Several members of the superfamilies of voltage-gated and ligand-regulated channel proteins have been characterized, and purification procedures described (e.g., 1–9, 26, 27). Membrane proteins are purified in the presence of detergents and it is necessary, therefore, to reconstitute the purified protein in a lipid bilayer for functional characterization. We have described the purification and reconstitution of the AChR (2, 20) and the voltage-gated sodium channel (3, 4).

The AChR is selected here to illustrate advantages of the reconstitution approach in obtaining fundamental insights into the modulation of channel activity by protein phosphorylation. The *T. californica* AChR contains up to nine potential phosphorylation sites, modified by different protein kinases and phosphatases (28). AChR phosphorylation modulates receptor assembly, clustering, and stabilization during development (28), as well as ion-channel activity. Fast-flux measurements on reconstituted vesicles (28) and single-channel recordings on reconstituted AChR in planar lipid bilayers (22) showed that AChR phosphorylation by protein kinase A modifies the kinetics of entry into and exit from the desensitized state and point to a key role of γ and δ subunits in AChR gating. The reconstitution approach allows specific investigation of the effects of chemical modification on single-channel properties, while excluding effects of other cellular modulators.

Design of Functional Channel Proteins

The identification of protein modules that determine specific properties of a channel protein may provide clues as to how amino acid sequence determines three-dimensional conformation and associated functional characteristics. Accordingly, we described a strategy for the identification of potential pore-lining segments of ligand-regulated and voltage-gated channel proteins and identified a plausible molecular blueprint for the pore-forming structure as a bundle of amphipathic α-helices, arranged such that charged or polar residues line a central hydrophilic channel (25). This structural motif is inferred from the occurrence of homologous protein domains or subunits arranged around a central pore, extensive sequence homology between members of superfamilies of channel proteins, and the identification of segments capable of forming membrane-spanning α-helical structures (25, 29).

The nonpolar nature of the cell membrane restricts potential conformations of a membrane protein: hydrophobic surfaces interact with the membrane interior and the aqueous pathway for ionic diffusion is considered lined with polar residues. Accordingly, plausible pore-lining segments include amphipathic α-helices, distinguishable from the amino acid sequence using secondary structure prediction algorithms (30). Segments are selected based on sequence conservation and the presence of functional residues compatible with observed ionic selectivity of the authentic protein. Helical modules must be greater than 20 residues in length to span the hydrophobic core of the bilayer (25, 30).

As an approach to modeling a pore-forming structure, we designed oligomeric proteins (14, 16) composed of four identical amphipathic α-helices tethered to a nine amino acid template (31). Attachment to the template promotes interaction between peptide modules and induces a four-helix bundle conformation (31).

Single-Channel Characteristics of Designed Channel Proteins

Monomeric peptides representing identified segments may be synthesized and their ability to form ionic channels tested in lipid bilayers. However, peptides self-assemble in the bilayer to form conductive oligomers of different sizes (11–13). The covalent attachment of channel-forming peptides to a template specifies oligomeric number. Reconstitution of oligomeric proteins in lipid bilayers yields conductance events of uniform amplitude and prolonged channel mean open time, indicating stability of the tethered α-helical bundle (14, 16).

This molecular design strategy has been used to design and synthesize functional channel proteins that represent specific segments of ligand-gated channels, the cation-selective AChR channel (14) and the anion-selective glycinergic receptor channel (32), and the voltage-gated calcium channel (16). Designed proteins emulate pore properties of corresponding target proteins, such as ionic selectivity, channel blocker binding sites, and stereospecific drug-binding sites (14, 16, 32). Accordingly, key functional elements of the pore-forming structure of authentic channel proteins are contained within bundles of α-helices with a specific amino acid sequence.

CaIVS2	I AM NILNM LFTGLFT VEM I LK
CaIVS3	DPWNVFDF LI V I GS II DV I LSE
CaIVS4	NSRISITFFRLFRVMRLIKLLSR
CaIVS5	Y VALLIVMLFF I YAVIGMQMFGK
IVS3-Random	S IDLP IF I V I VNG WVF SLDEID
NaIVS4	YRV I RLAR I GRILRL I KGAKGIR
TEMPLATE	KKKPGKEKG
	* * * *

FIG. 1   Amino acid sequences of predicted transmembrane segments S2–S5 from the fourth repeat of the DHP-sensitive calcium channel; sequences are conserved between isoforms of cardiac muscle and brain (29). IVS3-Random is a computer-generated randomized sequence with same amino acid composition as CaIVS3. The sequence of NaIVS4 is from rat brain (29). CaIVS2, CaIVS3, and CaIVS5 are used to generate four-helix bundle proteins by attachment of four identical peptide modules to the ε-amino groups of template lysines, indicated by an asterisk.

Here, we use peptides and four-helix bundle proteins representing sequences of the calcium channel to illustrate the general approach to determining single-channel properties of an identified sequence; emphasis is placed on the need for reproducibility, appropriate controls of sequence specificity, and the significance of membrane lipid composition.

Voltage-Gated Calcium Channel

The α_1 subunit of the dihydropyridine (DHP)-sensitive calcium channel forms a functional voltage-gated channel and contains binding sites for specific modulators [for review, see (33)]. The primary structure, which is similar to the sodium-channel α subunit suggests the occurrence of four internal repeats (I–IV) organized as pseudosubunits around a central pore (25, 29). Each repeat contains six segments (S1–S6), predicted to form membrane-spanning α-helical structures (25, 29). The loop connecting transmembrane segments S5 and S6 modulates ionic selectivity and blocker sensitivity of voltage-gated potassium and sodium channels and is considered to contribute to the pore lining (34). S3 segments contain conserved negatively charged and polar residues and, therefore, may also be implicated in lining the cation-selective pore of voltage-gated sodium and calcium channels (12, 16, 25). Figure 1 illustrates sequences of transmembrane segments from the calcium channel used to generate four-helix bundle proteins, as well as segments used as controls of sequence specificity.

Experimental Considerations

Formation of Lipid Bilayers at the Tip of Patch Pipettes

Bilayers are formed at the tip of patch pipettes by adjoining the hydrocarbon tails of two lipid monolayers initially formed at the air–water interface (18, 19). Compared to bilayers formed across an aperture in a Teflon partition that separates two aqueous compartments (17), bilayers formed at the tip of patch pipettes offer greater sensitivity and time resolution (18, 20). Pipettes are made from Microhematocrit Capillary Tubes (Fisher, Pittsburgh, PA) and pulled in a four-step procedure using an automated pipette puller (Model P-87; Sutter Instrument Co., San Rafael, CA). The tip size is adjusted by regulating the heating of the last pull and calibrated to yield 5–10 MΩ open resistance when the pipette is filled and immersed in the buffer described. Chambers for bilayer formation are lids from Eppendorf tubes (Applied Scientific, San Francisco, CA). The chambers have a capacity of approximately 200 μl and are used only once. Planar bilayer experiments are performed at 24 ± 2°C. Other conditions are as described (18–20).

Lipids

1-Palmitoyl-2-oleoyl-*sn*-glycero-3-phosphoethanolamine (POPE), 1-palmitoyl-2-oleoyl-*sn*-glycero-3-phosphocholine (POPC), 1-palmitoyl-2-oleoyl-*sn*-glycero-3-phosphoserine (POPS), and soybean lipids are from Avanti Biochemicals (Alabaster, AL). Cholesterol is from Sigma Chemical Co. (St. Louis, MO).

Electrical Recordings

Electrical recordings are carried out (18–21) using a patch-clamp system (List L-M EPC-7, Medical Systems Corp., NY). Constant voltage is applied from a DC source (Omnical 2001, W-P Instruments, New Haven, CT). The signal output from the clamp is stored on videocassette, using a VCR (Sony Betamax) equipped with a modified digital audioprocessor (Sony PCM 501ES, Unitrade, Philadelphia) and displayed on a storage oscilloscope. The two aqueous compartments are connected to the amplifier by two Ag/AgCl electrodes (In Vivo Metric Systems, Healdsburg, CA). The reference electrode is immersed in the chamber used for bilayer formation and the pipette electrode attached to a pipette holder. Ionic selectivity is determined from the reversal potential measured under a single salt concentration gradient. For experiments involving a concentration gradient across the membrane, the reference electrode is placed in a reservoir containing the same buffer as the pipette. Reservoir and bilayer chamber are then connected by a salt bridge (1% agar in 2 M KCl). Both electrodes are connected to the EPC-7 headstage, which is mounted directly on a micromanipulator. The experimental setup is enclosed within a Faraday cage which is sound insulated with foam padding and placed on a vibration-free table (Kinetic Systems, Roslindale, MA).

Data Processing

Records are filtered at 2–5 kHz with an 8-pole Bessel filter (Frequency Devices, Haverhill, MA) and digitized at 100 μsec per point using an Axon TL-1 interface (Axon Instruments, Burlingame, CA) connected to an Everex Step 386 computer (Everex, Fremont, CA). The pClamp 5.5 program package (Axon Instruments) is used for data processing. Conductance values are calculated from current histograms best fitted by the sum of two Gaussian distributions. Channel open (τ_o) and closed (τ_c) lifetimes are determined by exponential fits to probability density distributions of dwell times in the open and closed states. Reliable measurements of single-channel conductance and channel mean open and closed times must be obtained from multiple experiments, each with ≥300 openings in a continuous recording.

Reconstitution of Acetylcholine Receptor in Soybean Vesicles

Buffer Solutions

Homogenization buffer contains 100 mM NaCl, 5 mM EDTA, 5 mM EGTA, 0.02% NaN$_3$ (w/v), and 10 mM HEPES, pH 7.4. Prior to tissue homogenization, the buffer is supplemented with 5 mM iodoacetamide and 1 mM phenylmethylsulfonyl fluoride (PMSF). Other protease inhibitors may be used (e.g., aprotinine, leupeptine).

Buffer A: 100 mM NaCl, 0.02% NaN$_3$, 10 mM HEPES, pH 7.4.
Buffer B: Buffer A + 1.2% (w/v) sodium cholate.

Buffer C: Buffer B + 1 mg/ml soybean lipids.
Buffer D: Buffer C + 200 mM NaCl + 10 mM carbamylcholine.
Dialysis Buffer: 145 mM sucrose, 10 mM NaN$_3$, and 10 mM NaH$_2$PO$_4$
   or 10 mM HEPES, pH 7.4.

Buffered salt medium contains 0.5 M KCl or NaCl, 5 mM CaCl$_2$, 50 μM
4,4′-diisothiocyanatostilbene-2,2′-disulfonic acid (DIDS), 10 mM HEPES,
pH 7.4, filtered through a 0.2-μm nylon syringe filter.

Phosphorylation Buffer contains 10 mM MgCl$_2$, 0.1 mM EGTA, 20 mM
HEPES, pH 7.4.

Purification of AChR from Torpedo Californica

AChR from *T. californica* electric organ is purified by affinity chromatography (35, 36). Briefly, 200 g of frozen electroplax is homogenized with a Polytron and subjected to centrifugation at 5000 rpm in a Beckman JA-21 rotor for 10 min. The supernatant is decanted through eight layers of cheesecloth and centrifuged at 35,000 rpm in a Beckman Ti45 rotor for 30 min. The resulting pellet constituting the crude membrane fraction is resuspended in Buffer A. All steps are carried out at 4°C.

For solubilization, crude membranes are diluted to a final protein concentration of 2 mg/ml with Buffer A and sodium cholate is added to a final concentration of 1.2%. The suspension is stirred at 4°C for 20 min and centrifuged at 35,000 rpm in a Beckman Ti45 rotor for 50 min. The supernatant is applied to a bromoacetylcholine-derivatized affinity column (36) prepared from Affi-Gel 10 (Bio-Rad, Richmond, CA) and previously equilibrated with Buffer B. The affinity column is washed with 3 volumes of Buffer C, and AChR eluted with 10 mM carbamylcholine (Buffer D). AChR concentration is determined from the absorbance at 280 nm and purity tested by SDS–PAGE electrophoresis. Purified AChR is reconstituted in lipid immediately to prevent deterioration.

Reconstitution of Purified AChR in Soybean Vesicles

Soybean lipid (960 mg) is dissolved in chloroform, 240 mg of cholesterol added, and solvent evaporated under a stream of nitrogen. Traces of chloroform are eliminated by drying the lipid mixture in vacuum for 3 hr. The lipid film is resuspended in Buffer A (96 mg/ml soybean lipid, 24 mg/ml cholesterol) and sodium cholate added to a final concentration of 4%.

For reconstitution, 2 volumes of solubilized, purified AChR (≥1 mg/ml) are mixed with 1 volume of lipid solution. Liposomes containing AChR are formed on removal of detergent by dialysis for 16 hr against 500 volumes of Buffer A, followed by dialysis for 24 hr against 500 volumes of dialysis buffer. Reconstituted AChR is stored either at −80°C or in liquid nitrogen. The final AChR concentration is ≈4.0 μM and the lipid composition 32 mg/ml soybean lipid, 8 mg/ml cholesterol. At this protein/lipid ratio, binding sites for α-bungarotoxin are equally distributed between the interior and exterior of the reconstituted vesicle (20).

Incorporation of Reconstituted AChR in Planar Lipid Bilayers

Incorporation of the AChR channel in planar lipid bilayers may be accomplished by two different methods. Vesicles containing receptor may be fused to preformed bilayers; efficiency of incorporation is low, however, this approach is recommended for proteins sensitive to surface inactivation. Alternatively, bilayers may be formed from vesicle-derived monolayers, yielding a higher efficiency of incorporation. For the AChR, this is the method of choice; AChR-channel properties are independent of the pathway used for reconstitution (1, 2, 6, 20).

Planar Lipid Bilayers from Vesicle-Derived Monolayers

Planar lipid bilayers from vesicle-derived monolayers may be formed if the surface pressure of the monolayer is ≥25 dyne/cm (1, 20). In addition, two requirements for the functional integrity of the AChR must be taken into account: (i) cholesterol, which stabilizes the channel, and (ii) a monolayer cohesive pressure of 30 dyne/cm, which seems to reduce AChR aggregation in the membrane (1, 20).

A volume (1.0–5.0 μl) of reconstituted AChR is added to 200–250 μl of buffered salt medium, vortexed for 60 sec, and sonicated for 5 sec. The suspension is incubated for 20 min at room temperature and transferred to the bilayer chamber by gently delivering 20-μl drops. A lipid monolayer containing the AChR is spontaneously formed at the air–water interface within 2–4 min. Bilayers are assembled at the tip of patch pipettes by apposition of monolayers (18, 20). The membrane resistance, R_m, is 5 GΩ ≤ R_m ≤ 25 GΩ.

Fusion of Reconstituted Vesicles to Preformed Bilayers

Detailed descriptions of fusion protocols are published (1, 10, 37, 38). Here, we address basic considerations to improve efficiency. Membrane fusion results from the interaction of lipid vesicles with preformed planar bilayers, in the presence of negatively charged lipids and calcium ions and under osmotic conditions which induce swelling of the vesicles. The fusion rate may be controlled and, therefore, the efficiency of channel incorporation improved by variation of these parameters (1, 10, 37, 38). Another variable to be considered is the area of the lipid bilayer: larger bilayer areas produce higher rates of fusion events, however, the use of large-diameter patch pipettes reduces seal resistance, membrane stability, and signal-to-noise ratio.

Bilayers are formed at the tip of the patch pipettes by apposition of two monolayers derived from solutions of POPE/POPC, 4 : 1, 5 mg/ml in hexane. AChR-vesicles in dialysis buffer are added to the external bath solution (buffered salt medium) at a protein concentration of 0.1–1 μM. Fusion events occur 5–15 min after vesicle addition and are identified at an applied voltage, $V = 100$ mV, as abrupt, transient increases in ionic current. The occurrence of fusion events may be stopped by bath addition of 100–500 μM EGTA. In our hands, the efficiency of AChR-channel incorporation is 2–5%. Low efficiency is primarily a consequence of the use of solvent-free bilayers that exhibit a lower fusion rate than solvent-containing bilayers (10) and the small area of the pipette tip.

Phosphorylation of Purified AChR by cAMP-Dependent Protein Kinase

Chemical modification of the receptor is carried out in the reconstituted vesicles (22). Routinely, AChR (5–15 μg) is equilibrated with phosphorylation buffer supplemented with 1 μg of recombinant catalytic subunit of protein kinase A from murine heart. The reaction is started by addition of 10 nmol of ATP. For control samples, ATP, kinase A, or both are omitted in the reaction mixture. Reactions proceed at 22°C on a horizontal shaker and are terminated by addition of EDTA at a final concentration of 60 mM. Suspensions are centrifuged in an airfuge at 30 psi for 30 min. Pellets are resuspended either in 200–250 μl of buffered salt medium and transferred to bilayer chambers for electrical recordings or in 5–10 μl of dialysis buffer for fusion experiments.

For quantitation of phosphate incorporation, identical protocols are used, supplementing the phosphorylation mixture with [γ-^{32}P]ATP (22). Radiolabeled material is subjected to SDS–PAGE, protein bands are isolated, and radioactivity is counted in scintillation fluid (Filtron-X; National Diagnostics, Manbille, NJ). Stoichiometries are calculated based on an evenly distributed orientation of AChR in the reconstituted vesicles (20); only receptors with their cytoplasmic domain facing the exterior of the lipid vesicle are accessible to the kinase. Alternatively, phosphorylation of AChR may be performed on the purified, solubilized receptor. The advantage of this protocol is that all receptor molecules are accessible to chemical modification.

Reconstitution of Designed Channel-Forming Peptides and Proteins

Buffer Solutions

The aqueous compartments separated by the lipid bilayer contain the desired concentration of permeant ion, typically 50 mM BaCl$_2$ or 500 mM NaCl. For experiments using monovalent permeant ion, 1 mM CaCl$_2$ may be included to enhance bilayer stability. Solutions are buffered with 10 mM HEPES, pH 7.3, and filtered through a 0.2-μm nylon syringe filter.

Incorporation of Protein in Lipid Bilayers

Monolayers are formed from solutions of lipid, POPE/POPC, 4 : 1, 5 mg/ml in hexane. Peptide or protein is incorporated into bilayers by either of two approaches: purified protein (21) is extracted with lipid, POPE/POPC, 4 : 1, 5 mg/ml in hexane, by vortexing 1 min and sonicating for 10 sec in a water bath sonicator (Laboratory Supplies Co., Inc., Hicksville, NY) to achieve final protein/lipid ratios in the range of 1 : 1000. Lipid/protein mixtures are spread into monolayers at the air–water interface and bilayers formed as described above. Alternatively, peptide or protein may be dissolved in trifluoroethanol (TFE); TFE extracts are added to the aqueous subphase following formation of lipid bilayers (final concentration 300–600 nM). Spontaneous insertion into bilayers is observed after 5–15 min. Both methods of incorporation lead to similar channel recordings.

Illustrative Results

Modulation of AChR-Channel Activity by Protein Phosphorylation

Regulation of AChR-channel activity by protein phosphorylation is illustrated. The AChR is phosphorylated by the catalytic subunit of protein kinase A, which specifically modifies the γ and δ subunits (22, 28). To distinguish specific effects of this covalent modification on AChR-channel activity, single-channel properties of both phosphorylated and unmodified AChRs are determined. The studies are performed in the absence of acetylcholine (ACh) or in presence of low concentrations of ACh to minimize the extent of receptor desensitization (22).

Figure 2A depicts spontaneous single-channel currents recorded from lipid bilayers containing unmodified AChRs. AChR spontaneous activity is characterized by the infrequent occurrence of very brief openings (<1 msec) with a conductance of 44 pS in 0.5 M KCl. The open probability is 0.08%. Phosphorylation of AChRs increases the spontaneous channel activity (Fig.

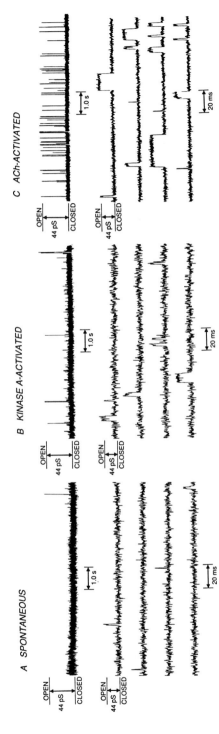

Fig. 2 Single-channel currents from AChR reconstituted in lipid bilayer. Spontaneous activity recorded at 50 mV in the absence of agonist from (A) unmodified and (B) phosphorylated AChR (0.3 mol Pi/mol receptor). (C) ACh-activated single-channel currents from unmodified AChR. ACh concentration is 1 μM and $V = 100$ mV. Records filtered at 1 kHz [From Ref. (22)].

2B) without affecting the single-channel conductance. Receptor phosphorylation leads to an increased channel open probability (0.5%), arising from an increment in the frequency of channel openings and a prolongation of residence time in the open state (22). Altered single-channel properties are clearly attributable to protein phosphorylation; omission of protein kinase A or ATP from the phosphorylation reaction mixture results in channel activity similar to that of unmodified AChRs. Further, observed activity is associated with activation of the receptor as indicated by its sensitivity to α-bungarotoxin, ruling out the possibility of an unspecific lipid-mediated channel-like activity (1, 39, 40).

Spontaneous activity of phosphorylated AChR resembles that of unmodified receptors activated by ACh (Fig. 2C): AChR-channel activity is characterized by the paroxysmal occurrence of events with short (<1 msec) and long (4–5 msec) open times, separated by quiescent periods reflecting a modest onset of receptor desensitization. Similarly, in the presence of low ACh concentrations, phosphorylated AChR showed a higher open probability than unmodified receptors. This increment arises primarily from a shortening of the characteristic long silent periods that separate bursts of channel openings. These results clearly demonstrate that covalent modification of the AChR, in the absence of cellular modulators, induces a significant change in single-channel properties.

Characterization of Designed Peptides and Proteins

Single-Channel Properties of Monomeric Peptides

Monomeric peptide CaIVS3 forms discrete channels of different amplitudes. Figure 3 (left) illustrates conductance events in symmeric 150 mM BaCl$_2$ with single-channel conductances of 14 and 21 pS (41). Channels recorded with CaIVS3 are the result of peptides self-assembling into conductive oligomers, indicated by the heterogeneity of γ and the brevity of channel mean open time. An important control of the sequence specificity of the activity observed with CaIVS3 is IVS3-Random, a peptide of the same amino acid composition as CaIVS3 and predicted helical conformation, but with a computer-generated randomized sequence (Fig. 1). IVS3-Random does not elicit single-channel events in lipid bilayers (Fig. 3, right), reflected in the current histogram which shows one band rather than discrete peaks corresponding to the closed and open states, thus confirming the sequence requirement for the activity observed (41).

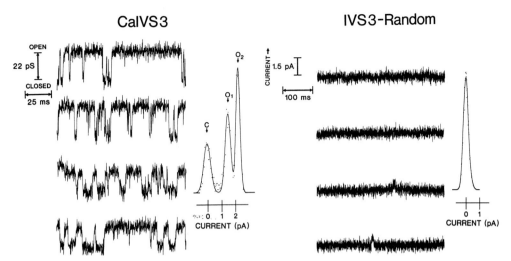

F IG. 3   Single-channel currents from lipid bilayers containing 22-residue peptides CaIVS3 (left) or IVS3-Random (right). Currents are recorded from POPE/POPC bilayers in symmetric 150 mM BaCl$_2$ at 100 mV. Filtered at 2.5 kHz (CaIVS3) or 2 kHz (IVS3-Random). The corresponding current histogram for CaIVS3 indicates the presence of multiple conductances; C, closed; O$_1$ and O$_2$, two distinct open states.

Sequence Specificity of Oligomeric Proteins

Attachment of four identical peptide modules with the sequence of CaIVS3 to a template (Fig. 1) generates the four-helix bundle protein T$_4$CaIVS3. Figure 4 depicts single-channel currents recorded in symmetric 500 mM NaCl, 1 mM CaCl$_2$ (Fig. 4, middle). Conductance events are uniform, consistent with a tetrameric array of pore-forming segments (16, 41). The single-channel conductance, γ, calculated from the corresponding current histogram is 11 pS. The brief channel mean open time reflects block by Ca^{2+} (16).

Extensive single-channel characterization of T$_4$CaIVS3 suggests that this structure represents a plausible molecular blueprint for the pore-forming element of voltage-gated calcium channels; T$_4$CaIVS3 mimics key pore properties of authentic channels, including ionic conductance, the presence of high (μM) and low (mM) affinity binding sites for Ca^{2+}, and selectivity for cations. Importantly, the modulation of authentic calcium channels by specific drugs is closely emulated by T$_4$CaIVS3 (16, 41).

For examining sequence specificity and evaluate the inferred functional significance of observed pore properties, a crucial control is the single-channel characterization of four-helix bundle proteins representing different

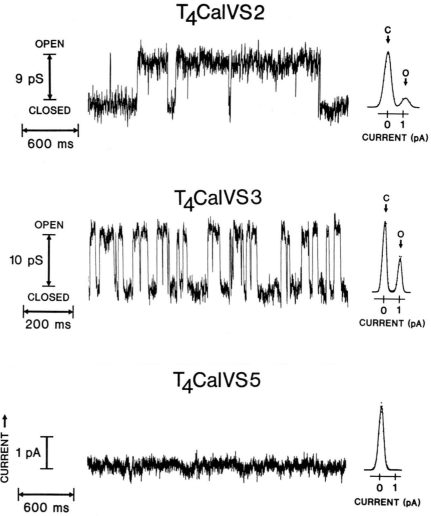

FIG. 4   Single-channel currents recorded from POPE/POPC bilayers containing four-helix bundle proteins $T_4CaIVS2$ (top), $T_4CaIVS3$ (middle), or $T_4CaIVS5$ (bottom). Currents are recorded in 500 mM NaCl in presence of 1 mM $CaCl_2$ at 100 or 120 mV ($T_4CaIVS2$). Filtered at 1 kHz. Corresponding current histograms were generated from segments of recordings lasting $\geq$30 sec. C, closed; O, open. For $T_4CaIVS2$, $\tau_{o1} = 5$ msec, $\tau_{o2} = 525$ msec; for $T_4CaIVS3$, $\tau_{o1} = 2$ msec, $\tau_{o2} = 23$ msec.

transmembrane segments. S2 segments contain conserved negatively charged and polar residues (Fig. 1) and a bundle of four IVS2 segments is predicted to form a central conductive pathway (25). Indeed, $T_4CaIVS2$ forms channels in symmetric 500 mM NaCl (Fig. 4, top), $\gamma = 9$ pS. However, $T_4CaIVS2$ does not reproduce pore properties of authentic calcium channels, including single-channel conductance and saturation for divalent cations, blockade by Ca^{2+} as indicated by the long mean open time (Fig. 4, legend), and sensitivity to specific calcium-channel modulators (41).

$T_4CaIVS5$ represents a very hydrophobic transmembrane segment not considered involved in lining an aqueous pore (25, 30). In accord with expectations, $T_4CaIVS5$ does not form distinct unitary conductance events (Fig. 4, bottom) as reflected in the current histogram (16). Erratic current fluctuations indicate that the protein is embedded in the bilayer. Thus, conductance properties of authentic calcium channels are emulated only by $T_4CaIVS3$, representing a specific sequence of the target protein.

Membrane Lipid Composition

Channel properties, both single-channel conductance and residence times in the open and closed states, may depend on membrane lipid composition (1, 20, 42, 43). This aspect may be directly investigated in lipid bilayers. In particular, the presence of charged phospholipids, e.g., phosphatidylserine (PS), influences the surface potential created by phospholipid head groups and the consequent interfacial accumulation of counterions. A cation-selective pore that senses the membrane surface charge will exhibit increasing single-channel conductance with increasing concentration of negatively charged phospholipid (42, 43).

Interaction of charged peptides with oppositely charged membranes may induce perturbations, reflected as changes in membrane conductance. Such current fluctuations have been interpreted as the generation of ion-conducting pores and even assigned biological relevance (44–47). A case in point is the positively charged S4 segment of voltage-gated channels (44). However, insertion of highly charged peptides in lipid bilayers is critically dependent on membrane lipid composition. This is illustrated in Fig. 5. Incorporation of a peptide representing an S4 segment of calcium or sodium channels (Fig. 1) in POPE/POPC bilayers supplemented with 30% POPS produces activity characterized by frequent transitions between low- and high-conducting states (shown for NaIVS4 in Fig. 5, top left). The corresponding current histogram reflects the predominant occurrence of two conducting states. Note, however, that activity is irregular, as indicated at higher time resolution. Further, the conductance does not mimic that of the authentic sodium

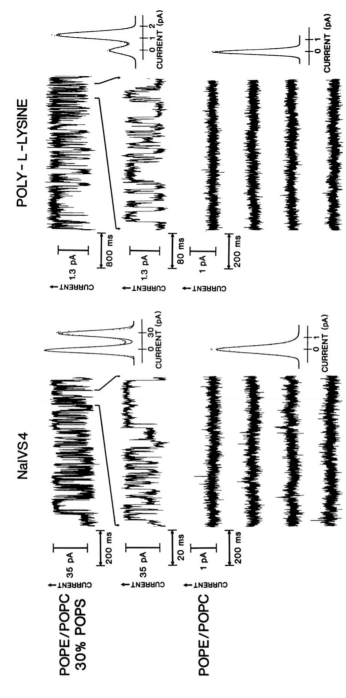

FIG. 5   Activity elicited by NaIVS4 (left) or poly-L-lysine (MW 3000, Sigma; right) in lipid bilayers composed of POPE/POPC (bottom) or in bilayers supplemented with 30% POPS (top). Recorded in 500 mM NaCl. Current histograms correspond to segments of recordings lasting ≥30 sec. Currents from POPE/POPC bilayers recorded at 100 mV and filtered at 1 kHz. Currents were from bilayers containing 30% POPS recorded at 50 mV; compressed and expanded records were filtered at 5 and 10 kHz (NaIVS4) or 1 and 2 kHz (poly-L-lysine).

channel, recorded under similar conditions. By contrast, reconstitution of NaIVS4 in POPE/POPC bilayers does not produce discrete transitions between current levels (Fig. 5, bottom left). Activity is characterized by transient, erratic changes in membrane conductance.

Very similar recordings are obtained after incorporation of poly-L-lysine in bilayers containing negatively charged lipid (Fig. 5, top right). Transitions between two current levels are evident and reflected in the corresponding current histogram. No such activity is observed in POPE/POPC bilayers (Fig. 5, bottom right). The highly charged peptides adopt an interfacial orientation allowing favorable interactions with negatively charged phospholipid head groups (48), leading to transient destabilization of the lipid bilayer. Accordingly, it is essential to discriminate between activity postulated to arise from a specific amino acid sequence and activity consequent to the interaction of oppositely charged interfaces that may bear no relevance to the biology of the system under study.

Signal Specificity: Detergents

Membrane proteins are purified in presence of detergents. Surfactants are notorious for their capacity to interact with membranes, perturb them, compromise their stability, and ultimately, solubilize them (1, 40). For reconstitution of channel proteins it is, therefore, of utmost importance to ascertain that residual detergent associated with the purified protein does not interfere with functional characterization. Specifically, detergents such as Triton X-100 form cation-selective conductive pathways in lipid bilayers (49). Other detergents, such as cholate and β-octylglucoside are readily removed from the purified protein and less effective in perturbing bilayers. Accordingly, careful selection of a detergent to be used in protein purification and its controlled removal prior to reconstitution in bilayers should assist in optimizing the fidelity of the system.

Advantages, Limitations, and Scope

Bilayers continue to provide valuable information on structure and activity of channel proteins. The reconstituted system provides a sensitive assay to identify protein components necessary for channel function, to investigate modulation of authentic ion channels in the absence of cellular components, and to characterize the pore properties of designed channel proteins. The biophysical and pharmacological characterization of the channel under study, and the realization of appropriate controls, should aid in identifying the channel activity relevant to the biology of the system.

However, we must sound a note of caution: bilayers are very sensitive to perturbants, particularly to surface-active molecules, including amphipathic peptides. Accordingly, specificity must be a major focus when assaying the channel activity of potential pore-forming peptides. Amphiphatic peptides may induce large, transient changes in membrane conductance (44–47) and elicit events of heterogeneous amplitude and lifetime (Fig. 5). An interfacial orientation may cause surface-adsorbed monomers to aggregate and disturb the bilayer by creating discrete, yet transient micellization loci that at low abundance do not compromise bilayer stability but tend to propagate and eventually may lead to membrane rupture (48). Amphiphilicity is necessary for membrane interaction and insertion, but not sufficient to generate discrete single-channel events. Sequence specificity is a requirement to reproduce functional attributes of authentic channel proteins (14, 16).

A wealth of information on structural elements underlying functional properties of ion channels has accrued from mutagenesis studies on cloned channel proteins, assayed after their expression in amphibian oocytes or cultured cells (23, 24). Recently, it was shown that lipid bilayers may broaden this endeavor: cloned channels (and mutants) produced in an *in vitro* translation system may be characterized after fusion of microsomes with lipid bilayers (50). This advance combines the power of recombinant DNA technology with the sensitivity of the lipid bilayer assay, expanding the paths toward elucidating molecular determinants of function.

Acknowledgments

We are indebted to our collaborators J. M. Tomich, T. Iwamoto, and M. S. Montal. This work was supported by grants from the National Institutes of Health (GM-42340 and MH-44638), the Office of Naval Research (N00014-89-J-1469), and the Department of the Army Medical Research (DAMD17-89-C-9032) and by a Research Scientist Award to M.M. from the Alcohol, Drug Abuse and Mental Health Administration (MH-00778). A.V.F.-M. is a postdoctoral fellow of NATO.

References

1. M. Montal, A. Darszon, and H. Schindler, *Q. Rev. Biophys.* **14,** 1 (1981).
2. P. Labarca, J. Lindstrom, and M. Montal, *J. Gen. Physiol.* **83,** 473 (1984).
3. R. P. Hartshorne, B. U. Keller, J. A. Talvenheimo, W. A. Catterall, and M. Montal, *Proc. Natl. Acad. Sci. U.S.A.* **82,** 240 (1985).

4. R. Hartshorne, M. Tamkun, and M. Montal, *in* "Ion Channel Reconstitution" (C. Miller, ed.), p. 337. Plenum Press, New York, 1986.

5. M. Montal, *J. Membr. Biol.* **98,** 101 (1987).

6. M. Montal, *in* "Techniques for the Analysis of Membrane Proteins" (C. I. Ragan and R. J. Cherry, eds.), p. 97. Chapman & Hall, London, 1986.

7. D. S. Duch, A. Hernandez, S. R. Levinson, and B. W. Urban, *J. Gen. Physiol.* **100,** 623 (1992).

8. R. Coronado and H. Affolter, *in* "Ion Channel Reconstitution" (C. Miller, ed.), p. 483. Plenum Press, New York, 1986.

9. R. Coronado, S. Kawano, C. J. Lee, C. Valdivia and H. H. Valdivia, *Methods Enzymol.* **207,** 699 (1992).

10. P. Labarca and R. Latorre, *Methods Enzymol.* **207,** 447 (1992).

11. M. S. P. Sansom, *Prog. Biophys. Mol. Biol.* **55,** 139 (1991).

12. S. Oiki, W. Danho, and M. Montal, *Proc. Natl. Acad. Sci. U.S.A.* **85,** 2393 (1988).

13. S. Oiki, W. Danho, V. Madison, and M. Montal, *Proc. Natl. Acad. Sci. U.S.A.* **85,** 8703 (1988).

14. M. Montal, M. S. Montal, and J. M. Tomich, *Proc. Natl. Acad. Sci. U.S.A.* **87,** 6929 (1990).

15. D. Langosch, K. Hartung, E. Grell, E. Bamberg, and H. Betz, *Biochim. Biophys. Acta* **1063,** 36 (1991).

16. A. Grove, J. M. Tomich, and M. Montal, *Proc. Natl. Acad. Sci. U.S.A.* **88,** 6418 (1991).

17. M. Montal, *Methods Enzymol.* **32,** 545 (1974).

18. B. A. Suarez-Isla, K. Wan, J. Lindstrom, and M. Montal, *Biochemistry* **22,** 2319 (1983).

19. T. Hamamoto and M. Montal, *Methods Enzymol.* **126,** 123 (1986).

20. M. Montal, R. Anholt, and P. Labarca, *in* "Ion Channel Reconstitution" (C. Miller, ed.), p. 157. Plenum Press, New York, 1986.

21. A. Grove, T. Iwamoto, M. S. Montal, J. M. Tomich, and M. Montal, *Methods Enzymol.* **207,** 510 (1992).

22. A. V. Ferrer-Montiel, M. S. Montal, M. Díaz-Muñoz, and M. Montal, *Proc. Natl. Acad. Sci. U.S.A.* **88,** 10213 (1991).

23. E. Neher, *Science (Washington, D.C.)* **256,** 498 (1992).

24. B. Sakmann, *Science (Washington, D.C.)* **256,** 503 (1992).

25. M. Montal, *FASEB J.* **4,** 2623 (1990).

26. V. Florio, J. Striessnig, and W. A. Catterall, *Methods Enzymol.* **207,** 529 (1992).

27. T. M. DeLorey and R. W. Olsen, *J. Biol. Chem.* **267,** 16747 (1992).

28. S. L. Swope, S. J. Moss, C. D. Blackstone, and R. L. Huganir, *FASEB J.* **6,** 2514 (1992).

29. S. Numa, *Harvey Lect.* **83,** 121 (1989).

30. J. Finer-Moore, J. F. Bazan, J. Rubin, and R. M. Stroud, *in* "Prediction of Protein Structure and the Principles of Protein Conformation" (G. D. Fasman, ed.), p. 719. Plenum Press, New York, 1989.

31. M. Mutter and S. Vuilleumier, *Angew. Chem. Int. Ed. Engl.* **28(5),** 535 (1989).

32. L. G. Reddy, T. Iwamoto, J. M. Tomich, and M. Montal, *J. Biol. Chem.* **268,** 14608 (1993).

33. P. Hess, *Annu. Rev. Neurosci.* **13,** 337 (1990).
34. C. Miller, *Science (Washington, D.C.)* **252,** 1092 (1991).
35. A. Bushan and M. G. McNamee, *Biochim. Biophys. Acta* **10,** 93 (1990).
36. A. Chak and A. Karlin, *Methods Enzymol.* **207,** 546 (1992).
37. F. S. Cohen, *in* "Ion channel Reconstitution" (C. Miller, ed.), p. 131. Plenum Press, New York, 1986.
38. W. Hanke, *in* "Ion Channel Reconstitution" (C. Miller, ed.), p. 141. Plenum Press, New York, 1986.
39. D. J. Woodbury, *J. Membr. Biol.* **109,** 145 (1989).
40. J. R. Silvius, *Annu. Rev. Biophys. Biomol. Struct.* **21,** 323 (1992).
41. A. Grove, J. M. Tomich, T. Iwamoto, and M. Montal, *Protein Sci.* **2,** 1918 (1993).
42. F. Gambale and M. Montal, *Biophys. J.* **53,** 771 (1988).
43. R. Latorre, P. Labarca, and D. Naranjo, *Methods Enzymol.* **207,** 471 (1992).
44. M. T. Tosteson, D. S. Auld, and D. C. Tosteson, *Proc. Natl. Acad. Sci. U.S.A.* **86,** 707 (1989).
45. K. Anzai, M. Hamasuna, H. Kadono, S. Lee, H. Aoyagi, and Y. Kirino, *Biochim. Biophys. Acta* **1064,** 256 (1991).
46. D. G. Reid, L. K. MacLachlan, C. J. Salter, M. J. Saunders, S. D. Jane, A. G. Lee, E. J. Tremeer, and S. A. Salisbury, *Biochim. Biophys. Acta* **1106,** 264 (1992).
47. P. Ghosh and R. M. Stroud, *Biochemistry* **30,** 3551 (1991).
48. M. Montal, *J. Membr. Biol.* **7,** 245 (1972).
49. H. van Zutphen, A. J. Merola, G. P. Brierley, and D. G. Cornwell, *Arch. Biochem. Biophys.* **152,** 755 (1972).
50. R. L. Rosenberg and J. E. East, *Nature (London)* **360,** 166 (1992).

Index